T0269054

This book deals with the various thermodynamic concepts used for the analysis of nonlinear dynamical systems. The most important concepts are introduced in such a way as to stress the interconnections with thermodynamics and statistical mechanics.

The basic ideas of the thermodynamic formalism of dynamical systems are explained in the language of physics. The book is an elementary introduction to the subject and no advanced mathematical preknowledge on the part of the reader is required. Among the subjects treated are probabilistic aspects of chaotic dynamics, the symbolic dynamics technique, information measures, the maximum entropy principle, general thermodynamic relations, spin systems, fractals and multifractals, expansion rate and information loss, the topological pressure, transfer operator methods, repellers and escape. The more advanced chapters deal with the thermodynamic formalism for expanding maps, thermodynamic analysis of chaotic systems with several intensive parameters, and phase transitions in nonlinear dynamics.

Written for students and scientists working in physics and related fields, this easily readable introduction can also be used as a textbook for special courses.

*Thermodynamics of chaotic systems*

**Cambridge Nonlinear Science Series 4**

**Titles in this series**

1 Weak chaos and quasi-regular patterns
*G. M. Zaslavsky, R. Z. Sagdeev, D. A. Usikov and A. A. Chernikov*
2 Quantum chaos: a new paradigm for nonlinear dynamics
*K. Nakamura*
3 Chaos, dynamics and fractals: an algorithmic approach to deterministic chaos
*J. L. McCauley*
4 Thermodynamics of chaotic systems: an introduction
*C. Beck and F. Schlögl*

# *Thermodynamics of chaotic systems*

## *An introduction*

**Christian Beck**

*School of Mathematical Sciences,*
*Queen Mary and Westfield College,*
*University of London*

*and*

**Friedrich Schlögl**

*Institute for Theoretical Physics,*
*University of Aachen*

CAMBRIDGE UNIVERSITY PRESS
Cambridge, New York, Melbourne, Madrid, Cape Town, Singapore, São Paulo

Cambridge University Press
The Edinburgh Building, Cambridge CB2 2RU, UK

Published in the United States of America by Cambridge University Press, New York

www.cambridge.org
Information on this title: www.cambridge.org/9780521433679



First published 1993
First paperback edition 1995
Reprinted 1997

*A catalogue record for this publication is available from the British Library*

*Library of Congress Cataloguing in Publication data*
Beck, Christian.
Thermodynamics of chaotic systems : an introduction / Christian
Beck and Friedrich Schlögl.
p. cm. — (Cambridge nonlinear science series : 4)
Includes bibliographical references.
ISBN 0-521-43367-3
1. Chaotic behavior in systems. 2. Thermodynamics.
3. Nonlinear theories. I. Schlögl, Friedrich. II. Title.
III. Series
Q172.5.C45B43 1993
536′.7 – dc20 92-33322 CIP

ISBN-13 978-0-521-43367-9 hardback
ISBN-10 0-521-43367-3 hardback

ISBN-13 978-0-521-48451-0 paperback
ISBN-10 0-521-48451-0 paperback

Transferred to digital printing 2005

# Contents


*Preface* xiii
*Introduction* xv

**PART I ESSENTIALS OF NONLINEAR DYNAMICS** 1

**1 Nonlinear mappings** 1

1.1 Maps and trajectories 1
1.2 Attractors 2
1.3 The logistic map 4
1.4 Maps of Kaplan–Yorke type 11
1.5 The Hénon map 14
1.6 The standard map 16

**2 Probability in the theory of chaotic systems** 20

2.1 Relative frequencies and probability 20
2.2 Invariant densities 22
2.3 Topological conjugation 28

**3 Symbolic dynamics** 32

3.1 Partitions of the phase space 32
3.2 The binary shift map 33
3.3 Generating partition 35
3.4 Symbolic dynamics of the logistic map 37
3.5 Cylinders 39
3.6 Symbolic stochastic processes 40

**PART II ESSENTIALS OF INFORMATION THEORY AND THERMODYNAMICS** 44

**4 The Shannon information** 44

4.1 Bit-numbers 44
4.2 The Shannon information measure 46
*4.3 The Khinchin axioms 47

**5 Further information measures** 50

5.1 The Rényi information 50
5.2 The information gain 51
*5.3 General properties of the Rényi information 53

**6 Generalized canonical distributions** 56

6.1 The maximum entropy principle 56
6.2 The thermodynamic equilibrium ensembles 59
6.3 The principle of minimum free energy 61
*6.4 Arbitrary *a priori* probabilities 63

**7 The fundamental thermodynamic relations** 65

7.1 The Legendre transformation 65
7.2 Gibbs' fundamental equation 68
*7.3 The Gibbs–Duhem equation 71
7.4 Susceptibilities and fluctuations 73
7.5 Phase transitions 75

**8 Spin systems** 78

8.1 Various types of spin models 78
8.2 Phase transitions of spin systems 82
8.3 Equivalence between one-dimensional spin systems and symbolic stochastic processes 83
8.4 The transfer matrix method 85

**PART III THERMOSTATISTICS OF MULTIFRACTALS** 88

**9 Escort distributions** 88

9.1 Temperature in chaos theory 88
9.2 Two or more intensities 90
*9.3 Escort distributions for general test functions 91

**10 Fractals** 94

10.1 Simple examples of fractals 94
10.2 The fractal dimension 97
10.3 The Hausdorff dimension 100
10.4 Mandelbrot and Julia sets 102
*10.5 Iterated function systems 109

**11 Multifractals** 114

11.1 The grid of boxes of equal size 114
11.2 The Rényi dimensions 115
11.3 Thermodynamic relations in the limit of box size going to zero 117
11.4 Definition of the Rényi dimensions for cells of variable size 123

***12 Bit-cumulants in multifractal statistics** 127

12.1 Moments and cumulants of bit-numbers 127
12.2 The bit-variance in thermodynamics 129
12.3 The sensitivity to correlations 131
12.4 Heat capacity of a fractal distribution 132

***13 Thermodynamics of finite volume** 136

13.1 Boxes of finite size 136
13.2 Thermodynamic potentials for finite volume 139
13.3 An analytically solvable example 143

PART IV DYNAMICAL ANALYSIS OF CHAOTIC SYSTEMS 146

**14 Statistics of dynamical symbol sequences** 146

14.1 The Kolmogorov–Sinai entropy 146
14.2 The Rényi entropies 149
14.3 Equivalence with spin systems 155
14.4 Spectra of dynamical scaling indices 156

**15 Expansion rate and information loss** 158

15.1 Expansion of one-dimensional systems 158
15.2 The information loss 160
*15.3 The variance of the loss 161
15.4 Spectra of local Liapunov exponents 164
15.5 Liapunov exponents for higher-dimensional systems 167
15.6 Stable and unstable manifolds 172

| | | |
|---|---|---|
| **16** | **The topological pressure** | 178 |
| 16.1 | Definition of the topological pressure | 178 |
| 16.2 | Length scale interpretation | 182 |
| 16.3 | Examples | 184 |
| *16.4 | Extension to higher dimensions | 187 |
| *16.5 | Topological pressure for arbitrary test functions | 188 |
| **17** | **Transfer operator methods** | 190 |
| 17.1 | The Perron–Frobenius operator | 190 |
| 17.2 | Invariant densities as fixed points of the Perron–Frobenius operator | 193 |
| 17.3 | Spectrum of the Perron–Frobenius operator | 196 |
| 17.4 | Generalized Perron–Frobenius operator and topological pressure | 197 |
| *17.5 | Connection with the transfer matrix method of classical statistical mechanics | 199 |
| **18** | **Repellers and escape** | 204 |
| 18.1 | Repellers | 204 |
| 18.2 | Escape rate | 207 |
| *18.3 | The Feigenbaum attractor as a repeller | 208 |
| **PART V** | **ADVANCED THERMODYNAMICS** | 211 |
| **19** | **Thermodynamics of expanding maps** | 211 |
| 19.1 | A variational principle for the topological pressure | 211 |
| 19.2 | Gibbs measures and SRB measures | 214 |
| 19.3 | Relations between topological pressure and the Rényi entropies | 218 |
| 19.4 | Relations between topological pressure and the generalized Liapunov exponents | 221 |
| 19.5 | Relations between topological pressure and the Rényi dimensions | 222 |
| **20** | **Thermodynamics with several intensive parameters** | 226 |
| 20.1 | The pressure ensemble | 226 |
| *20.2 | Equation of state of a chaotic system | 229 |
| 20.3 | The grand canonical ensemble | 233 |
| 20.4 | Zeta functions | 237 |
| *20.5 | Partition functions with conditional probabilities | 240 |

**21 Phase transitions** 243

21.1 Static phase transitions 243
21.2 Dynamical phase transitions 248
21.3 Expansion phase transitions 251
21.4 Bivariate phase transitions 254
*21.5 Phase transitions with respect to the volume 256
21.6 External phase transitions of second order 260
*21.7 Renormalization group approach 262
21.8 External phase transitions of first order 267

*References* 273

*Index* 282

# Preface

In recent years methods borrowed from thermodynamics and statistical mechanics have turned out to be very successful for the quantitative analysis and description of chaotic dynamical systems. These methods, originating from the pioneering work of Sinai, Ruelle, and Bowen in the early seventies on the thermodynamic formalism of dynamical systems, have been further developed in the mean time and have attracted strong interest among physicists. The quantitative characterization of chaotic motion by thermodynamic means and the thermodynamic analysis of multifractal sets are now an important and rapidly evolving branch of nonlinear science, with applications in many different areas.

The present book aims at an elucidation of the various thermodynamic concepts used for the analysis of nonlinear dynamical systems. It is intended to be an elementary introduction. We felt the need to write an easily readable book, because so far there exist only a few classical monographs on the subject that are written for mathematicians rather than physicists. Consequently, we have tried to write in the physicist's language. We have striven for a form that is readable for anybody with the knowledge of mathematics a student of physics or chemistry at the early graduate level would have. No advanced mathematical preknowledge on the part of the reader is required. On the other hand, we also tried to avoid serious loss of rigour in our presentation. Often we have worked out the subject matter in a pedestrian way, that is to say, we have often dealt first with a special case, doing the generalization later. Our main goal was to emphasize the interesting parallels between thermodynamic methods in nonlinear dynamics and conventional statistical mechanics, and to make these beautiful new applications of statistical mechanics accessible to a broad spectrum of readers from the physics community.

The sections marked with an asterisk * in the Contents can be omitted at a first reading. These deal with special topics that are not essential for the development of the following sections, and so their omission will not disrupt the logical thread of the book.

We would like to thank the editors of the Cambridge Nonlinear Science Series for expressing their interest in the subject and for the invitation extended (to F.S.) to write a book on it. Moreover, we are very grateful to T. Tél, P. Grassberger, and S. Grossmann for carefully reading the manuscript. Their kind advice and criticism helped us to make several improvements. This book also benefited from many discussions with colleagues and friends. In particular, we would like to thank D. H. Mayer, G. Roepstorff, G. Gürbüz, D. Graudenz, M. Ringe and Z. Kaufmann, for useful comments and encouragement. One of us (F.S.) is grateful for the kind hospitality of the National Tsing Hua University in Hsinchu, Taiwan, where he gave a series of lectures on related topics in 1991. Special thanks are directed to K. Y. Lin and C. C. Hsu, who initiated the invitation to these lectures. C.B. would like to thank P. Cvitanović, T. Bohr, and M. H. Jensen for the kind hospitality of the Niels Bohr Institute, Copenhagen, during summer 1992. He also gratefully acknowledges some helpful comments from Ya. Sinai and M. J. Feigenbaum.

Aachen, June 1992 C. Beck, F. Schlögl

# Introduction

A fundamental aspect of statistical physics is the following question: in which way can a physical system, described by deterministic evolution equations, exhibit quasi-stochastic behaviour? Until the end of the nineteenth century it was generally accepted that a necessary condition for unpredictable behaviour is a large number of degrees of freedom. For example, the movement of a single molecule in a gas is *random*, as we project down from a $10^{23}$-dimensional phase space onto a six-dimensional subspace when observing the trajectory. Today we know that even very low-dimensional, simple deterministic systems can exhibit an unpredictable, quasi-stochastic long-time behaviour. It has become common to call this phenomenon 'chaos'. The first system of this kind, namely the three-body problem of classical mechanics, was investigated by Henri Poincaré at the end of the nineteenth century (Poincaré 1892). Since then a large number of dynamical systems that exhibit chaotic behaviour have become known. Indeed, for nonlinear systems, chaos appears to be a generic rather than an exotic phenomenon.

The time evolution of a dynamical system is determined by a deterministic evolution equation. For continuous-time dynamical systems it is a differential equation

$$dx/dt = F(x), \tag{1}$$

and for discrete-time dynamical systems it is a recurrence relation

$$x_{n+1} = f(x_n) \tag{2}$$

that determines the time evolution of an initial state $x_0$. In this book we shall restrict ourselves to discrete-time dynamical systems, or more simply, to 'maps'. A map is an evolution law of type (2). In fact, maps reveal all the essentials of the known chaotic phenomena and are easier to treat, both analytically and numerically. Each state $x_n$ is characterized by a set of $d$ variables, which can be represented as a point in a $d$-dimensional space, the 'phase space' $X$. Given an initial point $x_0$, the sequence $x_0, x_1, x_2, \ldots$ obtained by iteration is called a 'trajectory'. The index $n$ is regarded as a discrete time variable. In principle, the trajectory is uniquely determined if $x_0$ is given with infinite precision. As a rule, however, a longer trajectory can be obtained only empirically, by a computer experiment, and cannot be predicted by analytical means.

There is yet another reason for unpredictable behaviour. A chaotic map is characterized by the fact that there is sensitive dependence on the initial conditions: a small uncertainty of the initial condition grows very fast (in fact, exponentially fast) with the time. This means that if the initial value $x_0$ is only known with a certain restricted precision then the behaviour of the system is already unpredictable and indeterminable after a short time.

Due to this unpredictability, the trajectory can only be described by statistical means. A common procedure then is to determine the relative frequency of the iterates $x_n$ in a certain subregion of the phase space, i.e., to investigate probability distributions of the iterates. For a generic chaotic map $f$ this probability distribution turns out to be extremely complicated:

(*a*) At some points in the phase space $X$ the probability density may diverge, often it even diverges at infinitely many points, i.e., there is a spectrum of singularities.
(*b*) Moreover, the subset of $X$ that attracts the trajectories often has a fractal structure.
(*c*) Another complicated structure can be observed with respect to the sensitive dependence on initial conditions: for initial data from some regions of the phase space the uncertainty in the specification of the initial data may grow much faster during the time evolution than for other regions.

A central task of the theory of chaotic systems then is to find adequate forms of description that are able to grasp the most important and characteristic features of the system. In this context analogous techniques to those used in conventional thermodynamics have been used that have turned out to be very successful for this purpose. Today the corresponding development is called the 'thermodynamic formalism of dynamical systems'. The mathematical foundations of this formalism go back to work by Sinai, Ruelle, and Bowen in the early seventies (Sinai 1972; Ruelle 1978a; Bowen 1975). In the mean time, the original concept has been further developed and applied to fractal sets in general. These recent developments are sometimes called the 'thermodynamics of multifractals'; applied to mappings they are the subject of this book. The principal idea of the thermodynamic formalism is to analyse the complex behaviour generated by nonlinear dynamical systems using methods borrowed from statistical mechanics. It should be stressed that the need to do thermodynamics for a nonlinear system does not arise from a very high dimension of the phase space as in classical statistical mechanics: the dimension $d$ of the phase space of a chaotic mapping usually is very low (typically not more than $d = 3$). Rather, thermodynamic tools are needed because of the complexity inherent in nonlinear systems. When seeking an adequate classification, we do not want to know all the details of the complicated probability distribution, but only a few important quantities such as, for example, the 'fractal dimension'. To analyse chaotic behaviour in a quantitative way typical thermodynamic notions are used, such as entropy, which is already familiar as a concept of general statistics and information theory. But other notions are also used such as temperature, pressure, and free energy; and it is not only for beginners that the question arises: what do these concepts have to do with the corresponding quantities in traditional thermodynamics and how far can the analogies to thermodynamics be extended; moreover, how can they be a helpful tool? To give answers to such questions is one of the aims of this book. We shall introduce the most important quantities and techniques used to analyse chaotic dynamical systems. We shall try to do this in a way that stresses the interconnections with thermodynamics and statistical mechanics.

The book is divided into five parts I–V. In *part I* 'Essentials of nonlinear dynamics' we give a short and elementary introduction to some basic notions and phenomena that occur in the theory of nonlinear dynamical systems. Of course, this part is not meant as a complete introduction to nonlinear dynamics: there are many excellent and more detailed textbooks on this field (e.g., Guckenheimer and Holmes 1983; Lichtenberg and Liebermann 1983; Schuster 1984; Hao 1984; Collet and Eckmann 1980; Arnol'd and Avez 1968; Zaslavsky 1985; Devaney 1986; Percival and Richards 1982). Rather, we just restrict ourselves to those subjects that are of most interest with respect to the thermodynamical analysis of chaotic systems. That is, we deal with attractors, invariant probability densities, partitions of the phase space, and the symbolic dynamics technique. Moreover, we present a short introduction to those maps that serve as standard examples in the following chapters, such as the binary shift map, the logistic map, maps of Kaplan–Yorke type, the Hénon map, and the standard map. Summarizing, part I provides the reader with the necessary prerequisites from nonlinear dynamics to understand the rest of the book.

In *part II* 'Essentials of information theory and thermodynamics' we develop the basic concepts of information theory and thermostatistics, again restricting ourselves to those concepts which are relevant to the thermodynamic analysis of chaotic systems. We introduce the Shannon information as well as further information measures, the Kullback information gain and the Rényi informations, and discuss general properties of these quantities. It is shown how the method of 'unbiased guess' or the 'maximum entropy principle' can be used to construct the generalized canonical probability distributions. We also deal with the various thermodynamic equilibrium ensembles, with susceptibilities and fluctuations, and with critical exponents. Moreover, we give a short introduction to spin systems and the transfer matrix method of statistical mechanics. In other words, part II yields the prerequisites from thermodynamics that are necessary for an understanding of the following chapters.

After this, in *part III* 'Thermostatistics of multifractals' we deal with thermodynamic tools used to analyse complicated, possibly fractal probability distributions. These distributions can, for example, be generated by chaotic mappings, i.e., they are given in terms of the natural invariant probability density and the attractor of the map. Nevertheless, the considerations of part III are quite generally applicable to *any* probability distribution, no matter how it is generated. To any such probability distribution one can attribute deduced distributions, which we call 'escort distributions'. These are classes of probability distributions that are constructed as a tool for scanning the structure of a given probability distribution in more detail. The escort distributions have the same formal structure as the generalized canonical distributions of thermodynamics. They yield the same basic thermodynamic relations, hence they are the common root of the various analogies to thermodynamics.

We present an elementary introduction to fractals and multifractals, i.e., fractals with a probability distribution on the fractal support. The Hausdorff dimension, and more generally the Rényi dimensions are introduced. They are obtained from the free energy of an appropriately defined partition function. By a Legendre transformation we pass to the spectra of crowding indices. We also describe an alternative approach in terms of bit-cumulants (chapter 12) and consider finite box size effects (chapter 13).

Whereas part III deals with static aspects, namely the analysis of fractals and invariant probability densities, in *part IV* 'Dynamical analysis of chaotic systems' the discussion is extended to dynamical aspects concerning the time evolution of the dynamical system. The general thermodynamic scheme will be worked out in various directions, corresponding to different sets of the elementary events for which the escort distributions are defined. Such events are, alternatively, the occurrence of a single iterate in a certain subset or 'cell' of the phase space, the occurrence of entire sequences of cells visited by the trajectories, or the occurrence of certain local expansion rates of small volume elements of the phase space. It should be clear that for chaotic dynamical systems different types of partition functions can be defined, dealing with different aspects of the system. Consequently, there are several types of 'free energies' associated with these partition functions.

We obtain the most important quantities that characterize a nonlinear system, such as the Rényi dimensions, the (dynamical) Rényi entropies, the generalized Liapunov exponents, and the topological pressure from the free energy densities in the thermodynamic limit. In statistical mechanics the name 'thermodynamic limit' is used for a well defined limit of infinite volume and infinite particle number. In connection with a map the name is used for a limit in which the size of cells of the phase space goes to zero and the number of iterations of the map goes to infinity.

Free energies dealing with dynamical aspects are the Rényi entropies, the generalized Liapunov exponents, and the topological pressure. We introduce these quantities in an elementary and physically motivated way. An important method for calculating the topological pressure, the transfer operator method, is discussed in chapter 17. This method can be regarded as the generalization of the transfer matrix method of conventional statistical mechanics. At the end of part IV we finally deal with repellers and the escape of trajectories from the neighbourhood of repellers. These considerations are relevant for the phenomenon of 'transient chaos', i.e., chaotic behaviour on a long but finite time scale.

*Part V*, 'Advanced thermodynamics', is devoted to a possible unification of the various thermodynamic concepts, i.e., of the various free energies, in one theory. This can easily be achieved for special classes of maps, the so called 'expanding' (or 'hyperbolic') maps. For these maps one type of free energy (for example, the topological pressure) is sufficient to deduce all the others (Rényi dimensions, Rényi entropies, generalized Liapunov exponents). We shall derive an important variational principle for the topological pressure, which corresponds to the principle of minimum free energy of traditional thermodynamics. This principle allows us to distinguish the natural invariant probability density of a map from other, less important invariant probability distributions. On the other hand, for nonexpanding (nonhyperbolic) maps, in general, the various free energies are independent quantities. Nevertheless, it is still possible to introduce generalized partition functions depending on several intensive parameters such that all the previously introduced free energies can be obtained as special values of a more general thermodynamic potential. We report on these analogues of the pressure ensemble and the grand canonical ensemble. Also an effective method for calculating the topological pressure, the so called 'zeta function approach', is dealt with in this context. Finally we deal with phase transitions of chaotic systems, which – just as in conventional statistical mechanics – correspond to nonanalytic behaviour of the free energy at a critical point. We classify the various phase transitions, and explain the mechanism that is generating them.

Part I · Essentials of nonlinear dynamics

# 1
# Nonlinear mappings

In this chapter we shall first give an elementary introduction to mappings and attractors, and consider several examples. Mappings can be regarded as discrete-time dynamical systems. Usually they are evaluated on a computer. Maps defined by nonlinear equations in many cases exhibit 'chaotic' behaviour and the attractors may have a 'fractal' structure. The quantitative analysis of these phenomena will be a main subject of this book. We shall introduce and discuss some of the most important standard examples of nonlinear mappings, namely the logistic map, maps of Kaplan–Yorke type, the Hénon map, and the standard map.

## 1.1 Maps and trajectories

Let us consider a *map*, or *mapping* in a $d$-dimensional space with appropriate coordinates (for example, Cartesian coordinates). The set of possible values of the coordinates is called the 'phase space' $X$. The map is given by

$$x_{n+1} = f(x_n), \tag{1.1.1}$$

where

$$x_n = (x_n^{(1)}, \ldots, x_n^{(d)}) \tag{1.1.2}$$

is a vector in $X$ and

$$f = (f^{(1)}, \ldots, f^{(d)}) \tag{1.1.3}$$

is a vector-valued function. The dynamical system is called 'nonlinear' if the function $f(x)$ is nonlinear. Only nonlinear maps can exhibit chaotic behaviour and are of interest in this book. We start with an initial point $x_0$ and iterate it step by step. Each point $x_n$ is called an *iterate*. In a computer experiment, the number of iteration steps is supposed to be very large – let us say of the order $10^4$ or larger. The sequence of iterates $x_0, x_1, x_2, \ldots$, which we call a *trajectory*, describes the motion of a point in the space $X$. Let us assume that each step from $x_n$ to $x_{n+1}$ takes the same time. Then the entire time is proportional to $n$, the total number of steps. We adopt the convention of calling the length of the trajectory the 'time'.

A trajectory may either become *periodic*, or stay *aperiodic* forever. In the first case after a certain number $K$ of iterations the iterates approach a sequence $x_K, x_{K+1}, \ldots, x_{K+L}$ satisfying

$$x_{K+L} = x_K. \tag{1.1.4}$$

The sequence $x_K, x_{K+1}, \ldots, x_{K+L-1}$ is called a *periodic orbit* or *cycle* of $f$. The smallest possible $L$ satisfying eq. (1.1.4) is called 'length' of the cycle. A periodic orbit of length $L$ equal to 1 is called a *fixed point* of the map $f$. A fixed point $x^*$ satisfies

$$x^* = f(x^*). \tag{1.1.5}$$

Of course, a periodic orbit of length $L$ can be regarded as a fixed point of the $L$-times iterated function

$$f^L(x) = f(f(\ldots f(x))) \qquad (L \text{ times}). \tag{1.1.6}$$

Hence we may restrict the discussion of periodic orbits to the discussion of fixed points.

## 1.2 Attractors

In nonlinear dynamics we are mainly interested in the long-time behaviour of the map for a generic initial value. This long-time behaviour is totally different for different kinds of maps. We distinguish between so called *Hamiltonian* dynamical systems and *dissipative* dynamical systems. A Hamiltonian system conserves the volume of an arbitrary volume element of the phase space during the time evolution. We may express this fact by means of the Jacobi determinant

$$U(x) = \det Df(x) \tag{1.2.1}$$

(by $Df(x)$ we denote the $d \times d$-matrix $\partial f^{(\beta)}/\partial x^{(\alpha)}$). Namely, a Hamiltonian system satisfies $U(x) = 1$ for all $x$. For a dissipative system, however, a small phase space volume either shrinks or expands, and this usually depends on the position $x$ in the phase space. Typical for a dissipative system is the fact that a large number of trajectories approach a certain subset $A$ of the phase space $X$ in the limit $n \to \infty$. The subset $A$ is called an *attractor*. There may be one or several attractors; but typically, in low-dimensional systems, the number of different attractors is very small. Indeed, in many cases there is just one attractor that attracts almost all trajectories.

One possible type of attractor is a *stable fixed point*: suppose $f$ is a differentiable function such that $Df(x)$ exists. The fixed point $x^*$ is called 'stable' if all $d$ eigenvalues $\eta_\alpha$ of the matrix $Df(x^*)$ satisfy $|\eta_\alpha| < 1$. In particular, for a one-dimensional map $f$, a fixed point is stable if $|f'(x^*)| < 1$. The stability means that a large number of trajectories is attracted by $x^*$: consider a point $x$ in a small neighbourhood of $x^*$, i.e., the distance $|x - x^*|$ is assumed to be small. $x$ is mapped onto $f(x)$. Expanding $f(x)$ in a Taylor series around $x^*$ we obtain in the one-dimensional case

$$f(x) \simeq x^* + (x - x^*)f'(x^*). \tag{1.2.2}$$

Hence the new distance $|f(x) - x^*|$ after one iteration step satisfies

$$|f(x) - x^*| \simeq |x - x^*|\,|f'(x^*)|, \tag{1.2.3}$$

i.e., it is smaller than the old distance $|x - x^*|$ provided $|f'(x^*)| < 1$. Thus a stable fixed point is characterized by the fact that its neighbourhood is contracted under the action of $f$. The generalization to the $d$-dimensional case is straightforward: in this case the condition $|\eta_\alpha| < 1$, $(\alpha = 1, 2, \ldots, d)$ guarantees that the vicinity of $x^*$ contracts in all directions.

More generally, the attractor of a map may also be a stable periodic orbit of length $L$. In the previous section we mentioned that a periodic orbit of length $L$ can be regarded as a fixed point of the $L$-times iterated function $f^L$. The periodic orbit of length $L$ is called stable if the corresponding fixed point of $f^L$ is stable. Let us denote the periodic orbit by $x_0, x_1, \ldots, x_{L-1}$. The stability condition means that all eigenvalues $\eta_\alpha$ of the matrix

$$Df^L(x_0) = \prod_{n=0}^{L-1} Df(x_n) \tag{1.2.4}$$

satisfy

$$|\eta_\alpha| < 1. \tag{1.2.5}$$

In this case the vicinity of the periodic orbit is contracting, and thus a large number of trajectories is attracted. If, on the other hand, at least one eigenvalue of $Df^L(x_0)$ satisfies $|\eta_\alpha| > 1$, the periodic orbit is called 'unstable'. An unstable periodic orbit is certainly not an attractor, because it repels trajectories. We notice that the simple existence of a fixed point does not mean that it is of relevance for the long-time behaviour of the mapping. Rather, it must be a stable fixed point in order to attract a large number of trajectories.

Besides stable periodic orbits, there are other types of attractors that correspond to *chaotic motion*. A dynamical system is said to be 'chaotic' if it possesses sensitive dependence on the initial conditions. This means that randomly chosen, very close initial values $x_0$ and $x_0' = x_0 + \delta x_0$ generate totally different trajectories in the long-time run. For simplicity, let us for the moment consider a one-dimensional map $f$. Let $x_n$ be the iterates of $x_0$, and $x_n'$ the iterates of $x_0'$. As a rule, for a chaotic map an exponential increase of the difference $|x_n - x_n'|$ is observed:

$$|x_n - x_n'| \approx |\delta x_0| \exp(\Lambda n) \qquad (|\delta x_0| \ll 1). \tag{1.2.6}$$

The average of the separation rate $\Lambda$ is called the *Liapunov exponent* of the map. Chaotic motion corresponds to a positive Liapunov exponent. In chapter 15 we shall treat Liapunov exponents in much more detail and shall then generalize this concept for higher-dimensional maps.

Chaotic attractors may have an extremely complicated structure (especially in dimensions $d \geqslant 2$) and are called *strange attractors* (Eckmann and Ruelle 1985; and references therein). In a sense, they are the subject of this book. For strange attractors there is at least one direction of the phase space where small distances expand on average. But in spite of that the strange attractor is confined to a finite phase space. Strange attractors often have a *fractal* structure. This means that we observe a complicated structure on arbitrary length scales, which can be described by a 'noninteger dimension'. This and further concepts related to fractals will be discussed in detail in chapter 10.

Let us now consider a few important standard examples of maps and their attractors.

## 1.3 The logistic map

This map has played an important role in the development of the theory of nonlinear dynamical systems (May 1976; Feigenbaum 1978). The phase space $X$ is one-dimensional. It is the section $[-1, 1]$ of the real axis. The map is defined by

$$f(x) = 1 - \mu x^2. \tag{1.3.1}$$

$\mu$ is a so called 'control parameter' with possible values $\mu \in [0, 2]$. Numerical results are presented in fig. 1.1: the iterates $x_n$ of the map are plotted as points in the plane $(\mu, x)$. The figure shows the attractor of this map as a function of $\mu$. We observe the following scenario: for $\mu < \mu_1 = \frac{3}{4}$ the attractor is a stable fixed point, namely the solution

$$x^* = \frac{1}{2\mu}[(1 + 4\mu)^{1/2} - 1] \tag{1.3.2}$$

of the fixed point equation (1.1.5). At $\mu = \mu_1 = \frac{3}{4}$ this fixed point suddenly loses its stability, because

$$|f'(x^*)| = 2\mu|x^*| > 1 \tag{1.3.3}$$

for $\mu > \mu_1$. Instead, a periodic orbit of length 2 (a fixed point of $f^2$) becomes stable (see fig. 1.2). This phenomenon is called a *pitchfork bifurcation*. Further increasing $\mu$, at $\mu_2 = \frac{5}{4}$ the period 2 orbit becomes unstable. Instead, a stable periodic orbit of length 4 is created (see fig. 1.3). In fact, one observes a whole cascade of period doublings $2^{k-1} \rightarrow 2^k$ at parameter values $\mu_k$. The sequence $\mu_k$ approaches a critical value $\mu_\infty = 1.401155189\ldots$. At this point we formally observe an orbit of period $2^\infty$. This is called the *Feigenbaum attractor*. Its complicated fractal structure will be analysed by thermodynamic means in later chapters.

Fig. 1.1 Attractor of the logistic map depending on $\mu$.

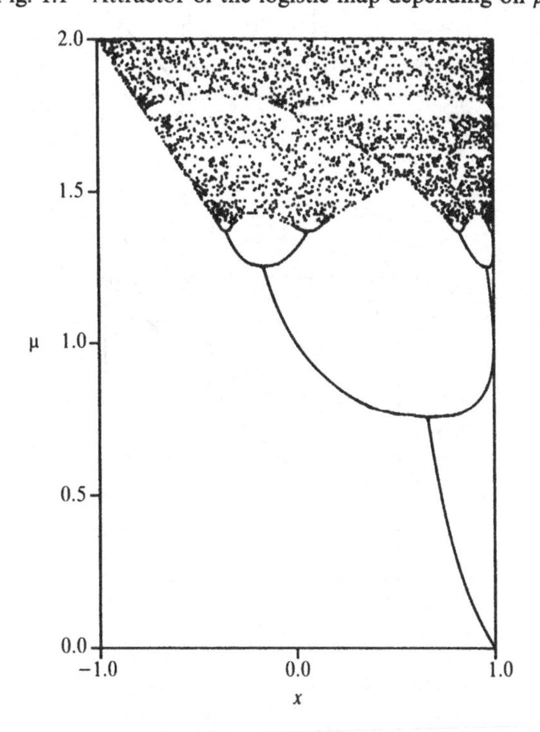

The Feigenbaum attractor marks the threshold between order and chaos. On further increasing $\mu$, trajectories behave chaotically – with the exception of certain 'windows', i.e., values of the parameter $\mu > \mu_\infty$ where again stable periodic behaviour is observed. For example, for $\mu = 1.75$ a stable period 3 is observed, which – as every stable orbit – again exhibits a period doubling scenario. The endpoint of the entire scenario is given by $\mu = 2$; for $\mu > 2$ generic orbits escape to infinity.

Feigenbaum (1978, 1979) and independently Grossmann and Thomae (1977) and Coullet and Tresser (1978) numerically discovered that certain ratios of bifurcation parameters approach a fixed value for $k \to \infty$. The

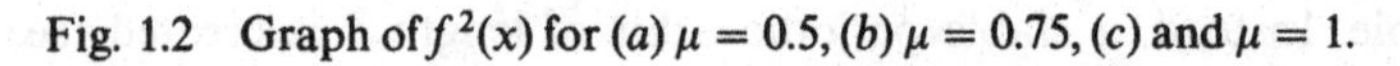

Fig. 1.2 Graph of $f^2(x)$ for (*a*) $\mu = 0.5$, (*b*) $\mu = 0.75$, (*c*) and $\mu = 1$.

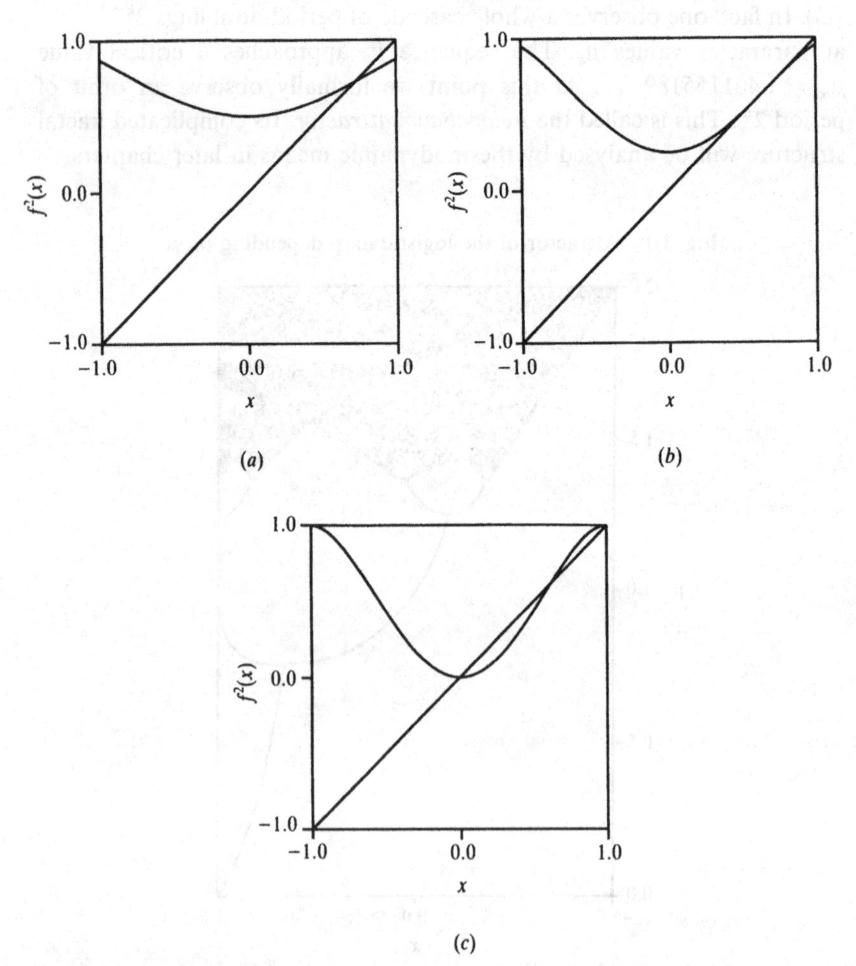

asymptotic ratio

$$\delta = \lim_{k\to\infty} \frac{\mu_k - \mu_{k-1}}{\mu_{k+1} - \mu_k} = 4.6692011\ldots \tag{1.3.4}$$

is called the *Feigenbaum constant*. Alternatively, one can also study the parameter values $\tilde{\mu}_k$ where $f$ has a superstable periodic orbit of length $2^k$. A periodic orbit of length $L$ of a one-dimensional map is called *superstable* if

$$f^{L\prime}(x_0) = 0. \tag{1.3.5}$$

Fig. 1.3 Graph of $f^4(x)$ for (*a*) $\mu = 1.18$, (*b*) $\mu = 1.25$, and (*c*) $\mu = 1.32$.

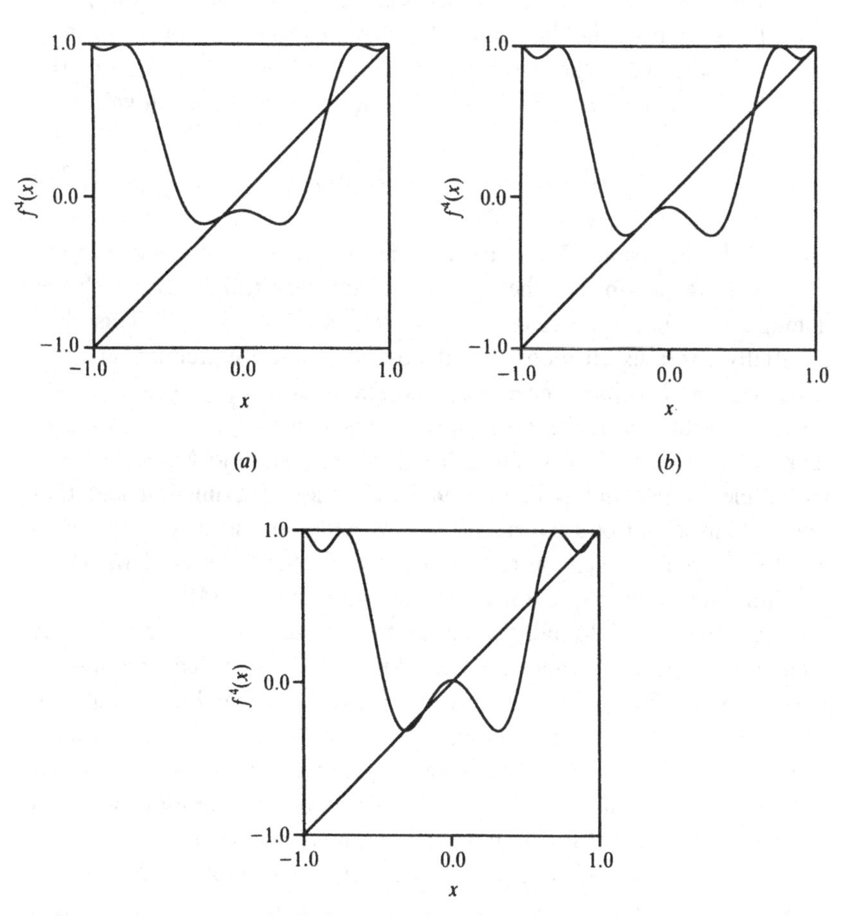

Here $x_0$ is an arbitrary orbit element. Due to the chain rule,

$$f^{L\prime}(x_0) = \prod_{n=0}^{L-1} f'(x_n). \tag{1.3.6}$$

Hence, for the logistic map a superstable orbit is a periodic orbit that contains the point 0, as $f'(x) = 0$ for $x = 0$ only. If we form the same ratios as in eq. (1.3.4) using the superstable parameter values $\tilde{\mu}_k$ instead of $\mu_k$, the same constant $\delta$ is approached in the limit $k \to \infty$:

$$\lim_{k\to\infty} \frac{\tilde{\mu}_k - \tilde{\mu}_{k-1}}{\tilde{\mu}_{k+1} - \tilde{\mu}_k} = \delta. \tag{1.3.7}$$

While the Feigenbaum constant $\delta$ describes scaling behaviour in the parameter space, there is also interesting scaling behaviour in the phase space. Let $d_k$ denote the diameters of the bifurcation forks in the vicinity of $x = 0$ at the superstable parameter values $\mu_k$. In the limit $k \to \infty$ the ratio of successive diameters $d_k$ and $d_{k+1}$ approaches a certain value:

$$|\alpha| = \lim_{k\to\infty} \frac{d_k}{d_{k+1}} = 2.50290787\ldots, \tag{1.3.8}$$

$\alpha = -|\alpha|$ is the second Feigenbaum constant. It is the merit of Feigenbaum to have shown that these constants are universal for entire classes of maps, i.e., they do not depend on details of the map $f$. This class essentially contains all maps $f$ that possess a single extremum at some value $x_{max}$ and behave near $x_{max}$ as $f(x) \sim |x - x_{max}|^z + \text{const}$, with $z = 2$. For other values of $z$ the above limits also exist but with different numerical values $\alpha(z)$, $\delta(z)$ (Vilela Mendes 1981; Hu and Mao 1982; van der Weele, Capel and Kluiving 1987). It is an experimental fact that Feigenbaum scenarios are observed in a variety of physical systems, which are described by much more complicated evolution laws than just a one-dimensional map (see, for example, Cvitanović (1984)).

Besides the period doubling scenario, the logistic map exhibits a variety of further interesting phenomena. One of them is the so called *intermittency* phenomenon. This kind of behaviour is closely related to a different mechanism leading from order to chaos, commonly called a *tangent bifurcation.* The logistic map exhibits a tangent bifurcation at the parameter value $\mu_T = \frac{7}{4}$. The phenomenon is easily understood if we plot the graph of three-times iterated function $f^3$ for parameter values $\mu > \mu_T$, $\mu = \mu_T$, and $\mu < \mu_T$ (see fig. 1.4). For $\mu > \mu_T$ there are three stable fixed points of $f^3$ near $x = 0, 1, 1 - \mu$, corresponding to a stable periodic orbit of length 3. At $\mu = \mu_T$ the graph of $f^3$ touches the diagonal. Formally, an unstable and a stable fixed point of $f^3$ collide. For $\mu < \mu_T$ the graph of $f^3$ no longer touches the diagonal. The map exhibits chaotic behaviour. However, it is chaotic behaviour of a special type: as soon as the trajectory reaches the vicinity of, e.g., $x = 0$, it stays almost constant in this vicinity for a large number of iterations of $f^3$. This can be easily understood from fig. 1.4(*c*): Successive iterations of $f^3$ correspond to a staircase line near $x = 0$. The almost constant behaviour near $x = 0$, often called the 'laminar phase' of intermittency, is interrupted by sudden chaotic outbursts. Here the iterate leaves the vicinity of $x = 0$ for a short time, until it enters a laminar phase again (see fig. 1.5). Intermittency is quite a general phenomenon in

Fig. 1.4 Graph of $f^3(x)$ for (*a*) $\mu = 1.76$, (*b*) $\mu = 1.75$, and (*c*) $\mu = 1.74$.

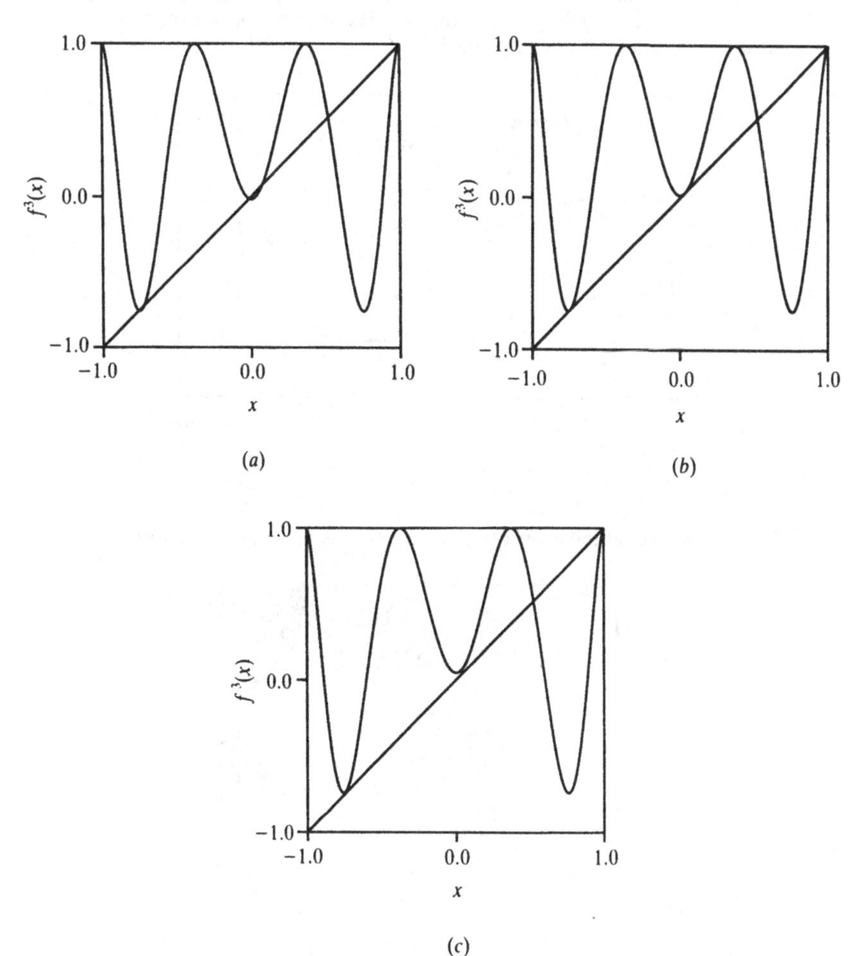

nonlinear dynamics. It is not restricted to the logistic map alone, and, in fact, it is observed in a variety of physical experiments (see, for example, Schuster (1984)).

Another interesting phenomenon is the *band splitting* phenomenon. It occurs if we approach the critical value $\mu_\infty$ of period doubling accumulation from above, i.e., from the chaotic regime. For the logistic map the first splitting takes place at the parameter value $\hat{\mu}_1 = 1.5436\ldots$. At this point the chaotic attractor splits into two chaotic bands (see fig. 1.6), i.e., on further decreasing $\mu$ there is an empty region between two separated parts of the attractor. The iterate alternates between both bands in a periodic way, however, inside each band the movement is chaotic. On further

Fig. 1.5 Intermittent behaviour of the logistic map for $\mu = 1.74999$. The iterates $x_{n+1} = f^3(x_n)$ of the three-times iterated logistic map are shown as a function of $n$. The initial value is $x_0 = 0.2$.

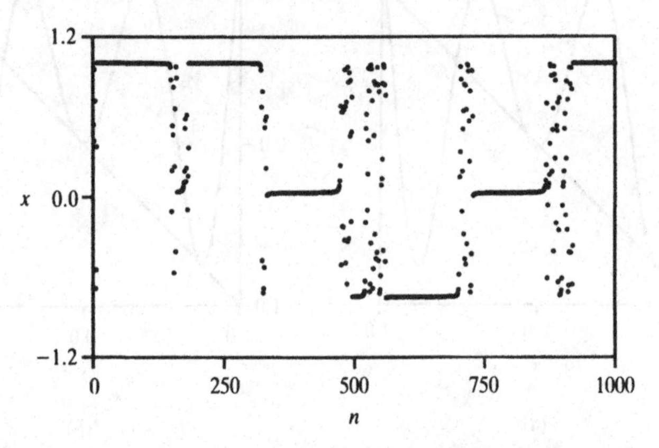


Fig. 1.6 Magnification of fig. 1.1 in the band splitting region.

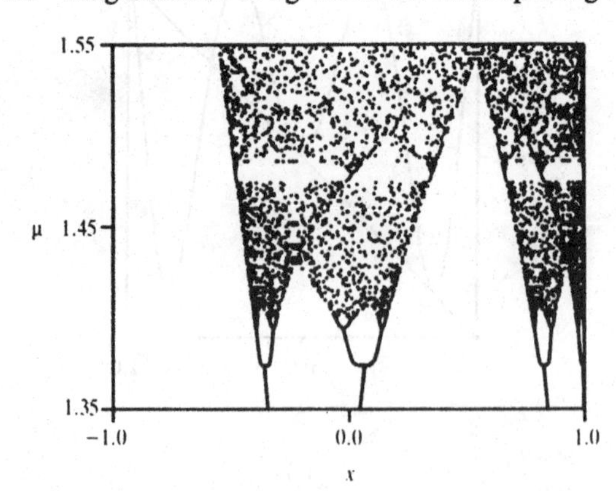

decreasing $\mu$, at $\hat{\mu}_2 = 1.43035\ldots$ the two bands split into four bands. Again there is a regular movement of the trajectory between the four bands, but inside each band the behaviour is chaotic. In fact, again a whole cascade of parameter values $\hat{\mu}_k$ exists where there is a splitting from $2^{k-1}$ to $2^k$ chaotic bands. The sequence of band splitting parameter values converges to the critical point $\mu_\infty$ of period doubling accumulation, and the ratio of band splitting parameter values again converges to the Feigenbaum constant

$$\lim_{k\to\infty} \frac{\hat{\mu}_k - \hat{\mu}_{k-1}}{\hat{\mu}_{k+1} - \hat{\mu}_k} = \delta. \tag{1.3.9}$$

The band splitting phenomenon has to do with the fact that at this point a certain iterate of the initial value $x_0 = 0$ (or, in general, of the point where the map has its extremum) falls onto an (unstable) fixed point $x^*$ of $f$. For example, for $\mu = \hat{\mu}_1$ we have for the third iterate of 0

$$f^3(0) = x^*. \tag{1.3.10}$$

Thus also

$$f^3(0) = f^4(0) = f^5(0) = \cdots = x^*, \tag{1.3.11}$$

whereas for $\mu < \hat{\mu}_1$

$$f^4(0) < f^3(0). \tag{1.3.12}$$

It turns out that the empty region between the two chaotic bands is just given by the interval $[f^4(0), f^3(0)]$. For the higher band splitting points $\hat{\mu}_k$ the analogous consideration applies if we replace the map $f$ by the $2^{k-1}$-times iterated map $f^{2^{k-1}}$. In general, parameter values for which the orbit of the initial value 0 falls onto an unstable fixed point are called *Misiurewicz points.*

## 1.4 Maps of Kaplan–Yorke type

Our next example is a class of two-dimensional dissipative maps, which we call maps of Kaplan–Yorke type (Kaplan and Yorke 1979; Kaplan, Mallet-Paret and Yorke 1984; Mayer and Roepstorff 1983; Beck 1990c). The maps are defined as

$$f: \begin{cases} x_{n+1} = T(x_n) \\ y_{n+1} = \lambda y_n + h(x_n). \end{cases} \tag{1.4.1}$$

Here $T$ is a one-dimensional map, $h$ is a smooth function, and $\lambda \in [0, 1]$

is a parameter. Kaplan and Yorke (1979) have introduced a special member of this class determined by

$$T(x) = 2x - \lfloor 2x \rfloor \tag{1.4.2}$$

$$h(x) = \cos \pi x. \tag{1.4.3}$$

Here $\lfloor 2x \rfloor$ denotes the integer part of the number $2x$. As a standard example, we choose for $T$ the logistic map with $\mu = 2$ and $h(x) = x$:

$$f: \begin{cases} x_{n+1} = 1 - 2x_n^2 \\ y_{n+1} = \lambda y_n + x_n. \end{cases} \tag{1.4.4}$$

Fig. 1.7 shows 2000 iterates of this map for $\lambda = 0.4$. We observe a very complicated pattern of points. The corresponding attractor is a *strange attractor* (for a precise mathematical definition of this concept see, for example, Eckmann and Ruelle (1985)). Typically, a strange attractor of a two-dimensional map has the following properties. For a generic point on the attractor there is one direction (the unstable direction) where the attractor consists of a 'connected line', whereas in the other direction (the stable direction) the line pattern is curled up in a complicated way; it has a *fractal structure* (see fig. 1.7). Moreover, the attractor is confined to a finite phase space. The motion of the iterates on the attractor is chaotic.

Fig. 1.7 Attractor of the map (1.4.4) for $\lambda = 0.4$.

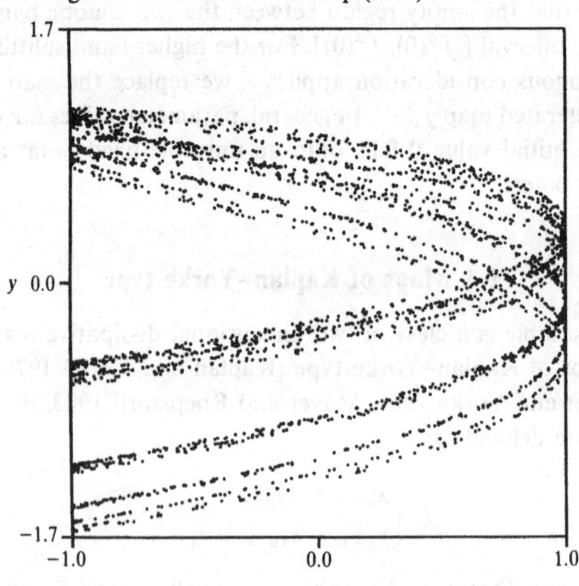

Looking at the figure, we see that there are some regions (near $x = \pm 1$) in the phase space where the relative frequency of the iterates is much greater than in other regions. Indeed, the density of the iterates of a dissipative system often possesses singularities, either at a finite number of points in the phase space, or even at an infinite number of points. This fact, together with the fractal structure of the attractor, yields the motivation to use thermodynamic tools.

A nice property of maps of Kaplan–Yorke type is that they have a direct physical meaning. This fact distinguishes them from other well known two-dimensional maps, such as the Hénon map (see next section). Consider a particle of mass 1 that moves in a liquid of viscosity $\gamma$. The velocity $v$ of the particle as a function of the time $t$ obeys

$$dv/dt = -\gamma v, \tag{1.4.5}$$

with the solution

$$v(t) = v(0)\exp(-\gamma t). \tag{1.4.6}$$

For a small particle, the motion is not sufficiently described by a mere friction term as done in eq. (1.4.5). The influence of the irregularly kicking liquid molecules on the particle will generate an irregular movement (a dynamical simulation of *Brownian motion*). To take this into account, in statistical mechanics an irregular *stochastic* force, the so called 'Langevin force' $L(t)$, is added to the right hand side of eq. (1.4.5)

$$dv/dt = -\gamma v + L(t). \tag{1.4.7}$$

In the Langevin theory (see, for example, van Kampen (1981), Nelson (1967)), the force $L(t)$ fulfils certain stochastic conditions, the most important being that the time average is zero and that the force is $\delta$-correlated. That means the autocorrelation function is

$$\langle L(t)L(t')\rangle = C\,\delta(t - t'), \tag{1.4.8}$$

where $\delta(t)$ denotes the Dirac delta function, and $\langle\cdots\rangle$ denotes the expectation value. In our case, let us assume that a more complicated kick force $L_\tau(t)$ is given in the following way. At discrete time points $n\tau$, with a fixed $\tau$ and integers $n = 0, 1, 2, \ldots$, the particle gets a kick of strength $x_n$:

$$L_\tau(t) = \sum_n x_n\,\delta(t - n\tau). \tag{1.4.9}$$

Moreover, we assume that the kick strength at time $(n+1)\tau$ is a deterministic function of the kick strength at time $n\tau$:

$$x_{n+1} = T(x_n). \tag{1.4.10}$$

By integration of eq. (1.4.7) from $\tau + \varepsilon$ to $\tau + 1 + \varepsilon$ (with $\varepsilon \to 0^+$) we obtain

$$v[(n+1)\tau + \varepsilon] = v(n\tau + \varepsilon)\exp(-\gamma\tau) + x_{n+1}. \tag{1.4.11}$$

In this equation the notation $v(n\tau + \varepsilon)$, $\varepsilon \to 0^+$, indicates that we consider the velocity $v$ at time $n\tau$ immediately after the kick. Defining $v(n\tau + \varepsilon) = y_n$, the equations of motion are equivalent to the following map of Kaplan–Yorke type:

$$x_{n+1} = T(x_n), \tag{1.4.12}$$

$$y_{n+1} = \lambda y_n + x_{n+1}, \tag{1.4.13}$$

where $\lambda = \exp(-\gamma\tau)$ is a fixed parameter.

If we slightly generalize and consider forces

$$L_\tau(t) = \sum_n h(x_{n-1})\,\delta(t - n\tau) \tag{1.4.14}$$

with some arbitrary function $h$, we end up with the general form (1.4.1) of the Kaplan–Yorke map. Hence such a map always has the above given physical interpretation. $h(x_{n-1})$ is the external momentum transfer at time $n\tau$ and $y_n$ is the velocity of the kicked particle at time $n\tau + \varepsilon$. For a generalization of the approach to more general nonlinear evolution equations rather than the linear equation (1.4.5), see Beck and Roepstorff (1987a) and Beck, Roepstorff and Schroer (1992).

## 1.5 The Hénon map

The Hénon map has played an important historical role, as it was the first published example of a two-dimensional mapping that possesses a strange attractor (Hénon 1976). Until Hénon wrote his article in 1976, strange attractors with a complicated fractal structure were commonly associated with three- or higher-dimensional continuous-time dynamical systems such as the Lorenz model (Lorenz 1963; Sparrow 1982) rather than with simple two-dimensional maps.

The Hénon map is given by

$$f: \begin{cases} x_{n+1} = 1 - a x_n^2 + y_n \\ y_{n+1} = b x_n. \end{cases} \tag{1.5.1}$$

$a$ and $b$ are parameters; a standard choice is $a = 1.4$, $b = 0.3$. For $b \to 0$ the Hénon map reduces to the logistic map, for finite $b$, however, it exhibits a much more complicated behaviour. The strange attractor obtained for $a = 1.4$ and $b = 0.3$ is shown in fig. 1.8. Qualitatively, we get a similar picture to that for the map (1.4.4) of Kaplan–Yorke type. However, there is an important difference between the two types of mapping. The Hénon map is an invertible mapping. Given $(x_{n+1}, y_{n+1})$, we can uniquely reconstruct $(x_n, y_n)$ via

$$f^{-1}: \begin{cases} x_n = b^{-1} y_{n+1} \\ y_n = x_{n+1} - 1 + a x_n^2 = x_{n+1} - 1 + a b^{-2} y_{n+1}^2. \end{cases} \tag{1.5.2}$$

This distinguishes the Hénon map from a map of Kaplan–Yorke type, where usually each tuple $(x_{n+1}, y_{n+1})$ has several preimages. For example, for the following map of Kaplan–Yorke type

$$\begin{cases} x_{n+1} = 1 - \mu x_n^2 \\ y_{n+1} = \lambda y_n + x_n \end{cases} \tag{1.5.3}$$

one always has the two preimages

$$\begin{cases} x_n = \pm[\mu^{-1}(1 - x_{n+1})]^{1/2} \\ y_n = \lambda^{-1}\{y_{n+1} \mp [\mu^{-1}(1 - x_{n+1})]^{1/2}\}. \end{cases} \tag{1.5.4}$$

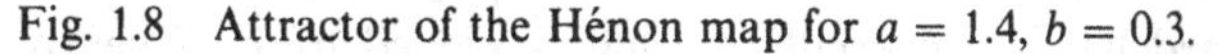
Fig. 1.8 Attractor of the Hénon map for $a = 1.4$, $b = 0.3$.

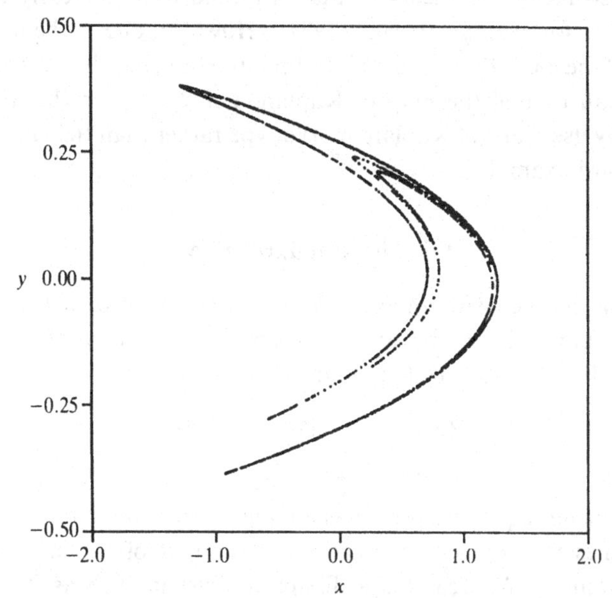

Mathematicians call a (uniquely) invertible differentiable map $f$ a *diffeomorphism.* The Hénon map is a diffeomorphism, the Kaplan–Yorke map is not.

Both mappings, the Hénon map given by (1.5.1) as well as the generalized Kaplan–Yorke map given by (1.5.3), exhibit period doubling scenarios for appropriate values of the parameters. In both cases, the Feigenbaum constants coincide with those of the one-dimensional logistic map (Collet and Eckmann 1980; Beck 1990c). In general, it turns out that many rigorous results can be proved much more easily for maps of Kaplan–Yorke type than for maps of Hénon type. This has to do with the fact that for maps of Kaplan–Yorke type the $x$-dynamics is 'autonomous', it is not influenced by the $y$-dynamics, whereas for Hénon type models it is.

For the Hénon map, a critical point of period doubling accumulations is observed for the parameter values $a = 1.058049\ldots$ and $b = 0.3$. If one changes the parameter $b$, one also has to change the parameter $a$ in order to stay at the accumulation point. The parameters interfere in a very complicated way. For the map (1.5.3) the behaviour is much simpler. It exhibits a period doubling scenario at the same parameter values $\mu$ as the logistic map, for arbitrary $|\lambda| < 1$. The parameters $\mu$ and $\lambda$ are decoupled. Whereas $\mu$ measures the nonlinearity, $\lambda$ is just responsible for the 'dissipation rate'.

The Hénon map also possesses a kind of physical interpretation in terms of a damped harmonic oscillator that is kicked by nonlinearly coupled impulses (see, for example, Heagy (1992)). However, due to the nonlinear coupling of the kicks this physical interpretation is somewhat less straightforward than that of the map of Kaplan–Yorke type. In this book we shall mainly use maps of Kaplan–Yorke type rather than the Hénon map as a standard example.

## 1.6 The standard map

We end this chapter with an example of a two-dimensional *Hamiltonian* map that exhibits chaotic behaviour for appropriate values of the control parameter. It is the so called standard map defined by

$$\phi_{n+1} = \phi_n + p_n - \kappa \sin \phi_n \tag{1.6.1}$$

$$p_{n+1} = \phi_{n+1} - \phi_n. \tag{1.6.2}$$

$\kappa$ is a parameter, $\phi_n$ is interpreted as an angle and taken modulo $2\pi$. One easily verifies that indeed the Jacobi determinant of this map has the constant value 1. Whereas for a dissipative system such as the logistic map or the Kaplan–Yorke map it is typical that there is just one attractor, the behaviour of a Hamiltonian map is totally different: each initial point $x_0$ (here $(\phi_0, p_0)$) generates a trajectory of its own. For some initial points the trajectory may behave in a chaotic way, for others it is regular.

Figure 1.9 shows a large number of iterates of the standard map for several initial values. We have chosen the parameter values $\kappa = 0.5, 0.95$, and 1.4. For small $\kappa$ the phase portrait appears to be quite regular and smooth. It reminds us of the phase portrait of an integrable Hamiltonian system (the reader should be aware of the fact that the dotted lines in fig. 1.9(*a*) would become closed curves if we invested a bit more computing time). For $\kappa = 0.95$ there is a complex pattern of chaotic, quasi-stochastic trajectories, but still there are also several closed lines (the so called 'invariant tori'). Each structure corresponds to just one initial point. We observe the typical phasc portrait of a nonintegrable Hamiltonian system. On further increasing $\kappa$, the stochastic behaviour becomes more and more dominant. None of the closed lines leading from the left to the right are present anymore. The mathematical theory for transition scenarios of this type is the KAM theory (see, for example, Lichtenberg and Liebermann (1983)). We shall not go into details here.

Fig. 1.9 Phase portrait of the standard map for (*a*) $\kappa = 0.5$.

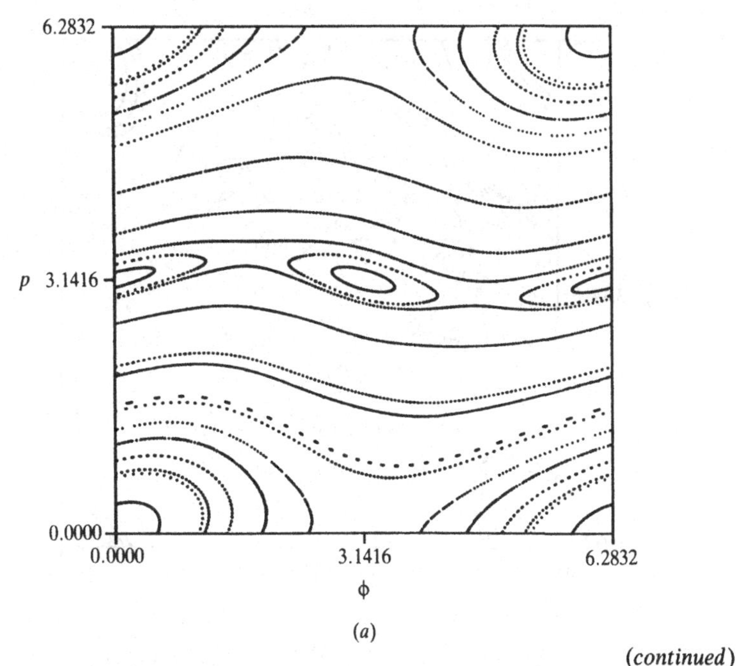


(*a*)

(*continued*)

Fig. 1.9 (*continued*) (*b*) $\kappa = 0.95$, and (*c*) $\kappa = 1.4$.

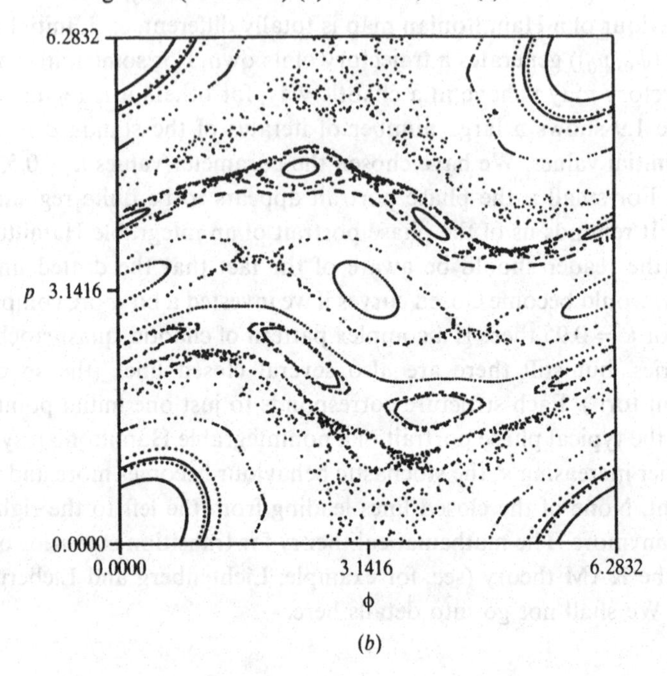


(*b*)

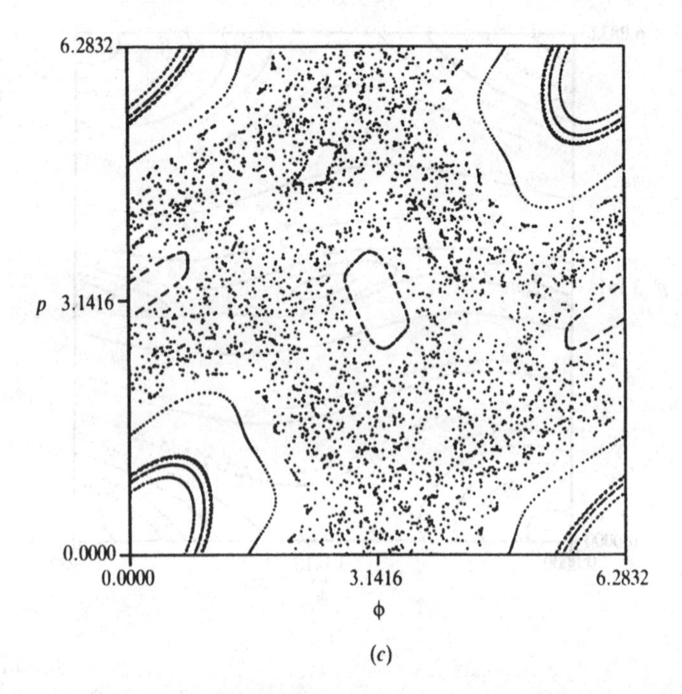


(*c*)

The standard map also has kind of a 'physical' interpretation: consider the following differential equation:

$$d^2\phi/dt^2 + \kappa \sin \phi \sum_{n=0}^{\infty} \delta(t - n\tau) = 0. \tag{1.6.3}$$

This equation is the equation of motion of a 'kicked rotator', i.e., of a plane pendulum under the influence of a force that is switched on at discrete time points $n\tau$ only. Let us integrate eq. (1.6.3). For $n\tau < t < (n+1)\tau$ we have $\ddot{\phi} = 0$. Hence $\dot{\phi}$ is a constant, which we call $p_{n+1}$ (a dot denotes the derivative with respect to time). It follows that

$$\phi(t) = (t - n\tau)p_{n+1} + \phi(n\tau). \tag{1.6.4}$$

Defining $\phi(n\tau) = \phi_n$, we obtain

$$\phi_{n+1} = \tau p_{n+1} + \phi_n. \tag{1.6.5}$$

To get a recurrence relation for $p_n$, we integrate eq. (1.6.3) from $n\tau - \varepsilon$ to $n\tau + \varepsilon$, where $\pm\varepsilon$ again indicates respectively that the time is taken immediately after and before the kick:

$$\int_{n\tau-\varepsilon}^{n\tau+\varepsilon} dt\, (d^2\phi/dt^2) + \kappa \sin \phi_n = 0 \qquad (\varepsilon \to 0) \tag{1.6.6}$$

As $\dot{\phi}(n\tau + \varepsilon) = p_{n+1}$ and $\dot{\phi}(n\tau - \varepsilon) = p_n$, we obtain

$$p_{n+1} - p_n + \kappa \sin \phi_n = 0. \tag{1.6.7}$$

Substituting eq. (1.6.7) into eq. (1.6.5) we end up with

$$\phi_{n+1} = \phi_n + (p_n - \kappa \sin \phi_n)\tau. \tag{1.6.8}$$

Hence, eq. (1.6.8) and eq. (1.6.5) are equivalent to the standard map for the choice $\tau = 1$. Thus we can interpret $\phi_n$ as the angle, and $p_n$ as the momentum of a kicked rotator.

We have mentioned the example of the standard map in order to demonstrate in which way the complex behaviour of a Hamiltonian system is different from that of a dissipative system. We refer a reader interested in more details to some standard literature on Hamiltonian systems (Percival and Richards 1982; Lichtenberg and Liebermann 1983; MacKay, Meiss and Percival 1984; Escande 1985). In the following chapters we will mainly deal with dissipative systems.

# 2
# Probability in the theory of chaotic systems

The iterates $x_n$ of a deterministic map are not random events but are uniquely determined by the initial value $x_0$. In spite of that, as already mentioned in the introduction, it is useful to consider probability distributions of the iterates. In this chapter, we shall be concerned with the use of probability in the theory of chaotic systems. We shall introduce and discuss important notions such as invariant density, ergodicity, and topological conjugation.

## 2.1 Relative frequencies and probability

Before we begin to analyse chaotic dynamical systems by means of probabilistic method, let us quite generally consider arbitrary random events and their probabilities. We would like to give an *empirical definition of probabilities.* Let us generally start with an experiment with $R$ different possible results $i$. The set of the $R$ results is called a 'sample set' or a 'complete disjunction' of random events $i$. The second one of these names expresses the defining properties: first, that the experiment always yields one event $i$ of the set; secondly, the events are disjoint in the sense that they exclude one another; only one result is possible. Now we consider a sequence of $n$ such experiments, always under the same conditions and statistically independent. If in this sequence the result $i$ occurs $n_i$ times, we call the ratio $n_i/n$ the *relative frequency* $H_i$ of the event $i$. If in the limit of very large $n$ this frequency takes a fixed value

$$\lim_{n\to\infty} H_i = p_i, \tag{2.1.1}$$

we call this limiting value the *probability* of observing the event $i$.

Now let us proceed to chaotic systems. Suppose we have a dissipative chaotic $d$-dimensional mapping, for example, the map of Kaplan–Yorke type described in section 1.4. For a fixed initial value $x_0$ the iterates of a deterministic map are not random events. Nevertheless, for a chaotic system they are unpredictable in practice. The long-time behaviour of a chaotic system can be systematically described by statistical means only. For this purpose we divide the $d$-dimensional phase space $X$ into $d$-dimensional cells. The 'cells' are disjoint subsets of the phase space $X$ which together cover it completely. We label the cells by a subscript $i = 1, 2, \ldots, R$, where $R$ is the total number of cells. After a certain number of iterations of a certain initial value $x_0$, the number of iterates found in cell $i$ may be called $n_i$. The total number of iterates is

$$n = \sum_{i=1}^{R} n_i, \tag{2.1.2}$$

where the summation runs over all cells. We suppose that if the iteration time $n$ is very long, the relative frequencies

$$H_i = n_i/n \tag{2.1.3}$$

become practically independent of $n$. Notice that this assumption is not identical with the corresponding assumption of the existence of the limit in eq. (2.1.1) made by the empirical definition of probabilities. There the $n$ experiments were statistically independent events. This, however, is not true for the $n$ iterates. Our basic assumption is that the limit of (2.1.3) exists in the sense that we shall only consider dynamical systems that fulfil this condition. For sufficiently large $n$, we call the relative frequency $H_i$ simply the probability $p_i$ of finding an iterate of the trajectory in cell $i$. The experiment, mentioned in the empirical definition, is the computer calculation of a trajectory. A result or possible outcome of the experiment is to find the iterate in the cell $i$. In general the probabilities may depend on the initial value $x_0$. This dependence does not occur if the map is *ergodic*. We shall come back to this in the next section.

The introduction of probabilities leads us to a comparison with *statistics in thermodynamics* where thermodynamic states of a system are described by probability distributions over 'microstates' $i$, that means, by the entire set of all $p_i$. As a probability distribution can always be approximated by relative frequencies of an ensemble of realizations, a probability distribution is often called an 'ensemble'. The microstates $i$ are either dynamical states of the system in the phase space of classical mechanics or pure states in the quantum mechanical Hilbert space. In thermodynamics only macroscopic quantities are observed which characterize global features of the probability distributions. The central problem there is to find adequate probability distributions which express the expectation about the question of which microstate $i$ the system would be found in if the corresponding observation were possible. This observation is, of course, not possible in a realistic macroscopic system. Nevertheless, such an observation is not only impossible but, moreover, not interesting. What are interesting in thermodynamics are macroscopically observable quantities and a prediction about the macroscopic global behaviour, based on knowledge of an earlier macroscopic observation and of the microdynamical laws. The probability distribution is a theoretical tool to find laws for such predictions.

Also in the *theory of dynamical systems* the interest is not primarily directed to the question in which 'microstate', that is to say, in which cell $i$, the system actually is. The interesting properties are not immediately given by these microstates. Rather, the central aim is to find adequate global features to characterize the behaviour of the system. The situation in chaos theory is different from thermodynamics in so far as the probability distributions are known in the form of relative frequencies in a computer experiment. Thermodynamic probabilities, however, are an expression of an expectation. Hence, a method to find these probabilities is a prognostic one, and serves to find predictions about the future behaviour of a system based on preceding observations. In a numerical experiment we go the other way round. The relative frequencies – say the probabilities – are given experimental data which have to be analysed afterwards. Notwithstanding these differences, it is not surprising that the question of finding adequate characteristic features by statistical methods leads us to formal analogies in both theories.

## 2.2 Invariant densities

In section 2.1 we were dealing with relative frequencies obtained by iterating a map for a single initial value $x_0$. Let us now do something different. We consider an ensemble of different initial values $x_0$. The relative frequency of initial values in a certain subset $A$ of the phase space can be interpreted as the probability $\mu_0(A)$ of having an initial value $x_0$ in $A$. The quantity $\mu_0$ is called a *probability measure* of $A$. We may write

$$\mu_0(A) = \int_A dx\, \rho_0(x). \qquad (2.2.1)$$

The function $\rho_0(x)$ defined on the entire phase space is called a *probability density*. Notice the difference between a probability *measure* and its probability *density*: a measure is a function of a set, a density is a function of the coordinates $x$. The function $\rho_0(x)$ may sometimes be quite singular and contain distribution-like ingredients such as Dirac's delta function $\delta(x)$. For this reason one often uses instead of eq. (2.2.1) the equivalent, mathematically well defined notation

$$\mu_0(A) = \int_A d\mu_0(x). \tag{2.2.2}$$

*Remark*: This notation is traditionally used if the integral is interpreted as a Stieltjes integral, whereas the notation (2.2.1) corresponds to a 'usual' integral. The two types of integrals are equivalent if we allow $\rho_0(x)$ to be a 'distribution' in the sense of Laurent Schwartz (not to be confused with our use of the word 'distribution' as an abbreviation for 'probability distribution'). The Schwartz distributions are generalizations of the usual functions in the sense that they include, for instancc, the Dirac delta function $\delta(x)$ or 'derivatives' of $\delta(x)$.

Given a map $f$, we are interested in the time evolution of the ensemble of trajectories corresponding to the ensemble of different initial values $x_0$. Let us denote by $\mu_n$ the probability distribution of the iterates after $n$ iterations of the map $f$. In other words, $\mu_n(A)$ is the probability of finding an iterate $x_n$ in the subset $A$ of the phase space. The corresponding density is denoted by $\rho_n$:

$$\mu_n(A) = \int_A dx\, \rho_n(x). \tag{2.2.3}$$

Because of conservation of probability, the following condition is trivially fulfilled for arbitrary subsets $A$:

$$\mu_{n+1}(A) = \mu_n(f^{-1}(A)). \tag{2.2.4}$$

Here $f^{-1}(A)$ is the *preimage* of $A$, i.e., the set of all points that are mapped onto $A$ by one iteration step. Eq. (2.2.4) simply means that the relative frequency of iterates $x_{n+1}$ in the subset $A$ must be equal to the relative frequency of iterates $x_n$ in the subset $f^{-1}(A)$.

We are especially interested in invariant probability distributions, i.e., distributions that do not change under the action of $f$. This means that we are looking for special probability measures that satisfy the additional condition

$$\mu_{n+1}(A) = \mu_n(A). \tag{2.2.5}$$

A probability measure $\mu_n$ satisfying both eq. (2.2.4) and eq. (2.2.5) is called an *invariant* measure, and the corresponding density function $\rho(x)$ an *invariant density*. It describes an ensemble of points whose density does not change in time, a stationary state. It should be clear that individual points certainly evolve in time under the action of $f$. What is constant is just their probability distribution.

According to eq. (2.2.5), the invariant measure no longer depends on time. Therefore, we can suppress the index $n$ and denote $\mu_n$ by $\mu$. Putting this into eq. (2.2.4) we obtain the result that an invariant measure satisfies

$$\mu(A) = \mu(f^{-1}(A)) \tag{2.2.6}$$

for arbitrary subsets $A$ of the phase space. The corresponding invariant density $\rho$ satisfies

$$\int_A \mathrm{d}x\, \rho(x) = \int_{f^{-1}(A)} \mathrm{d}x\, \rho(x). \tag{2.2.7}$$

We may introduce expectation values of an arbitrary integrable test function $Q(x)$ with respect to the invariant density $\rho$:

$$\langle Q \rangle = \int_X \mathrm{d}x\, \rho(x) Q(x). \tag{2.2.8}$$

$\langle Q \rangle$ is called the *ensemble average* of the observable $Q$. The integration runs over the entire phase space $X$. The mean value $\langle Q \rangle$ is an average with respect to space, each element of the phase space is weighted with a certain probability. Equivalently, in agreement with the notation of eq. (2.2.1) and eq. (2.2.2) we can write

$$\langle Q \rangle = \int_X \mathrm{d}\mu(x)\, Q(x). \tag{2.2.9}$$

As a consequence of the supposed invariance property of the probability distribution, we may also write

$$\langle Q \rangle = \int_X \mathrm{d}\mu(x)\, Q(x) = \int_X \mathrm{d}\mu(f^{-1}(x))\, Q(x) = \int_X \mathrm{d}\mu(x)\, Q(f(x)). \tag{2.2.10}$$

In the last step we substituted $f(x)$ for $x$ in the integral. We notice that the invariance property of the measure is equivalent to the fact that the ensemble average of an arbitrary observable $Q(x)$ is equal to the ensemble average of $Q(f(x))$. In other words, expectation values of observables in the invariant ensemble are invariant under the action of $f$.

The *time average* of the observable $Q$ with respect to a certain trajectory is different from the ensemble average and is defined by

$$\bar{Q} = \lim_{N \to \infty} \frac{1}{N} \sum_{n=0}^{N-1} Q(x_n). \tag{2.2.11}$$

This time average is an average for a single initial value $x_0$.

**Ergodicity** A map together with an invariant measure $\mu$ is called 'ergodic' if for any integrable test function $Q(x)$ the time average is equal to the ensemble average: for arbitrary $x_0$ up to a set of $\mu$-measure 0, we have the equality

$$\bar{Q} = \langle Q \rangle. \tag{2.2.12}$$

The ergodicity also means that the time average does not depend on $x_0$ (up to some exceptional initial values of $\mu$-measure 0).

> *Remark*: Mathematicians usually like to give a more abstract, but equivalent definition of ergodicity, namely the property that every $f$-invariant set has either $\mu$-measure 1 or 0. Then eq. (2.2.12) is a consequence of a theorem, the famous Birkhoff ergodic theorem (Cornfeld, Fomin, and Sinai 1982).

In general, it is very difficult to prove ergodicity. So far only a few examples of maps have been shown to be ergodic. The logistic map is known to be ergodic for certain values of the parameter $\mu$ (for example, $\mu = 2$ (Ulam and von Neumann 1947), and more generally for the Misiurewicz points (Collet and Eckmann 1980) (see section 1.3)). This is also true for the corresponding map of Kaplan–Yorke type with $\lambda < 1$ (Beck 1990c).

An ergodic map has the remarkable property that we can obtain an invariant measure of the map just by iterating a single initial value and producing a *histogram*, i.e., counting the relative frequency of the iterates in a certain subset $A$ of the phase space. This is easily seen by choosing for $Q(x)$ the so called *characteristic function* $\chi_A(x)$ of the subset $A$. The function $\chi_A(x)$ is defined to be 1 for all points $x$ in $A$, and it is zero otherwise. Eqs. (2.2.9) and (2.2.11) yield

$$\lim_{N \to \infty} \frac{1}{N} \sum_{n=0}^{N-1} \chi_A(x_n) = \int_X \mathrm{d}x\, \rho(x) \chi_A(x) = \int_A \mathrm{d}x\, \rho(x) = \mu(A). \tag{2.2.13}$$

One remark is in order. Usually there may exist several invariant measures for an ergodic map, but only one is really important, in the sense that if we iterate a randomly chosen initial point, the iterates will be distributed according to this measure 'almost sure'. This measure is called the *natural invariant measure.* It is the measure of physical interest, whereas other invariant measures do not have a 'physical meaning'. As an example let us assume that $f(x)$ has an unstable fixed point $x^*$. Then the Dirac delta function $\delta(x - x^*)$ is formally a normalized invariant density, but it is relevant for very special initial values $x_0$ only, namely the preimages of $x^*$, and thus unimportant.

The density function $\rho(x)$ corresponding to the natural invariant measure is called the *natural invariant density.* In the physical literature this density is sometimes introduced as the following limit

$$\rho(x) = \lim_{N \to \infty} \frac{1}{N} \sum_{n=0}^{N-1} \delta(x - x_n). \tag{2.2.14}$$

Here $\delta(x)$ denotes the $d$-dimensional Dirac delta function. It should, however, be clear that this limit has to be regarded as a limit in the weak sense: eq. (2.2.14) can, in general, be used if an integral over $x$ is taken afterwards. A possible way to give a meaning to the notation of eq. (2.2.14) is just eq. (2.2.13).

Only in very few cases can the natural invariant density be determined analytically. Let us consider one of these analytically solvable examples in the following. Further examples will be discussed in sections 2.3 and 17.2.

**The tent map** A simple example is the one-dimensional map

$$f(x) = \begin{cases} 2x & \text{for } x \leqslant \frac{1}{2} \\ 2(1-x) & \text{for } x > \frac{1}{2}. \end{cases} \tag{2.2.15}$$

The phase space is the unit interval $X = [0, 1]$. The map is called the 'symmetric triangular map' or simply the 'tent map'. It is obvious from fig. 2.1 that $f$ conserves the total length of arbitrary intervals. The lengths $l_0, l_1$ of the preimages of an arbitrary interval of length $l_2$ satisfy

$$l_0 + l_1 = l_2. \tag{2.2.16}$$

Hence, the constant probability distribution $\rho(x) = 1$ is an invariant probability density (in fact, the natural invariant density). It should be clear that, in general, for other maps, the invariant density can be quite a complicated function. In general, it is *not* the *uniform distribution,* where all events have equal probability.

**Mixing** An even stronger property than ergodicity is the *mixing property*. A map $f$ is called 'mixing' if an arbitrary smooth initial probability density $\rho_0(x)$ converges to the natural invariant density $\rho(x)$ under successive iterations

$$\lim_{n\to\infty} \rho_n(x) = \rho(x) \qquad \text{for all smooth } \rho_0(x). \tag{2.2.17}$$

Mixing implies ergodicity. The reverse, however, is not true.

An equivalent way to define the mixing property is possible via *correlation functions*. Suppose we have an ergodic map $f$. We may then choose two arbitrary integrable test functions $\phi_1(x)$, $\phi_2(x)$ and define the (generalized) correlation function of the map $f$ by

$$C(\phi_1, \phi_2; n) = \lim_{J\to\infty} \frac{1}{J} \sum_{j=0}^{J-1} \phi_1(x_{j+n})\phi_2(x_j) - \langle\phi_1\rangle\langle\phi_2\rangle. \tag{2.2.18}$$

Here

$$\langle\phi_i\rangle = \lim_{J\to\infty} \frac{1}{J} \sum_{j=0}^{J-1} \phi_i(x_j) = \int_X d\mu(x)\phi_i(x) \qquad (i = 1, 2) \tag{2.2.19}$$

denotes the expectation value of the observable $\phi_i$. The map $f$ is mixing if for arbitrary $\phi_1$, $\phi_2$

$$\lim_{n\to\infty} C(\phi_1, \phi_2; n) = 0, \tag{2.2.20}$$

i.e., the (generalized) correlation function asymptotically decays to zero. In other words, mixing means asymptotic statistical independence.

Fig. 2.1 The conservation of interval lengths by the tent map: $l_0 + l_1 = l_2$.

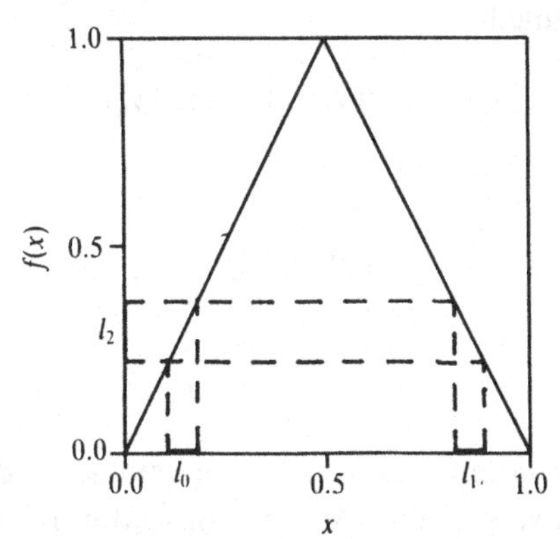

Examples of maps that are both ergodic and mixing are the tent map, the logistic map for $\mu = 2$, as well as the corresponding map of Kaplan–Yorke type with $|\lambda| < 1$ (see, for example, Beck (1990c) for a proof).

*Remark*: For the mixing property it is essential that eq. (2.2.20) is fulfilled for *arbitrary* test functions $\phi_1, \phi_2$ that are integrable with respect to the measure $\mu$. It is not sufficient to show that eq. (2.2.20) is valid for some test functions. A popular but, however, too restricted, choice is correlation functions based on the test functions $\phi_1(x) = \phi_2(x) = x$. The asymptotic decay of this special correlation function is a necessary, but not sufficient, criterion for the mixing property.

## 2.3 Topological conjugation

A map can be given in different coordinates. Sometimes it is useful to change coordinates such that the transformed map is simpler or has other advantages. Generally such a transformation to new coordinates is called a *topological conjugation*. It connects a map with an equivalent one. Two maps $f(x)$ and $g(y)$ are called topologically conjugated if they can be transformed into each other by means of a so called conjugating function $\phi$. It just yields the coordinates $x$ as a function of $y$

$$x = \phi(y). \tag{2.3.1}$$

Let us assume that a unique inverse $\phi^{-1}(x)$ exists. If the map

$$x_{n+1} = f(x_n) \tag{2.3.2}$$

is transformed into the topologically conjugated map

$$y_{n+1} = g(y_n), \tag{2.3.3}$$

the following is fulfilled:

$$x_{n+1} = \phi(y_{n+1}) = f(\phi(y_n)). \tag{2.3.4}$$

This means

$$\phi(g(y)) = f(\phi(y)) \tag{2.3.5}$$

or

$$g = \phi^{-1} \circ f \circ \phi, \tag{2.3.6}$$

where $\circ$ denotes the composition of two functions: $f \circ \phi(y) = f(\phi(y))$.

Let us consider a very simple example. In section 1.3 we introduced the logistic map

$$x_{n+1} = 1 - \mu x_n^2 \qquad \mu \in [0, 2]. \tag{2.3.7}$$

Many authors, however, do not call the map (2.3.7) a 'logistic map', but use the following map

$$y_{n+1} = r y_n (1 - y_n). \tag{2.3.8}$$

Here the control parameter $r$ takes values in $[0, 4]$. Eq. (2.3.8) apparently differs from eq. (2.3.7). The difference, however, is just produced by a different choice of coordinates.

Let us determine the conjugating function. We have

$$f(x) = 1 - \mu x^2 \tag{2.3.9}$$

and

$$g(y) = r y (1 - y). \tag{2.3.10}$$

Putting the ansatz

$$\phi(y) = a y + b \tag{2.3.11}$$

into eq. (2.3.5) and comparing equal powers of $y$, we obtain

$$\phi(y) = \left( y - \frac{1}{2} \right) \frac{1}{\mu} [(1 + 4\mu)^{1/2} + 1] \tag{2.3.12}$$

and

$$r = (1 + 4\mu)^{1/2} + 1. \tag{2.3.13}$$

We notice that for this simple example the conjugating function $\phi$ is linear in $y$. In the following, we shall consider an important and nontrivial example, where $\phi$ is a nonlinear function.

**The Ulam map** This is the logistic map for the special value $\mu = 2$. The Ulam map has several simplifying properties, which are not present for the general logistic map. For instance, it is topologically conjugated to the tent map by means of a smooth function $\phi$. This allows us to treat it analytically. In particular, it allows us to determine the natural invariant density $\rho(x)$. The transformation

$$x = -\cos \pi y = \phi(y), \tag{2.3.14}$$

which is unique in both directions, provided $y \in [0, 1]$, transforms the Ulam map

$$f(x) = 1 - 2x^2 \tag{2.3.15}$$

into a topologically conjugated map $g(y)$ that satisfies eq. (2.3.5). We obtain

$$-\cos \pi g(y) = 1 - 2\cos^2 \pi y = -\cos 2\pi y. \tag{2.3.16}$$

This equation is indeed fulfilled by the tent map $g(y)$ given by eq. (2.2.15).

Generally, the natural invariant density $\rho_x(x)$ of a map $f(x)$ is connected with the corresponding density $\rho_y(y)$ of a topologically conjugated map $g(y)$ by

$$\rho_y(y) = \rho_x(x)|\det D\phi(y)| \tag{2.3.17}$$

$(x = \phi(y))$ provided $\phi$ is uniquely invertible. Here $\det D\phi(y)$ denotes the

Fig. 2.2 Histogram of the iterates of the logistic map for (*a*) $\mu = 2$, (*b*) $\mu = 1.7495$, (*c*) $\mu = 1.7$, (*d*) $\mu = 1.5437$. In case (*b*) there is intermittent behaviour, case (*d*) corresponds to a Misiurewicz point.

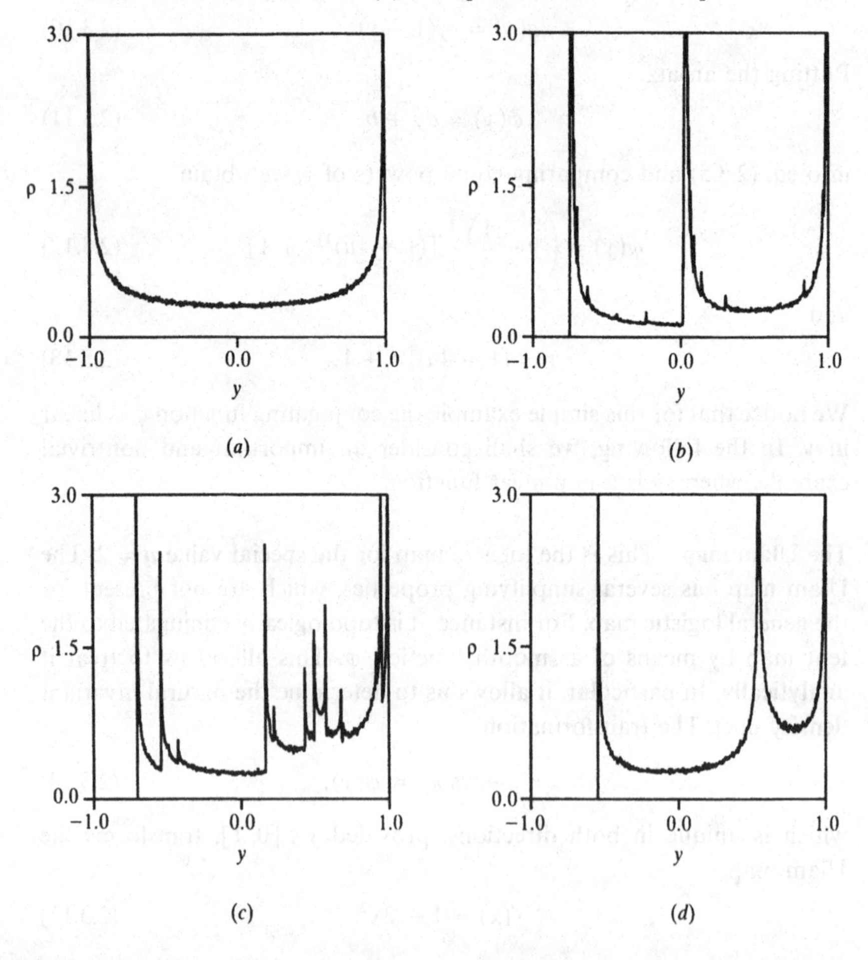

determinant of the Jacobi matrix of $\phi(y)$. In the one-dimensional case $\det D\phi(y)$ is simply $d\phi/dy$. Eq. (2.3.17) just expresses the conservation of probability

$$\rho_y(y)\,dy = \rho_x(x)\,dx. \tag{2.3.18}$$

For the tent map $\rho_y(y) = 1$. Differentiating

$$y = \frac{1}{\pi} \arccos(-x), \tag{2.3.19}$$

we obtain for the natural invariant density of the Ulam map the analytic expression

$$\rho_x(x) = \frac{1}{\pi(1 - x^2)^{1/2}}. \tag{2.3.20}$$

This is indeed remarkable, as no analytic expression for $\rho(x)$ is known for other values of $\mu$ than 2.

Fig. 2.2(*a*) shows a histogram of 50 000 iterates of the special logistic map with $\mu = 2$ for a single initial value $x_0$ obtained in a computer experiment. The distribution is in good agreement with the theoretical prediction (2.3.20). Histograms for other values of $\mu$ are shown in fig. 2.2(*b*)–(*d*). Here no analytic expression for $\rho_x(x)$ is known.

# 3
# Symbolic dynamics

To apply a statistical description to mappings, we need a partition of the phase space into subsets. Different methods will be discussed in this chapter. Which one is the best is dependent on the map and the problem under consideration. Given an appropriate partition, we can analyse chaotic behaviour in terms of symbol sequences generated by the trajectories. The symbol sequences describe the time evolution of the trajectories in a coarse-grained way.

## 3.1 Partitions of the phase space

The easiest way to partition a $d$-dimensional phase space $X$ is to choose $d$-dimensional cubes of equal size. It is common to call these cubes *boxes.* The length $\varepsilon$ of the side of the cube is called the *size* of the box. The entire partition is called a *grid.* We label the boxes by an index $i$ that runs from 1 to $R$, where $R$ is the total number of the boxes. The boxes do not overlap and they cover the entire phase space $X$.

Sometimes, however, it is more advantageous to partition the phase space into cells $A_i$ of different sizes and maybe of different shapes. Again each cell is labelled by an index $i$. The cells are 'disjoint'; i.e., they do not have a point in common. We may write this in the form

$$A_i \cap A_j = \varnothing, \tag{3.1.1}$$

that means the intersection is the empty set $\varnothing$. The cells cover the entire phase space $X$. This can be written in the form

$$\bigcup_{i=1}^{R} A_i = X, \tag{3.1.2}$$

that means the union of all cells is $X$. Generally, we shall call any partition of the phase space into subsets with these properties a partition into *cells*. If the cells are cubes of equal size, then we shall call them 'boxes'.

Let us now iterate a certain initial value $x_0$ with a map $f$. The point $x_0$ will be in some cell. We call its index $i_0$. The first iterate $x_1$ is in cell $i_1$, the second iterate $x_2$ in cell $i_2$, and so on. By this method we attribute to each initial value $x_0$ a symbol sequence

$$x_0 \to i_0, i_1, i_2, \ldots, i_n, \ldots. \tag{3.1.3}$$

This sequence of the symbols $i_n$ is called a 'symbolic dynamical sequence', or simply a 'symbol sequence'. It describes the trajectory in a coarse-grained way. The mapping (3.1.3) from the phase space to the symbol space is called a 'symbolic dynamics'. It attributes to each value $x_0$ a symbol sequence. As the size of the starting cell $i_0$ is finite and as it contains many initial values $x_0$, a given symbol sequence of finite length $N$ can be associated with many different sequences $x_0, x_1, \ldots, x_N$ of iterates. On the other hand, not all symbol sequences may be allowed in general. This depends on the map considered.

## 3.2 The binary shift map

As a preparation of a more general concept, we shall present an example of a map for which the partition of the phase space into two cells already yields quite a complete description. The map is given by

$$f(x) = \begin{cases} 2x & \text{for } x \in [0, \frac{1}{2}] \\ 2x - 1 & \text{for } x \in [\frac{1}{2}, 1]. \end{cases} \tag{3.2.1}$$

$f$ is called the 'binary shift map', it has some similarity with the tent map (we shall discuss this relationship in more detail at the end of this section). The phase space is the unit interval $X = [0, 1]$ with endpoints identified. A useful partition consists of the two intervals $A_1 = [0, \frac{1}{2})$ and $A_2 = [\frac{1}{2}, 1)$, where the square brackets denote a closed interval and the parentheses an open interval. We may call $A_1$ the 'left' and $A_2$ the 'right' interval. Let us choose an arbitrary initial value $x_0 \in [0, 1)$ and iterate it with the map $x_{n+1} = f(x_n)$. Alternatively we may write

$$f(x) = 2x - \lfloor 2x \rfloor \tag{3.2.2}$$

where the symbol $\lfloor 2x \rfloor$ denotes the integer part of $2x$. This means, $f(x)$ is obtained by writing $2x$ as a decimal number and deleting the part which stands in front of the decimal point. Instead of the decimal representation it is more useful to introduce the dual representation of the real number

$x_0$ by writing

$$x_0 = \sum_{k=1}^{\infty} i_k 2^{-k}. \tag{3.2.3}$$

The binary digits $i_k$ take the values 0 or 1. The iterates can be written as

$$x_n = \sum_{k=1}^{\infty} i_{k+n} 2^{-k}, \tag{3.2.4}$$

i.e., at each iteration step the first digit is 'thrown away' and the following digits are shifted to the left. $f$ acts as a shift in the binary representation. In general such an operation is called a 'Bernoulli shift'. The symbol sequence $i_0, i_1, i_2, \ldots$ provides us with a coarse-grained description of the trajectory: for $i_n = 0$ we may write the symbol $L$. It means that $x_n$ is 'left' (i.e., in $A_1$); the symbol $R$ or $i_n = 1$ means that $x_n$ is 'right' (i.e., in $A_2$). The binary shift map is distinguished by a very special property: notwithstanding that the two cells do not change their size during the sequence of iteration steps, the following holds: by the sequence $i_0, i_1, \ldots, i_N$ of 0 and 1, corresponding to 'left' and 'right', the initial value $x_0$ is determined more and more precisely for increasing $N$. From an infinite string $i_0, i_1, \ldots$ of symbols, we can uniquely determine the initial value $x_0$. For this reason the partition of the phase space into the two intervals 'left' and 'right' $\{A_1, A_2\}$ is called a 'generating partition'. We shall discuss this concept in more generality in the next section.

The binary shift map (3.2.1) is topologically conjugated to the tent map

$$g(y) = 1 - 2|y - \tfrac{1}{2}| \tag{3.2.5}$$

introduced at the end of section 2.2. The conjugating function $\phi$, however, is by no means trivial. Rather, it is quite a complicated fractal function. Without proof we mention that $\phi$ is given as follows:

$$y = \phi(x) = \sum_{k=1}^{\infty} b_k 2^{-k}. \tag{3.2.6}$$

Here the $b_k$ depend on $x$ and take on the values 0 or 1; they are related to the dual expansion of the number $x$ in the following way. If

$$x = \sum_{k=1}^{\infty} a_k 2^{-k} \qquad (a_k = 0 \text{ or } 1) \tag{3.2.7}$$

then the $b_k$ are obtained by the following recurrence relation:

$$b_0 = 0, \qquad b_{k+1} = |b_k - a_{k+1}| \qquad k = 0, 1, 2, 3, \ldots \tag{3.2.8}$$

Fig. 3.1 shows a graph of the function $\phi(x)$. It is a bijective function: each value $y$ corresponds uniquely to a value $x$ and vice versa. We obtain quite a complicated self-similar set of points. Although the conjugating function $\phi$ is neither differentiable nor continuous, the binary shift map and the tent map have many important properties in common. For both maps, the absolute value of the derivative has the constant value 2, from which it follows that the uniform distribution is the natural invariant density (see end of section 2.2). Moreover, for both maps all possible combinations of symbols $R$ and $L$ can occur as symbol sequences, there are no restrictions on the symbolic dynamics. Finally, from an infinite string of symbols one can uniquely determine the initial value $x_0$. For the binary shift map, this is easily done by writing down the dual expansion, i.e., by using eq. (3.2.3). For the tent map, the procedure to find $x_0$ is slightly more complicated: here one first has to change coordinates by means of the conjugating function $\phi$, and then eq. (3.2.3) can be used.

## 3.3 Generating partition

In general, we call a partition consisting of a fixed number $R$ of cells $A_i$ of fixed size a 'generating partition' if the infinite symbol sequence

Fig. 3.1 The function $y = \phi(x)$ conjugating the binary shift map and the tent map.

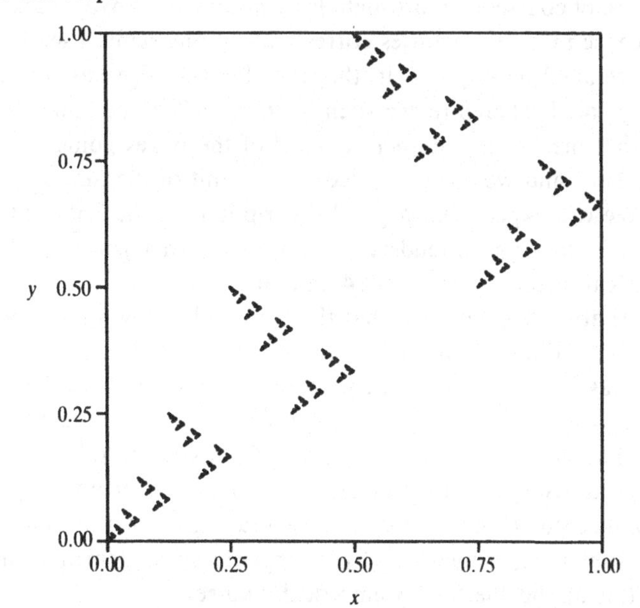

$i_0, i_1, i_2, \ldots$ uniquely determines the initial value $x_0$. This means that the mapping from the phase space to the symbol space can be uniquely reversed. Not only does the initial value $x_0$ determine the symbol sequence, but the reverse is also true. Only for special maps is this possible (Cornfeld *et al.* 1982). In general, the cells of a generating partition have different sizes and in a phase space with $d > 1$ they can also have different shapes.

It should be clear that the existence of a generating partition does not automatically imply the equivalence of the dynamics with that of a Bernoulli shift, as was the case for the special examples of the previous section. For a Bernoulli shift all sequences of symbols 0, 1 (or, in general, of symbols $1, 2, \ldots, R$) are allowed, and the symbols are statistically independent for random initial values. In general, however, these properties may not be satisfied, but still the partition can be a generating one. The only relevant property of a generating partition is that from an infinite symbol string we can uniquely reconstruct the initial value $x_0$. This may sometimes happen in a very complicated way, much more complicated than just writing down the dual expansion.

*A priori*, given some map $f$, it is usually not clear if a generating partition exists, and even when it does exist, how can we find it? If no generating partition is known, the best thing we can do is to use a grid of cubic boxes of equal size. The description of the trajectory by the symbol sequence of the boxes is, of course, only a rough description if the boxes are relatively large, because many trajectories correspond to one symbol sequence. To obtain a detailed description, in this case the box size has to be taken sufficiently small. Therefore, we shall be interested in the limit of $\varepsilon$ going to zero, that means the number $R \sim \varepsilon^{-d}$ of the boxes going to infinity. On the other hand we are interested in the limit of the time $N$ going to infinity. We can expect that a good description can be obtained only if $N$ goes to infinity and $\varepsilon$ tends to zero in a well defined way. We shall discuss this in detail in chapters 14 and 20.

We have already mentioned that there are only a few special maps for which a generating partition is known. The use of such a partition, however, has the advantage that it allows us some analytical treatment in typical cases. Therefore such maps are of special interest for the theoretical aspects of nonlinear dynamics. Compared to this, the description of a map using a grid of boxes of equal size is generally applicable and advantageous if the iterates of a map are given experimentally, i.e., by a computer experiment or by a physical measurement. The grid, moreover, is applicable to any empirical data set.

## 3.4 Symbolic dynamics of the logistic map

The logistic map

$$f(x) = 1 - \mu x^2 \quad (3.4.1)$$

introduced in section 1.3, will now be considered from the point of view of symbolic dynamics.

First we look at the map for the special parameter value $\mu = 2$, i.e., at the *Ulam map.* As a partition of the phase space $X = [-1, 1]$ we now choose the two intervals $\tilde{A}_1 = [-1, 0)$ and $\tilde{A}_2 = [0, 1)$. Again we introduce the symbol $L$ for 'left' if an iterate $x_n$ is in $\tilde{A}_1$, and the symbol $R$ for 'right' if it is in $\tilde{A}_2$.

It was shown in section 2.3 that the map is topologically conjugated to the tent map by the transformation

$$x = -\cos \pi y. \quad (3.4.2)$$

This transformation maps the phase space $Y = [0, 1]$ of the tent map onto the phase space $X = [-1, 1]$ of the Ulam map. The generating partition $\{A_1, A_2\}$ of $Y$ corresponds to the generating partition $\{\tilde{A}_1, \tilde{A}_2\}$ of $X$. To a given $y_0$ there uniquely corresponds an initial value $x_0$. The dual representation of $y_0$ can be uniquely related to a sequence of symbols $L$ or $R$ in the space $Y$ (see previous section). The symbol sequences in the space $X$ and in the space $Y$ coincide. Any sequence of the symbols $L$ or $R$ is allowed, due to the conjugacy with the Bernoulli shift.

The situation completely changes for *other values of the parameter* $\mu$ keeping the same partition $\tilde{A}_1 = [-1, 0)$, and $\tilde{A}_2 = [0, 1)$. Typically, in this case some sequences $i_0, i_1, \ldots$ never occur; they are 'forbidden'. Let us consider as an example the parameter value $\mu = 1.8$. On a pocket calculator one can easily verify that the following finite strings of symbols never occur:

*LLL*

*LLRR*

*LLRLRL*

*LLRLRRR*

*LLRLRRLRL*

...

It turns out that for all symbol sequences up to a given length, there are certain rules – a 'grammar' (Collet and Eckmann 1980; Alekseev and Yakobsen 1981; Grassberger 1988). This grammar determines which

Table 3.1. *Superstable orbits of the logistic map of period K ⩽ 7.*

| Period | Type | Parameter $\mu$ |
|---|---|---|
| 2 | *RC* | 1 |
| 4 | *RLRC* | 1.3107026 |
| 6 | *RLRRRC* | 1.4760146 |
| 7 | *RLRRRRC* | 1.5748891 |
| 5 | *RLRRC* | 1.6254137 |
| 7 | *RLRRLRC* | 1.6740661 |
| 3 | *RLC* | 1.7548777 |
| 6 | *RLLRLC* | 1.7728929 |
| 7 | *RLLRLRC* | 1.8323150 |
| 5 | *RLLRC* | 1.8607825 |
| 7 | *RLLRRRC* | 1.8848036 |
| 6 | *RLLRRC* | 1.9072801 |
| 7 | *RLLRRLC* | 1.9271477 |
| 4 | *RLLC* | 1.9407998 |
| 7 | *RLLLRLC* | 1.9537059 |
| 6 | *RLLLRC* | 1.9667732 |
| 7 | *RLLLRRC* | 1.9771796 |
| 5 | *RLLLC* | 1.9854243 |
| 7 | *RLLLLRC* | 1.9918142 |
| 6 | *RLLLLC* | 1.9963761 |
| 7 | *RLLLLLC* | 1.9990957 |

'words', that is to say which sequences, can be formed by the alphabet of the symbols. For the partition $\{A_1, A_2\}$ used here, the alphabet only contains two elements, $L$ and $R$. To characterize the grammar in total, an infinite set of allowed or forbidden symbol sequences has to be listed. Of course this grammar changes with the parameter $\mu$.

**Stable periodic orbits** Symbolic dynamics is also a useful tool to classify superstable periodic orbits. As we have already mentioned in section 1.3, for the logistic map a superstable orbit is characterized by the fact that the point 0 is part of the periodic orbit. If we denote this point separating the left interval from the right interval by the symbol $C$, a superstable orbit of length $K$ generates a symbol sequence of the following type:

$$C * \underbrace{\quad \cdots \quad}_{K-1 \text{ symbols } L \text{ or } R} * C. \tag{3.4.3}$$

Such a symbol sequence generated by the special initial value $x_0 = 0$ is often called a *kneading sequence*. Table 3.1 shows all possible superstable orbits of the logistic map with $K \leqslant 7$. The orbits are classified according to their kneading sequence. It has been proved that for entire classes of maps depending on one parameter the symbolic dynamics of the table is 'universal': if two orbits of this table are observed, then all orbits between them must also occur for certain values of the parameter $\mu$. There are no 'jumps' in the table (Metropolis, Stein and Stein 1973; Derrida, Gervois and Pomeau 1979; Collet and Eckmann 1980).

*Remark*: The symbolic analysis of dynamical systems is certainly a useful tool in higher dimensions as well, although much less rigorous results are known in this case. For a numerical investigation of the Hénon map see, for example, Grassberger, Kantz and Moenig (1989).

## 3.5 Cylinders

Let us return to a map exhibiting chaotic behaviour. Suppose we have chosen some partition $A_1, \ldots, A_R$ of the phase space. Each initial value $x_0$ generates a symbol sequence $i_0, i_1, i_2, \ldots$. For a given finite symbol sequence $i_0, \ldots, i_{N-1}$ of length $N$ let us denote by $J[i_0, \ldots, i_{N-1}]$ the set of initial values $x_0$ that generate this sequence. If the sequence is forbidden, we have $J[i_0, \ldots, i_{N-1}] = \varnothing$, i.e., the set is empty. Obviously the sets $J[i_0, \ldots, i_{N-1}]$ induce another partition of the phase space $X$. They satisfy

$$\bigcup_{i_0, \ldots, i_{N-1}} J[i_0, \ldots, i_{N-1}] = X \tag{3.5.1}$$

since each point in the phase space $X$ can be an initial value $x_0$. Moreover,

$$J[i_0, \ldots, i_{N-1}] \cap J[j_0, \ldots, j_{N-1}] = \varnothing \tag{3.5.2}$$

with $i_n \neq j_n$ for at least one $n \in \{0, \ldots, N-1\}$, because each $x_0$ determines the sequence $i_0, \ldots, i_{N-1}$ uniquely, and therefore different sequences must belong to different initial values. In the mathematical literature such a set $J[i_0, \ldots, i_{N-1}]$ is called an *N-cylinder* (at least for appropriate classes of maps). The reader should not be confused by the word 'cylinder'. This is just a technical term, which has turned out to be adequate for the mathematical analysis of chaotic maps. By no means has an $N$-cylinder the shape of an ordinary cylinder of classical geometry. For the example of the binary shift map, the $N$-cylinders are $2^N$ equally spaced intervals of length $2^{-N}$ (see Fig. 3.2(*a*)). The same is true for the tent map (Fig. 3.2(*b*)). The only difference compared to the binary shift map is the encoding of the cylinders, i.e., the way in which we identify the symbol strings with the intervals (compare Fig. 3.2(*a*) and (*b*)). The $N$-cylinders of the Ulam map, obtained by topological conjugation, are shown in Fig. 3.2(*c*). The encoding coincides with that of the tent map, but now the cylinders have different sizes. Generally, the edges of the $N$-cylinders in the phase space can be constructed by taking the union of all preimages of the edges of the partition $A_1, \ldots, A_R$ under the map $f^1, f^2, \ldots, f^{N-1}$. If the partition $A_1, \ldots, A_R$ is a generating one, it immediately follows from the definition that in this case the size of all $N$-cylinders must shrink to zero for $N \to \infty$, in order to determine the precise initial value $x_0$ from the sequence $i_0, i_1, i_2, \ldots$.

## 3.6 Symbolic stochastic processes

So far we have only dealt with the question of whether a certain symbolic sequence is allowed or forbidden. But it is intuitively clear that among

Fig. 3.2 The first three generations of the $N$-cylinders for (*a*) the binary shift map, (*b*) the tent map, and (*c*) the Ulam map.

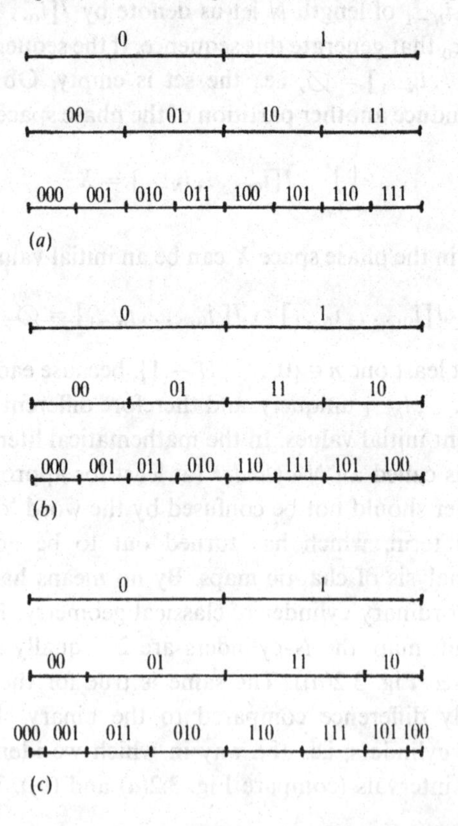

the allowed symbolic sequences some will occur more frequently than others. Hence, we can attribute to each sequence $i_0, \ldots, i_N$ a certain probability $p(i_0, \ldots, i_N)$ that it is observed. This means that we do not consider a single initial value $x_0$ but an entire ensemble of initial values $x_0$ distributed according to some probability density $\rho_0(x_0)$. Provided that it exists one usually chooses the natural invariant density for the function $\rho_0(x_0)$. As a consequence one has

$$p(i_0, \ldots, i_N) = \mu(J[i_0, \ldots, i_N]), \tag{3.6.1}$$

where $\mu$ is the natural invariant measure. The hierarchy of all probabilities $p(i_0, \ldots, i_N)$ with $N = 0, 1, 2, \ldots$ defines a *stochastic process* (see, for example, Doob (1953) and van Kampen (1981)). We can always write

$$p(i_0, \ldots, i_{N-1}, i_N) = p(i_N \mid i_0, \ldots, i_{N-1}) p(i_0, \ldots, i_{N-1}), \tag{3.6.2}$$

where the conditional probability $p(i_N \mid i_0, \ldots, i_{N-1})$ is the probability of the event $i_N$ provided we have observed the sequence $i_0, \ldots, i_{N-1}$ before. The symbolic stochastic process is called a *Markov chain* if

$$p(i_N \mid i_0, \ldots, i_{N-1}) = p(i_N \mid i_{N-1}). \tag{3.6.3}$$

This means the conditional probability does not depend on the entire history $i_0, \ldots, i_{N-1}$, but on the last event only.

A slightly different concept, of utmost interest for the theory of nonlinear dynamical systems, is that of a *topological Markov chain.* The latter is defined by the property

$$p(i_N \mid i_0, \ldots, i_{N-1}) = 0 \tag{3.6.4}$$

$$\text{if and only if } p(i_N \mid i_{N-1}) = 0 \text{ or } p(i_{N-1} \mid i_0, \ldots, i_{N-2}) = 0.$$

The meaning of this statement is the following. There are two ways in which the transition probability $p(i_N \mid i_0, \ldots, i_{N-1})$ to cell $i_N$ can be zero: either it is not possible to reach cell $i_N$ from cell $i_{N-1}$, or the sequence $i_0, \ldots, i_{N-1}$ is already forbidden.

In general, the stochastic process generated by a map $f$ will be neither a Markov chain nor a topological Markov chain but a complicated non-Markovian process. Moreover, the character of the process will depend on the partition chosen. But for some maps (the so called expanding maps; see section 15.6) a partition indeed exists that makes the corresponding stochastic process a topological Markov chain. Such a partition is called a *Markov partition.*

Technically, one usually defines a Markov partition by a topological property of the partition, that is to say essentially by the fact that – at least in the one-dimensional case – edges of the partition are mapped again onto edges. In fact, for a one-dimensional single-humped map a partition is a Markov partition if and only if the image of any of its elements is again an element or a union of elements of the partition (Bowen 1970). For higher-dimensional maps, one has to distinguish between expanding and contracting directions in the phase space (see Cornfeld *et al.* (1982) for more details).

In certain cases the stochastic process generated by the map is not only a topological Markov chain but even a conventional Markov chain in the sense of eq. (3.6.3) (Adler and Weiss 1967; Cornfeld *et al.* 1982). This depends on the measure $\mu$ that is chosen in eq. (3.6.1): it has to be a so called *maximum entropy measure* (Parry 1964; Bowen 1978; Walters 1981; Taylor 1991). We shall deal with simple examples in section 17.5.

The partition $\{A_1, A_2\}$ of the binary shift map, introduced in section 3.2, is not only a generating partition but even a Markov partition. This can be seen from the fact that the edges $\{0, \frac{1}{2}, 1\}$ of the partition are mapped again onto edges; namely onto $\{0, 1\}$, which is a subset of the set $\{0, \frac{1}{2}, 1\}$. We may also verify this by checking the validity of eq. (3.6.4). In this very simple case there are no forbidden sequences at all, hence all transition probabilities are nonzero, which means that condition (3.6.4) is trivially fulfilled. Moreover, if we choose the natural invariant measure for the initial values, not only a topological Markov chain but even a conventional Markov chain is obtained: as the natural invariant density $\rho(x)$ is 1 and as each cylinder has equal size, we have $p(i_0, \ldots, i_{N-1}) = 2^{-N}$ for all $N$. Hence,

$$p(i_N \mid i_0, \ldots, i_{N-1}) = \frac{p(i_0, \ldots, i_N)}{p(i_0, \ldots, i_{N-1})} = \frac{1}{2}. \tag{3.6.5}$$

This conditional probability does not depend on the history at all. Hence, eq. (3.6.3) is trivially fulfilled. The same consideration applies to the tent map as well as to the Ulam map, provided we choose the natural invariant density (2.3.20) and the partition $\tilde{A}_1 = [-1, 0)$, $\tilde{A}_2 = [0, 1)$. On the other hand, for the logistic map with $\mu = 1.8$ the partition $\{\tilde{A}_1, \tilde{A}_2\}$ is not a Markov partition, for obvious reasons. The edges $\{-1, 0, 1\}$ of the partition $\{\tilde{A}_1, \tilde{A}_2\}$ are not again mapped onto edges: for example, we have

$$f(1) = -0.8 \notin \{-1, 0, 1\}. \tag{3.6.6}$$

Thus one expects a complicated non-Markovian process to be generated.

Usually, for a generic chaotic map one does not know whether a Markov partition exists. And even if it does exist, there is no simple way to find it. In general, the cells of a Markov partition are not of equal size, and in a space with $d > 1$ they are not simple geometrical objects such as, for instance, cubes. In fact, often they have quite a complicated structure, for example, fractal boundaries. This means that we can either choose a simple partition, obtaining a complicated stochastic process, or a complicated partition, obtaining a simple stochastic process.

Part II · Essentials of information theory and thermodynamics

# 4
# The Shannon information

In this chapter we shall return to the general framework of probability theory and shall introduce an information measure. The aim of such a measure is to distinguish between probability distributions with higher and lower qualities of prediction. An information measure is a measure of the knowledge an observer gains with respect to the occurrence of an event if he knows the probability distribution and nothing more. The measure is important for the question of how to obtain probabilities when there is some knowledge of the system. In later chapters we shall use the measure to obtain the basic relations of thermodynamics, which, however, are also valid in the general theory of statistics.

## 4.1 Bit-numbers

We may regard a bit-storage unit of a computer as a switch with two possible positions. A set of $A$ of such switches can take on

$$N = 2^A \tag{4.1.1}$$

different states and thus give $N$ different bit patterns. In particular, each of these states can be used to select a particular one of the $N$ numbers $0, 1, 2, \ldots, N-1$. This can be done, for instance, by writing a number $m$ as a binary number of length $A$ in the form

$$m = \sum_{k=0}^{A-1} s_k 2^k \tag{4.1.2}$$

where $s_k$ is 0 or 1. Then we can represent each $s_k$ by the position of the $k$th switch. Eq. (4.1.1) says that we need

$$A = \frac{\ln N}{\ln 2} \tag{4.1.3}$$

bits to select one of $N$ events. It is convenient to use ln 2 as a unit and to define

$$b = \ln N \tag{4.1.4}$$

as the '*bit-number*'.

Now let $p_i$ be the probability of an event $i$ of a sample set of $R$ events $i = 1, 2, \ldots, R$. We call the set of all $p_i$, which in general may take on different values, a probability distribution, or simply a 'distribution', and we shall often denote it by $p$. All numerical quantities occurring in digital devices are rational numbers. Hence, if we simulate probabilities of certain events on a computer, we can suppose that each probability $p_i$ can be represented by the ratio of two positive integers $N_i$ and $N$, i.e.,

$$p_i = N_i/N \tag{4.1.5}$$

with

$$N = \sum_{i=1}^{R} N_i. \tag{4.1.6}$$

Such integers $N_i$, $N$ indeed occur if we construct compound events.

Let us first consider a simple example as an illustration. Let a pool of $N = 10$ lotto cards with figures $1, 2, \ldots, 10$ be given. The probability of drawing a particular one of these cards is equal to $1/N = 0.1$ for all cards. Now we separate the cards into three subsets labelled $i$. The subset $i = 1$ contains all figures $\leqslant 5$, the subset $i = 2$ contains the remaining figures $\leqslant 7$, the subset $i = 3$ contains all figures $> 7$. The number of cards in the subsets is $N_1 = 5$, $N_2 = 2$, $N_3 = 3$. The compound probability $p_i$ of drawing a card of the subset $i$ is $N_i/N$ (i.e., 0.5, 0.2, 0.3 respectively).

The generalization of this construction of compound probabilities is the following. We start with a large sample set of $N$ 'elementary' events $\alpha$ of *equal probability*. This set is divided into subsets $i = 1, 2, \ldots, R$ which have no elements in common. Each subset is a composition of $N_i$ 'elementary' events. The probability of finding an elementary event $\alpha$ in the $i$th subset is then given by the $p_i$ of eq. (4.1.5) and is the probability of the compound event $i$. This event is represented by the subset $i$ and means finding some arbitrary 'elementary' event $\alpha$ lying in the subset. The minimum bit-number necessary to select one elementary event $\alpha$ out of the large sample set is ln $N$. It has to be independent of whether we select first the subset $i$ in which $\alpha$ lies and then select $\alpha$ out of this subset, or whether we select $\alpha$ directly from the large sample set. Let us denote the bit-number needed for selecting the subset $i$ by $b_i$. Then the statement is

$$b_i + \ln N_i = \ln N. \tag{4.1.7}$$

Hence we obtain the result that

$$b_i = -\ln p_i \tag{4.1.8}$$

is the bit-number an observer is missing in order to know whether the event $i$ will occur, if he only knows the probability $p_i$ of this event. For convenience, we shall exclude events with probability zero. In practice this is not a serious restriction.

## 4.2 The Shannon information measure

Since in a long series of observations each event $i$ occurs with the relative frequency $p_i$, the mean value of $-b_i$ is

$$I(p) = \sum_{i=1}^{R} p_i \ln p_i. \tag{4.2.1}$$

This quantity is a function of the distribution $p$. We may also include events with probability zero, because the limit of $p_i \ln p_i$ for $p_i$ going to zero is zero. Such events do not contribute to $I(p)$ at all. As generally $b_i$ measures a lack of knowledge and thus $-b_i$ a knowledge, $I(p)$ was introduced by Shannon and Weaver (1948) and Wiener (1949) as an information measure in communication theory, in connection with the task of decoding enciphered messages during World War II. This was the beginning of the information theory. To distinguish $I(p)$ from other information measures, it is called the 'Shannon information'. It can be regarded as a measure for the knowledge of the observer about the question of which event of the sample set is to be expected, if he knows only the distribution $p$. As $0 \leqslant p_i \leqslant 1$, the quantity $I(p)$ is always negative with the maximum value zero belonging to optimum knowledge. The negative information measure

$$S(p) = -I(p) \tag{4.2.2}$$

is called the *Shannon entropy*. It is a measure of the lack of knowledge about the above formulated question. It is always positive.

The quantity $S(p)$ was already well known in the thermodynamic statistics of the last decades of the last century. It is the thermal entropy in the special case that the events $i$ are dynamical microstates in the Gibbs phase space of classical mechanics or pure quantum states in the Hilbert space of quantum mechanics. However, it was restricted to the case that the distribution $p$ corresponds to a macroscopic thermal equilibrium. In conventional units of energy and temperature we have to multiply the expression on the right hand side of eq. (4.2.1) by the Boltzmann constant $k_B$. In the following, however, we shall always use appropriate units in which the Boltzmann constant is set equal to one.

$I(p)$ takes its maximum value 0 for a 'pure state', that means for a distribution in which only one microstate – say $j$ – is realized

$$p_i = \delta_{ij}. \tag{4.2.3}$$

Here we have used the Kronecker symbol $\delta_{ij}$, which is 1 if $i = j$ and 0 if $i \neq j$.

$I(p)$ takes its minimum value $-\ln R$ if all events $i = 1, \ldots, R$ have equal probability

$$p_i = 1/R. \tag{4.2.4}$$

This distribution is called *uniform distribution.* Here no event is distinguished and knowledge about its occurrence indeed has a minimum.

## *4.3 The Khinchin axioms

To demonstrate the distinguished role of the Shannon information, it is worthwhile to show that it can uniquely be derived from reasonable requirements on the properties of an information measure. These are the following axioms given by Khinchin (1957):

(I)
$$I(p) = I(p_1, \ldots, p_R). \tag{4.3.1}$$

This means, $I(p)$ is a function of the probabilities $p_i$ only. It is not dependent on any other properties of the events $i$.

The second axiom,

(II)
$$I\left(\frac{1}{R}, \frac{1}{R}, \ldots, \frac{1}{R}\right) \leqslant I(p), \tag{4.3.2}$$

means $I(p)$ has to take its minimum value for the uniform distribution of eq. (4.2.4).

The next axiom requires that $I(p)$ remains unchanged if the sample set is enlarged by a new event with probability zero:

(III)
$$I(p_1, \ldots, p_R) = I(p_1, \ldots, p_R, 0). \tag{4.3.3}$$

To formulate the last axiom, we consider the composition of two subsystems $\Sigma^{\mathrm{I}}$ and $\Sigma^{\mathrm{II}}$ to one system $\Sigma$. We assume that the microstates of the two physical systems $\Sigma^{\mathrm{I}}$ and $\Sigma^{\mathrm{II}}$ form the two sample sets $i$ and $j$, respectively. We consider the composition of the two systems as a new system $\Sigma$ with the combined microstates $(i, j)$. In the special case that the two subsystems $\Sigma^{\mathrm{I}}, \Sigma^{\mathrm{II}}$ are statistically independent, the probability distribution of the compound system factorizes with respect to the subsystems:

$$p_{ij} = p_i^{\mathrm{I}} p_j^{\mathrm{II}}, \tag{4.3.4}$$

where $p^{\mathrm{I}}, p^{\mathrm{II}}$ are the probability distributions of the subsystems. Then $I$ should become additive with respect to the subsystems:

$$I(p) = I(p^{\mathrm{I}}) + I(p^{\mathrm{II}}). \tag{4.3.5}$$

In general, however, the two subsystems will be correlated and therefore only

$$p_{ij} = Q(j\,|\,i) p_i^{\mathrm{I}} \tag{4.3.6}$$

will hold, where $Q(j\,|\,i)$ is the conditional probability that the subsystems $\Sigma^{\mathrm{II}}$ is in state $j$ if the subsystem $\Sigma^{\mathrm{I}}$ is in state $i$. Then the axiom requires that

$$\text{(IV)} \qquad I(p) = I(p^{\mathrm{I}}) + \sum_i p_i^{\mathrm{I}} I(Q\,|\,i), \tag{4.3.7}$$

where

$$I(Q\,|\,i) = \sum_j Q(j\,|\,i) \ln Q(j\,|\,i) \tag{4.3.8}$$

is the *conditional information* of the probability distribution $Q(j\,|\,i)$ of the events $j$ for fixed $i$. The meaning of this axiom is that the information has to be independent of how it is collected, whether directly for the compound system or successively for the subsystems.

Now we shall show how these axioms uniquely determine the Shannon information. We are free to consider a special situation, namely that the events (or 'microstates') $j$ of system II are 'elementary' events, in the sense that they all have the same probability, and that the events $i$ of system I are compound events, in the following way. We partition the sample set of $N$ 'elementary' events $j$ into $R$ subsets containing $N_i$ elements, as explained in the text before eq. (4.1.7). We define each subset to be a compound event and denote it by the index $i$. In this special case we have

$$Q(j\,|\,i) = 1/N_i \tag{4.3.9}$$

if $j$ is in the subset $i$. Otherwise $Q(j\,|\,i)$ is zero. Consequently,

$$p_{ij} = 1/N \quad \text{or} \quad 0 \tag{4.3.10}$$

due to eq. (4.3.6) with

$$p_i^{\mathrm{I}} = N_i/N. \tag{4.3.11}$$

For the moment we may consider the special case that also all $N_i$ coincide, i.e., $N_i = r$ for all $i$. This means $p_i^{\mathrm{I}}$ is the uniform distribution of $R$ events, and $N$ is equal to $Rr$. For any uniform distribution, due to axiom I, the information measure is a function $f$ of the number of states only. This means

$$I(p) = f(N), \tag{4.3.12}$$

$$I(p^{\mathrm{I}}) = f(R), \tag{4.3.13}$$

$$I(Q \mid i) = f(r). \tag{4.3.14}$$

As

$$\sum_{i=1}^{R} p_i^{\mathrm{I}} = 1, \tag{4.3.15}$$

axiom IV yields

$$f(Rr) = f(R) + f(r), \tag{4.3.16}$$

thus

$$f(R) = -c \ln R, \tag{4.3.17}$$

where $c$ is some constant. Axioms II and III yield

$$I\left(\frac{1}{R}, \ldots, \frac{1}{R}, 0\right) \geqslant I\left(\frac{1}{R+1}, \ldots, \frac{1}{R+1}\right). \tag{4.3.18}$$

which implies the statement

$$f(R) \geqslant f(R+1), \tag{4.3.19}$$

i.e., $f(R)$ is a monotonically decreasing function of $R$. Therefore $c > 0$. Now we return to an arbitrary distribution $p^{\mathrm{I}}$ of the form (4.3.11). As the nonzero $p_{ij}$ and $Q(j \mid i)$ still form uniform distributions, axiom IV yields

$$f(N) = I(p^{\mathrm{I}}) + \sum_{i=1}^{R} p_i^{\mathrm{I}} f(N_i). \tag{4.3.20}$$

With the help of eq. (4.3.17) this can be written as

$$I(p^{\mathrm{I}}) = c \sum_{i=1}^{R} p_i^{\mathrm{I}} \ln p_i^{\mathrm{I}}. \tag{4.3.21}$$

The positive constant $c$ is undetermined, it is chosen to be 1 by convention.

# 5
# Further information measures

In this chapter we shall consider further information measures of probability distributions. They cannot replace the Shannon information but in spite of that they are also called 'informations'. They measure other features of probability distributions and will turn out to be important for all the following chapters of this book.

## 5.1 The Rényi information

It is noteworthy that the distinctive property of the Shannon information $I(p)$, that it becomes additive for independent subsystems, does not occur as one of the Khinchin axioms. Indeed, if we replaced Khinchin axiom IV by this requirement, we would not obtain the Shannon information in a unique way. It is easy to see that then another information measure, the so called *Rényi information* (Rényi 1960, 1970),

$$I_\beta(p) = \frac{1}{\beta - 1} \ln \sum_{i=1}^{r} (p_i)^\beta, \tag{5.1.1}$$

fulfils the changed axioms as well, namely Khinchin axioms I, II, III, and the additivity of independent systems, but not Khinchin axiom IV. The parameter $\beta$ is an arbitrary real number; $r$ is the number of all nonempty states $i$, that means the number of all events $i$ of the sample set with $p_i \neq 0$.

As the Rényi information $I_\beta$ plays a distinguished role in statistical mechanics (Schlögl 1980) and in nonlinear dynamics (Hentschel and Procaccia 1983), we shall analyse some of its features. For $\beta = 0$ we obtain

$$I_0(p) = -\ln r, \tag{5.1.2}$$

i.e., $|I_0(p)|$ grows logarithmically with the number of nonempty states.

A power expansion with respect to $\varepsilon = \beta - 1$ shows that for $\beta \to 1$ one has in first order of $\varepsilon$

$$\sum_{i=1}^{r} p_i^{1+\varepsilon} = \sum_{i=1}^{r} p_i \exp(\varepsilon \ln p_i) \approx \sum_{i=1}^{r} p_i(1 + \varepsilon \ln p_i)$$

$$= 1 + \varepsilon \sum_{i=1}^{r} p_i \ln p_i. \tag{5.1.3}$$

Hence,

$$\lim_{\varepsilon \to 0} I_{1+\varepsilon}(p) = \lim_{\varepsilon \to 0} \frac{1}{\varepsilon} \ln\left(1 + \varepsilon \sum_{i=1}^{r} p_i \ln p_i\right) = \sum_{i=1}^{r} p_i \ln p_i \tag{5.1.4}$$

or

$$\lim_{\beta \to 1} I_\beta(p) = I(p). \tag{5.1.5}$$

Thus $I_1$ is identified with the Shannon information, thereby rendering $I_\beta$ differentiable with respect to $\beta$. For finite $r$ we always have differentiability, only in the limit $r \to \infty$ may nonanalytic behaviour with respect to $\beta$ occur (see chapter 21).

## 5.2 The information gain

Let us assume that a probability distribution $p^0$ with $p_i^0 \neq 0$ over a sample set is given. Moreover, let us assume that an observer is led to a changed distribution $p$ for this sample set by a new piece of information, say by a new observation. What is the value of the corresponding information transfer measured in bit-numbers? The change of the bit-number for the event $i$ of the sample set is

$$b_i^0 - b_i = \ln(p_i/p_i^0). \tag{5.2.1}$$

As $b_i$ is the missing knowledge of the observer with respect to the event $i$, the decrease of $b_i$ describes the increase of knowledge. The corresponding sign is chosen in eq. (5.2.1). On average, the statistical weight of the result $i$ is determined by the new probability $p_i$, and we obtain the mean value of the difference in eq. (5.2.1) as

$$K(p, p^0) = \sum_{i=1}^{r} p_i \ln \frac{p_i}{p_i^0}. \tag{5.2.2}$$

$K(p, p^0)$ is called the *information gain* or *Kullback information* (Kullback

1951a,b). This measure plays an important role in information theory as well. We shall mention some of its remarkable features.

First of all, it is never negative:

$$K(p, p^0) \geqslant 0. \tag{5.2.3}$$

This is a consequence of the general inequality

$$\ln x \geqslant 1 - x^{-1}, \tag{5.2.4}$$

which holds for all $x > 0$. Choosing $x = p_i/p_i{}^0$ we obtain

$$K(p, p^0) = \sum_{i=1}^{r} p_i \ln(p_i/p_i{}^0) \geqslant \sum_{i=1}^{r} p_i(1 - p_i{}^0/p_i) = 1 - 1 = 0. \tag{5.2.5}$$

Secondly, $K(p, p^0)$ takes on its minimum value zero if and only if the two distributions $p$ and $p^0$ are identical. This can be seen from a stronger version of eq. (5.2.4), namely the fact that the equality holds for $x = 1$ only:

$$\ln x \begin{cases} = 1 - x^{-1} & \text{for } x = 1 \\ > 1 - x^{-1} & \text{otherwise.} \end{cases} \tag{5.2.6}$$

Hence, in eq. (5.2.5) we have $K(p, p^0) = 0$ if and only if all the ratios $p_i{}^0/p$ are equal to 1.

The special case where $p^0$ is the uniform distribution

$$p_i{}^0 = 1/R, \tag{5.2.7}$$

leads us to the Shannon information, because then

$$K(p, p^0) = I(p) + \ln R. \tag{5.2.8}$$

$K(p, p^0)$ is a convex function of the variables $p_i$, because

$$\frac{\partial^2 K}{\partial p_i \, \partial p_j} = \frac{1}{p_i} \delta_{ij} \geqslant 0. \tag{5.2.9}$$

Therefore $I(p)$ is a convex function of $p$ as well. The entropy

$$S(p) = -I(p), \tag{5.2.10}$$

on the other hand, is concave.

NB: A convex function $f(x)$ fulfils

$$f(c_1 x + c_2 x') \leqslant c_1 f(x) + c_2 f(x') \tag{5.2.11}$$

for all $x, x'$ and $c_1, c_2 \geqslant 0$, $c_1 + c_2 = 1$, whereas for a concave function $f(x)$

$$f(c_1 x + c_2 x') \geqslant c_1 f(x) + c_2 f(x'). \tag{5.2.12}$$

For twice differentiable functions $f$, convexity is equivalent to $f''(x) \geqslant 0$, and concavity to $f''(x) \leqslant 0$, for all $x$.

## *5.3 General properties of the Rényi information

An important general property of the Rényi information is that it is a monotonically increasing function of $\beta$:

$$I_\beta(p) \leqslant I_{\beta'}(p) \qquad \text{for } \beta < \beta' \tag{5.3.1}$$

for arbitrary probability distributions $p$.

Hence, in particular $I_\beta$ provides us with an upper ($\beta > 1$), respectively lower ($\beta < 1$), bound on the Shannon information ($\beta = 1$).

To prove eq. (5.3.1), we differentiate $I_\beta(p)$ with respect to $\beta$. The result can be written in the form

$$\frac{\partial I_\beta}{\partial \beta} = \frac{1}{(1-\beta)^2} \sum_{i=1}^{r} P_i \ln \frac{P_i}{p_i}, \tag{5.3.2}$$

where we call the probabilities

$$P_i = \frac{p_i^\beta}{\sum_{j=1}^{r} p_j^\beta} \tag{5.3.3}$$

*escort probabilities.* They will be analysed in more detail in chapter 9. Notice that $\sum P_i = 1$. The sum occurring on the right hand side of eq. (5.3.2) can be interpreted as Kullback information gain, which is always nonnegative. Hence $I_\beta$ is monotonically increasing with respect to $\beta$.

It is interesting that further general inequalities can be proved for $I_\beta$. In contrast to eq. (5.3.1), the following one provides us with a lower bound (Beck 1990a) on $I_\beta$ in terms of $I_{\beta'}$:

$$\frac{\beta - 1}{\beta} I_\beta(p) \geqslant \frac{\beta' - 1}{\beta'} I_{\beta'}(p) \tag{5.3.4}$$

for $\beta' > \beta$ and $\beta\beta' > 0$. To prove eq. (5.3.4), let us remember that the function $x^\sigma$ is convex for $\sigma > 1$ and concave for $0 \leqslant \sigma \leqslant 1$. Hence, for arbitrary $a_i$

$$\left(\sum_i a_i\right)^\sigma \geqslant \sum_i a_i^\sigma \qquad \text{for } \sigma > 1, \tag{5.3.5}$$

$$\left(\sum_i a_i\right)^\sigma \leqslant \sum_i a_i^\sigma \qquad \text{for } 0 \leqslant \sigma \leqslant 1. \tag{5.3.6}$$

We now choose $a_i = p_i^\beta$ and $\sigma = \beta'/\beta$. Let $\beta' > \beta > 0$. Then we have $\sigma > 1$, thus

$$\left(\sum_i p_i^\beta\right)^{\beta'/\beta} \geqslant \sum_i p_i^{\beta'} \qquad \text{for } \beta' > \beta > 0. \tag{5.3.7}$$

On the other hand, if $\beta < \beta' < 0$, we have $0 \leqslant \sigma = \beta'/\beta < 1$; thus

$$\left(\sum_i p_i^\beta\right)^{\beta'/\beta} \leqslant \sum_i p_i^{\beta'} \qquad \text{for } \beta < \beta' < 0. \tag{5.3.8}$$

Taking these inequalities to the power $1/\beta'$ we get in *both* cases

$$\left(\sum_i p_i^\beta\right)^{1/\beta} \geqslant \left(\sum_i p_i^{\beta'}\right)^{1/\beta'} \qquad \text{for } \beta' > \beta,\ \beta'\beta > 0. \tag{5.3.9}$$

The condition $\beta'\beta > 0$ just indicates that $\beta'$ and $\beta$ have the same sign. Taking the logarithm of the terms on both sides of eq. (5.3.9) indeed yields eq. (5.3.4).

Notice that in the above proof we did not need to assume that $I_\beta$ is differentiable with respect to $\beta$. In particular, this also holds in the *thermodynamic limit* of $r$ going to infinity, where nonanalytic behaviour with respect to $\beta$ can occur. This will be of importance in chapter 21 on phase transitions. Also the inequality (5.3.1) can be proved without the assumption of differentiability (Beck 1990a).

Further inequalities can easily be derived. For example, the function

$$\Psi(\beta) = (1-\beta)I_\beta = -\ln \sum_{i=1}^{r} p_i^\beta \tag{5.3.10}$$

is monotonically increasing and concave in $\beta$:

$$\Psi(\beta) \leqslant \Psi(\beta') \qquad \beta' > \beta \tag{5.3.11}$$

$$\frac{\partial^2}{\partial \beta^2}\Psi(\beta) \leqslant 0. \tag{5.3.12}$$

To prove eq. (5.3.11), we notice that

$$p_i^\beta \geqslant p_i^{\beta'} \tag{5.3.13}$$

for arbitrary $\beta' > \beta$, hence

$$-\ln \sum_i p_i^\beta \leqslant -\ln \sum_i p_i^{\beta'}. \tag{5.3.14}$$

Thus $\Psi(\beta)$ is monotonically increasing. The concavity will be proved in section 7.4.

In the case of differentiability all the above inequalities can also be written in differential form:

$$\frac{\partial}{\partial \beta} I_\beta \geqslant 0, \tag{5.3.15}$$

$$\frac{\partial}{\partial \beta} \frac{\beta - 1}{\beta} I_\beta \leqslant 0, \tag{5.3.16}$$

$$\frac{\partial}{\partial \beta} (1 - \beta) I_\beta \geqslant 0, \tag{5.3.17}$$

$$\frac{\partial^2}{\partial \beta^2} (1 - \beta) I_\beta \leqslant 0. \tag{5.3.18}$$

# 6
# Generalized canonical distributions

In traditional thermodynamics of a physical system, the canonical distribution is the probability distribution of microstates that is used to describe a thermodynamic equilibrium state if nothing but the temperature is given. When further quantities like pressure or chemical potentials are given as well, more detailed probability distributions are used, which are called generalized canonical distributions. In this chapter we shall deduce these distributions in a way that is not restricted to physical thermodynamics but is applicable to general statistics.

## 6.1 The maximum entropy principle

In thermodynamic statistics, different types of probability distribution are useful to describe a thermal equilibrium state; they are the 'microcanonical', 'canonical', 'grand canonical' distributions and the 'pressure ensemble'. We shall discuss these in a unified way as 'generalized canonical distributions'. They result from the so called 'maximum entropy principle' that will be described in the following. In information theory the application of this principle is sometimes called the method of *unbiased guess.* This method is connected with a transparent interpretation of the principle, first given by E. T. Jaynes in 1957 (Jaynes 1957, 1961a,b, 1962). We shall start with this approach.

Again a sample set of microstates $i = 1, 2, \ldots, R$ may be given. We consider the following situation. A random quantity $\tilde{M}$ takes on a certain value $M_i$ in each microstate $i$. The probabilities $p_i$ of the microstates are not known. What, however, is known is the mean value

$$\langle \tilde{M} \rangle = \sum_{i=1}^{R} p_i M_i. \tag{6.1.1}$$

For simplicity we shall denote this by $M$. In thermodynamics, for instance, $M$ may be the mean value $E$ of the energy $E_i$ in the microstates $i$. Now the question is: which probability distribution $p$ shall we assume? Before we get to this problem, we generalize it a bit by assuming that the mean values of not only one random quantity $\tilde{M}$ but of several such quantities are given. We call these quantities $\tilde{M}^\sigma$. They are labelled by the superscript $\sigma$. Hence, we have more than one condition

$$M^\sigma = \sum_{i=1}^{R} p_i M_i^\sigma, \tag{6.1.2}$$

which the distribution $p$ has to fulfil. Moreover, $p$ always has to satisfy the additional condition of normalization

$$\sum_{i=1}^{R} p_i = 1. \tag{6.1.3}$$

We suppose that the number of the quantities $M^\sigma$ is appreciably smaller than $R$. As a rule, in thermodynamics this number is not more than a few, whereas $R$ is an enormously large number.

Among all possible distributions $p$ that fulfil the requirements of eqs. (6.1.2) and (6.1.3), we have to distinguish the $p$ that comprises no unjustified prejudice. This is expressed by the distinction that for this $p$ the Shannon information $I(p)$ takes on its minimum value. For infinitesimal variations $\delta p_i$ that fulfil the conditions (6.1.2) and (6.1.3) in the form

$$\sum_{i=1}^{R} M_i^\sigma \, \delta p_i = 0, \tag{6.1.4}$$

$$\sum_{i=1}^{R} \delta p_i = 0 \tag{6.1.5}$$

we have to require

$$\delta I(p) = \sum_{i=1}^{R} (1 + \ln p_i) \, \delta p_i = 0. \tag{6.1.6}$$

As eq. (6.1.5) shows, we can omit the 1 in the sum of eq. (6.1.6). Moreover, we can multiply eqs. (6.1.4) and (6.1.5) by arbitrary factors $\beta_\sigma$ and $-\Psi$, and obtain by summation of these equations

$$\sum_{i=1}^{R} (\ln p_i - \Psi + \beta_\sigma M_i^\sigma) \, \delta p_i = 0. \tag{6.1.7}$$

Here and in all that follows, a repeated Greek super- and subscript in a product indicates a summation over this index (in our case $\sigma$). In eq. (6.1.7) the number of brackets that can be made zero by an appropriate choice of the 'Lagrange multipliers' $\beta_\sigma$ and $\Psi$ is equal to the number of such multipliers. The $\delta p_i$ which then remain in the sum of eq. (6.1.7) are not restricted, because the number of the restricting conditions (6.1.4) and (6.1.5) is equal to the number of Lagrange multipliers. Therefore the remaining brackets in the sum of eq. (6.1.7) also have to be zero. This means all brackets are zero and we are led to

$$P_i = \exp(\Psi - \beta_\sigma M_i^{\sigma}). \tag{6.1.8}$$

For our purpose it is convenient to denote this particular probability distribution by the capital letter $P$. The Lagrange mutlipliers have to be appropriately chosen to reproduce the required mean values $M^\sigma$ and to fulfil the normalization of $P$. The result $P$ of the unbiased guess is what we call a *generalized canonical distribution* or a *Gibbs distribution*.

$\Psi$ is called a 'generalized free energy'. The coefficients $\beta_\sigma$ are called 'intensities', and the mean values $M^\sigma$ 'extensities'. The reason for these names will become clear when we give their physical interpretation in the following sections.

If no mean value $M^\sigma$ is given, the unbiased guess yields the uniform distribution of the microstates; all $P_i$ have the same value. This, of course, is only reasonable if no microstate $i$ is preferable with respect to any other one. We say, the microstates are of *equal a priori probability*. It depends on the special situation and on the character of the system under consideration whether this can be assumed or not. The equal *a priori* probability of the microstates is a necessary condition for the applicability of Jaynes' unbiased guess method.

As already mentioned, the expression $-I(P)$ corresponding to eq. (4.2.1) was identified with the thermal entropy in thermodynamics long before any information measure was known. Therefore $-I(P)$ was called 'entropy' by Shannon. It will be denoted by the letter $S$. For the generalized canonical distribution we obtain

$$S = -\Psi + \beta_\sigma M^\sigma. \tag{6.1.9}$$

The normalization

$$\sum_{i=1}^{R} P_i = 1 \tag{6.1.10}$$

yields

$$\Psi = -\ln Z, \tag{6.1.11}$$

with the so called *partition function*

$$Z = \sum_{i=1}^{R} \exp(-\beta_\sigma M_i^\sigma). \tag{6.1.12}$$

## 6.2 The thermodynamic equilibrium ensembles

In statistical thermodynamics we have to distinguish between classical mechanics and quantum mechanics. In classical mechanics the microstates are cells in the Gibbs phase space of all position coordinates and their canonical conjugated momenta. They are called 'canonical dynamical coordinates'. It is a fundamental assumption that cells of equal volume in this space are of equal *a priori* probability. This supposition is supported by the Liouville theorem of classical mechanics. It says that in the Gibbs phase space the points representing the dynamical states of an ensemble of identical systems move like particles of an incompressible fluid, that means their point density remains unchanged. Notwithstanding that the supposition is in agreement with the Liouville theorem, it is not at all a logical consequence of it. It is a basic assumption. In quantum mechanics the microstates are 'pure' quantum states in which all observables that are simultaneously determinable have well defined values. It is again a fundamental assumption that these pure quantum states are of equal *a priori* probability.

Depending on the set of macroscopic thermal variables interpreted as mean values $M^\sigma$, the unbiased guess yields different types of distributions. In thermodynamics the following ensembles are most commonly used.

**The canonical distribution** This corresponds to the special case of the distribution of eq. (6.1.8) with only one mean value $M^\sigma$ being given, namely the mean energy $E$ of a physical system in thermal equilibrium. Conventionally it is written in the form

$$P_i = \exp[\beta(F - E_i)]. \tag{6.2.1}$$

Here the only $\beta_\sigma$ is denoted by $\beta$, and $T = (k_B\beta)^{-1}$ is identified with the absolute *temperature*; $k_B$ is the Boltzmann constant, which we shall set equal to 1 in the following. The quantity

$$F = (1/\beta)\Psi(\beta) \tag{6.2.2}$$

is called the *Helmholtz free energy*.

$$Z(\beta) = \exp(-\beta F) = \sum_{i=1}^{R} \exp(-\beta E_i) \tag{6.2.3}$$

is the partition function of the canonical ensemble. The Shannon entropy of probability theory of this ensemble

$$S = \beta(E - F) \tag{6.2.4}$$

is identified with the macroscopic entropy, which can be defined in an operational way, that means by macroscopic processes in which equilibrium states are changing in a certain controlled way.

**The pressure ensemble** This corresponds to the case where not only the energy $\tilde{E}$ but also the volume $\tilde{V}$ of the system is a random quantity of type $\tilde{M}^\sigma$. This, for instance, corresponds to the picture in which a gas or a fluid is enclosed in a vessel with a movable piston. Then the volume is a fluctuating quantity in the sense that it can take different values $V_i$ in different microstates $i$. The corresponding distribution in thermal equilibrium with given mean values $E$ and $V$ is called the 'pressure ensemble'. It is conventionally written in the form

$$P_i = \exp[\beta(G - E_i - \Pi V_i)], \tag{6.2.5}$$

where $\Pi$ is identified with the pressure, $T(=1/\beta)$ again with temperature, and $G$ is called *Gibbs free energy*. In the notation of eq. (6.1.9), now $G$ is identical to $\beta^{-1}\Psi$, and we obtain

$$S = \beta(E + \Pi V - G). \tag{6.2.6}$$

**The grand canonical ensemble** This is used to describe the thermal equilibrium of a system of many particles for which the number $N_i$ of particles fluctuates. For instance, such a system is realized by a gas in an open vessel. The concept of the phase space then has to be extended in such a way that different microstates can belong to different particle numbers $N_i$. The conventional form of writing down the distribution is

$$P_i = \exp[\beta(\Omega - E_i - \Pi V_i + \mu N_i)]. \tag{6.2.7}$$

Again $T = 1/\beta$ is the temperature, $\mu$ is called the *chemical potential*, and $\Omega$ has no special name.

We obtain

$$S = \beta(E + \Pi V - \mu N - \Omega). \tag{6.2.8}$$

This is the entropy of a system of particles consisting of one species only. If there is more than one species of particles, the product $\mu N$ has to be replaced by the sum $\mu_\alpha N^\alpha$ over all species labelled by the index $\alpha$.

**The thermodynamic limit** As will be explained in more detail, for macroscopic thermal systems the fluctuations of the random quantities $M^\sigma$, as a rule, are extremely small compared to the mean values in the thermal equilibrium states. Then, in practice, it does not matter for the equations of $S$ whether the quantities $M^\sigma$ enter in the description as fluctuating random quantities $\tilde{M}^\sigma$ or in the form of sharply valued parameters $M^\sigma$. The change from small systems to macroscopic systems is described by the so called 'thermodynamic limit', in which particle numbers of a molecular system and the volume become extremely large, whereas densities and intensities are kept fixed. In this limit the relations between the macroscopic thermal variables coincide for the different ensembles. We only have to require that $S$ has the same value in the different descriptions. That means we have to equate the expressions of eqs. (6.2.4), (6.2.6), and (6.2.8). This yields the equation

$$G = F + \Pi V = \Omega + \mu N, \tag{6.2.9}$$

which connects the various 'free energies' of the various thermodynamic ensembles.

**The microcanonical distribution** If all $M^\sigma$, including the energy, are introduced as sharp parameters rather than as fluctuating quantities, we are led to the 'microcanonical distribution'. The occurring microstates $i$ are then restricted to a subspace of the state space, the so called 'energy shell'. This consists of all microstates with the same energy $E$. Then, from the method of unbiased guess, $P$ is the uniform distribution in this subspace.

## 6.3 The principle of minimum free energy

The maximum entropy principle starts with given extensities $M^\sigma$. As we shall discuss in section 7.2, the intensities are determined by the thermal contact of the system with the surroundings (the 'thermal bath'). Therefore, in practice, more often a certain set of intensities $\beta_\sigma$ is given rather than extensities. In this case another form of the principle is more convenient.

We restrict from the beginning to normalized probabilities $p_i$. This means the variations always fulfil the equation

$$\sum_{i=1}^{R} \delta p_i = 0. \tag{6.3.1}$$

The principle of maximum entropy $S$ to given mean values $M^\sigma$ yields

$$\delta S = 0. \tag{6.3.2}$$

$$\delta M^\sigma = \sum_{i=1}^{R} M_i^\sigma \, \delta p_i = 0. \tag{6.3.3}$$

For fixed values of the intensities $\beta_\sigma$ we may write

$$\delta(S - \beta_\sigma M^\sigma) = 0. \tag{6.3.4}$$

For any probability distribution $p$ let us define the generalized free energy to be

$$\Psi[p] = \beta_\sigma M^\sigma - S = \sum_{i=1}^{R} (p_i \beta_\sigma M_i^\sigma + p_i \ln p_i). \tag{6.3.5}$$

It is also defined if $p$ is not the generalized canonical distribution; i.e., it is also defined for nonequilibrium and even for nonthermal states. In the new formulation the variational principle requires $\Psi[p]$ to have a minimum in the space of all probability distributions. This minimum is adopted for the canonical distribution

$$P_i = \exp(\Psi - \beta_\sigma M_i^\sigma), \tag{6.3.6}$$

and is given by

$$\Psi = -\ln \sum_{i=1}^{R} \exp(-\beta_\sigma M_i^\sigma). \tag{6.3.7}$$

In this form we call the variational principle the 'principle of minimum free energy'.

Depending on which intensities are given, $\Psi$ is a specific type of free energy. The equilibrium is distinguished by a minimum of this free energy. If only the inverse temperature $\beta$ is given and the volume $V$ is not a fluctuating quantity but a sharp parameter, $\beta^{-1}\Psi$ is the Helmholtz free energy $F$. If temperature and pressure are given, $\beta^{-1}\Psi$ is the Gibbs free energy $G$. We are led to more types of free energies, corresponding to the set of the given intensities.

It may also be that some $\beta_\sigma$ and some $M^\tau$ are given as independent thermal variables. Then the equilibrium is distinguished by an extremum of the thermodynamic potential for which the given thermal variables are the 'natural variables'. We shall explain this notion in section 7.2.

## *6.4 Arbitrary *a priori* probabilities

It may happen that some previous experience or any other kind of information tells us that – without any further knowledge – we have to start with *a priori* probabilities $p_i^0$ that are nonuniform for the different microstates. Then the unbiased guess method has to be extended to minimizing the information gain $K(p, p^0)$ instead of the Shannon information $I(p)$. Eq. (6.1.6) has to be replaced by

$$\delta K(p, p^0) = \sum_{i=1}^{R} (\ln p_i - \ln p_i^0 + 1)\, \delta p_i = 0 \tag{6.4.1}$$

and we obtain

$$p_i = p_i^0 \exp(\phi - \gamma_\sigma M_i^\sigma). \tag{6.4.2}$$

In particular, it may be that $p^0$ is a generalized canonical distribution, and that some $\tilde{M}^\sigma$ are newly observed differences to the mean value of random quantities $\tilde{M}^\tau$ which already occurred in the 'uncorrected' *a priori* distribution $p^0$.

In this case, the information gain is closely related to a fundamental quantity in thermodynamics, the *availability* or *exergy* (see, for example, Schlögl (1989)). To demonstrate this, let us consider a system which is put into a heat bath whose thermal state is characterized by the intensities $\beta_\sigma^0$. If the system were in equilibrium with the bath, it would have the same value of the intensities as the bath. This means its state would be given by the distribution

$$P_i^0 = \exp(\Psi^0 - \beta_\sigma^0 M_i^\sigma). \tag{6.4.3}$$

We consider, however, the situation that the system is not in equilibrium with the bath and that its state is given by another distribution $p_i$. The corresponding information gain (the entropy produced by the change from the state $p$ to the state $P^0$) is

$$\begin{aligned} K(p, P^0) &= \sum_{i=1}^{R} p_i(\ln p_i - \ln P_i^0) \\ &= \sum_{i=1}^{R} [p_i \ln p_i - P_i^0 \ln P_i^0 - (p_i - P_i^0) \ln P_i^0] \\ &= \sum_{i=1}^{R} [p_i \ln p_i - P_i^0 \ln P_i^0 - (p_i - P_i^0)(\Psi^0 - \beta_\sigma^0 M_i^\sigma)]. \end{aligned} \tag{6.4.4}$$

With the differences

$$\Delta S = S - S^0, \qquad \Delta M^\sigma = M^\sigma - M^{\sigma 0}, \tag{6.4.5}$$

we can write

$$K(p, P^0) = -\Delta S + \beta_\sigma{}^0 \Delta M^\sigma \geqslant 0. \tag{6.4.6}$$

In the special case that the bath is defined by the temperature $T^0$ and the pressure $\Pi^0$, we obtain the result that the so called availability or exergy defined by

$$A = \Delta E + \Pi^0 \Delta V - T^0 \Delta S \tag{6.4.7}$$

is always positive (Schlögl 1967). In the general case that the bath is defined by the intensities $\beta_\sigma{}^0$, the availability is

$$A = T^0 K(p, P^0) \geqslant 0. \tag{6.4.8}$$

*Remark*: The availability is the maximum available work that can be gained by the change from a thermal state $P$ of a system to the state $P^0$ which it would take on if it were in equilibrium with the surrounding thermal bath. It is considered to be a measure for the tendency of the system to approach equilibrium with the bath. The importance of the availability in thermodynamics was emphasized by Gibbs (see, for example, Hatsopoulos and Keeman (1965)). Also if the bath is replaced by an environment with intensities $\beta_\sigma$ changing sufficiently slowly with time, this quantity is considered to be the adequate measure for the tendency of the system to follow the thermal state of the environment. In this context the availability has gained considerable interest in the applied thermodynamics of nonequilibria during the last decades (see, for example, Salamon, Nulton and Ihrig (1984) and Salamon, Nulton and Berry (1985)).

# 7
# The fundamental thermodynamic relations

In this chapter we shall deduce and discuss fundamental relations of thermodynamics, which are consequences of the generalized canonical distributions. They are valid in a general statistical framework, not only in the classical application of conventional thermodynamics. We shall deal with Legendre transformations, discuss the Gibbs fundamental equation, and then turn to a few somewhat more special subjects of conventional thermodynamics that will turn out to possess analogues in the theory of nonlinear systems. These are the Gibbs–Duhem equation, the relations between fluctuations and susceptibilities, and phase transitions.

## 7.1 The Legendre transformation

A certain transformation plays a distinguished role in the thermodynamic formalism, and will therefore be explained in the following. We quite generally consider a function $F$ of a variable $\beta$ and define the variable $M$ by

$$M = \mathrm{d}F/\mathrm{d}\beta. \tag{7.1.1}$$

Let us assume that $M(\beta)$ is a monotonic function of $\beta$. In other words, we assume that $F(\beta)$ is concave or convex. We may then express $F$ as a function of $M$

$$F = G(M). \tag{7.1.2}$$

The function $G(M)$, however, does not allow us uniquely to reconstruct the function $F(\beta)$, because $F(\beta - \beta_0)$ with any constant $\beta_0$ has the same derivative at the point $\beta - \beta_0$ as $F(\beta)$ has at the point $\beta$. In Fig. 7.1 the function $F(\beta)$ is represented by a solid line, and $F(\beta - \beta_0)$ by a dashed line. They are identical curves but shifted. The pair of values $M$ and $G$ can either belong to the function $F(\beta)$ at the point $\beta$ or to the function

$F(\beta - \beta_0)$ at the point $\beta - \beta_0$. Thus it does not determine $F(\beta)$ uniquely (the registration of the mileage as a function of the velocity does not say when the journey started).

A unique reconstruction of $F(\beta)$, however, is possible by means of the so called 'Legendre transform' of $F$, defined as follows:

$$L(M) = -F(\beta) + \beta M. \tag{7.1.3}$$

(The sign of $F$ in the definition of $L$ is unimportant. We choose it in a way that yields a certain symmetry in the resulting equations.) We obtain

$$\frac{\mathrm{d}L}{\mathrm{d}M} = -\frac{\mathrm{d}F}{\mathrm{d}\beta}\frac{\mathrm{d}\beta}{\mathrm{d}M} + M\frac{\mathrm{d}\beta}{\mathrm{d}M} + \beta. \tag{7.1.4}$$

Due to definition (7.1.1) this means

$$\mathrm{d}L/\mathrm{d}M = \beta. \tag{7.1.5}$$

This relation implicitly yields $M$ as a function of $\beta$ and finally $F(\beta)$ via eq. (7.1.3). In Fig. 7.1 the geometrical construction is demonstrated. It justifies the name *contact transformation*, which is sometimes also used for the Legendre transformation.

The transformation can be generalized to more than one variable $\beta_\sigma$ as follows:

$$M^\sigma = \partial F/\partial \beta_\sigma, \tag{7.1.6}$$

$$L(M) + F(\beta) = \beta_\sigma M^\sigma, \tag{7.1.7}$$

$$\partial L/\partial M^\sigma = \beta_\sigma. \tag{7.1.8}$$

Eq. (7.1.7) is written in a form that illustrates the symmetry of the

Fig. 7.1 The Legendre transformation (contact transformation): $L(M) = -F + \beta M$ is constructed by the tangent touching the curve $F(\beta)$ at point $\beta$.

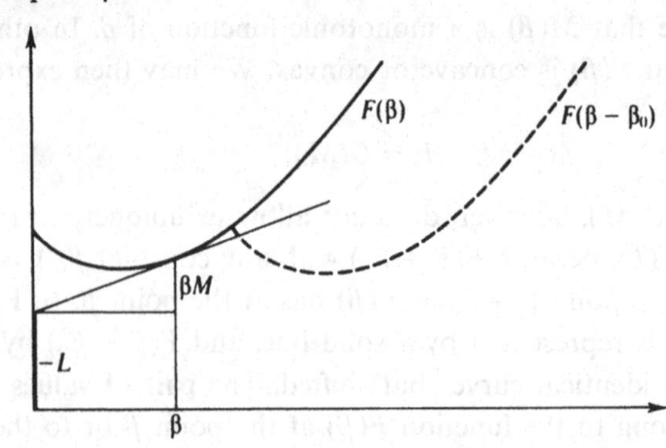

Legendre transformation under an exchange of $F$ and $L$. The last equation implicitly determines $M^{\sigma}(\beta)$. From this it is possible to construct $F(\beta)$ with the help of eq. (7.1.7).

For the above consideration we needed the condition that both the functions $F(\beta)$ and $L(M)$ are differentiable. In thermodynamics and other fields of statistics this may not always be the case. For example, if a system exhibits a phase transition, there is a critical point where thermodynamic functions depend on a control parameter in a nonanalytical way (see section 7.5).

Nevertheless, it is possible to generalize the concept of a Legendre transformation to the case of *nondifferentiable functions* (see, for example, Griffiths (1972) and Freidlin and Wentzell (1984)). In this general case the functions $F(\beta)$ and $L(M)$ are related by the variational principle

$$F(\beta) = \min_{M}[\beta M - L(M)], \tag{7.1.9}$$

$$L(M) = \min_{\beta}[\beta M - F(\beta)]. \tag{7.1.10}$$

In the first equation the minimum has to be taken with respect to the variation of $M$, in the second one with respect to the variation of $\beta$. Let us denote by $M(\beta)$, respectively by $\beta(M)$, the value where the minimum is taken on. We have

$$F(\beta) = \beta M(\beta) - L(M(\beta)), \tag{7.1.11}$$

$$L(M) = \beta(M)M - F(\beta(M)). \tag{7.1.12}$$

If $F$ and $L$ are differentiable, the minima are determined by the conditions

$$\frac{\mathrm{d}}{\mathrm{d}\beta}F(\beta) = M(\beta) + \beta\frac{\mathrm{d}M}{\mathrm{d}\beta} - \frac{\mathrm{d}L}{\mathrm{d}M}\frac{\mathrm{d}M}{\mathrm{d}\beta} = 0 \tag{7.1.13}$$

$$\frac{\mathrm{d}}{\mathrm{d}M}L(M) = \beta(M) + M\frac{\mathrm{d}\beta}{\mathrm{d}M} - \frac{\mathrm{d}F}{\mathrm{d}\beta}\frac{\mathrm{d}\beta}{\mathrm{d}M} = 0. \tag{7.1.14}$$

These equations are solved by

$$\mathrm{d}L/\mathrm{d}M = \beta, \tag{7.1.15}$$

$$\mathrm{d}F/\mathrm{d}\beta = M. \tag{7.1.16}$$

Hence we notice that in the case of differentiable functions the variational principle of eqs. (7.1.9) and (7.1.10) reduces to the standard form (7.1.1), (7.1.3) and (7.1.5).

## 7.2 Gibbs' fundamental equation

Let us consider the information gain from a generalized canonical distribution $P$, i.e.,

$$P_i = \exp(\Psi - \beta_\sigma M_i^\sigma) \tag{7.2.1}$$

to a neighbouring one $P + \delta P$, which is obtained by a variation $\delta\beta_\sigma$ of the intensities $\beta_\sigma$

$$\begin{aligned} K(P + \delta P, P) &= \sum_i (P_i + \delta P_i)[\ln(P_i + \delta P_i) - \ln P_i] \\ &= -\delta S - \sum_i \delta P_i \ln P_i \\ &= -\delta S + \beta_\sigma \,\delta M^\sigma \geqslant 0. \end{aligned} \tag{7.2.2}$$

As with any information gain, it is never negative. If the variations $\delta P_i$ are infinitesimally small, we can write

$$\left(\frac{\partial S}{\partial M^\sigma} - \beta_\sigma\right)\delta M^\sigma + \frac{1}{2}\frac{\partial^2 S}{\partial M^\sigma\,\partial M^\tau}\,\delta M^\sigma\,\delta M^\tau \leqslant 0. \tag{7.2.3}$$

As the infinitesimal variations $\delta M$ are arbitrary and take on both signs, we obtain two important relations. The first one is the statement that the term linear in $\delta M$ has to be zero

$$\partial S/\partial M^\sigma = \beta_\sigma. \tag{7.2.4}$$

The second one is the negativity of the second order term. By use of eq. (7.2.4) this can be written in the form

$$\frac{\partial^2 S}{\partial M^\sigma\,\partial M^\tau}\,\delta M^\sigma\,\delta M^\tau = \frac{\partial\beta_\sigma}{\partial M^\tau}\,\delta M^\sigma\,\delta M^\tau = \delta M^\sigma\,\delta\beta_\sigma \leqslant 0. \tag{7.2.5}$$

The consequences of this second relation will be discussed in section 7.4. Here let us just analyse the first relation (7.2.4). It implies that the entropy

$$S(M) = -\langle \ln P\rangle = -\Psi + \beta_\sigma M^\sigma \tag{7.2.6}$$

is the Legendre transform of $\Psi(\beta)$. This yields on the other hand

$$\partial\Psi/\partial\beta_\sigma = M^\sigma. \tag{7.2.7}$$

For the grand canonical ensemble with fluctuating volume, the general equation

$$\mathrm{d}S = \beta_\sigma\,\mathrm{d}M^\sigma \tag{7.2.8}$$

takes on the form

$$dE = T\,dS - \Pi\,dV + \mu\,dN. \qquad (7.2.9)$$

In phenomenological thermodynamics this is called the *Gibbs fundamental equation*. The quantities $S$, $V$, and $N$ are called *extensive parameters* or *extensities* because they increase proportionally with the amount of the substance if we compare homogeneous systems of different size but of the same chemical composition under the same external thermal conditions. The coefficients occurring in the total differential of the energy $E$ as a function of the extensities are $T, \Pi, \mu$. They are constant for the above systems of different size. Therefore they are called *intensive parameters* or *intensities*. In fact, as a rule, the intensities do not depend on the volume, whereas the extensities do. The intensities $\Pi$, $\mu$ differ from the coefficients $\beta_\sigma$ in eq. (7.2.8) only by the factor $1/T$. Indeed, usually the coefficients $\beta_\sigma$ in eq. (7.2.8) also are called intensities. We can describe the equilibrium state of a system alternatively by the entire set of extensities or by the entire set of intensities, or by a mixed set of variables. In this context all these variables are called 'thermal variables'. $\beta_\sigma$ and $M^\sigma$ are said to be 'thermally conjugated' to each other.

The intensive parameters have an important property: usually they coincide for systems in thermal equilibrium if the systems are in unrestricted thermal contact. Such a contact allows the exchange of energy, of the particle number, or of other extensities $M^\beta$. To give an example let us consider two systems $\Sigma'$ and $\Sigma''$ in thermal contact. The energies of the systems are $E'$ and $E''$, respectively. The systems can exchange energy in such a way that the total energy

$$E = E' + E'' \qquad (7.2.10)$$

remains constant. The two systems together form a compound system. In equilibrium it is described by a generalized canonical distribution where the inverse temperature $\beta = 1/T$ occurs as the same factor for both energies

$$\beta E_i = \beta(E_i' + E_i''). \qquad (7.2.11)$$

So far we have neglected an interaction energy $W$ between the two systems. As a rule, in macroscopic thermodynamics this can be done, because $W$ is a negligible surface contribution, whereas $E', E''$ are proportional to the volume. We may, however, include $W$ and write

$$E = E' + E'' + W. \qquad (7.2.12)$$

Also in this case, we have the same factor $\beta$ for both $E_i'$ and $E_i''$.

We should notice that the conventional phenomenological thermodynamic description is based on the fundamental assumption that the temperature is equal for systems in equilibrium if an unrestricted exchange of heat is possible (the 'Zeroth Law' of thermodynamics). Generally all intensities conjugated to conservative quantities coincide in equilibrium if their exchange is not restricted. In particular, this is valid for the pressure if the systems are separated by movable walls. Moreover, it is valid for the chemical potential $\mu_\alpha$ of each chemical component $\alpha$ if the exchange of the corresponding particles is allowed.

The defining equation of the Helmholtz free energy

$$F = E - TS \tag{7.2.13}$$

corresponds to a Legendre transformation from $E(S, V, N)$ to $F(T, V, N)$. With the help of $F$, the Gibbs fundamental equation takes on the form

$$\mathrm{d}F = -\Pi\,\mathrm{d}V + \mu\,\mathrm{d}N - S\,\mathrm{d}T. \tag{7.2.14}$$

This means the following. If the variables $V, N, T$ that occur in this equation as differentials are chosen as independent variables, we obtain

$$\partial F/\partial V = -\Pi, \qquad \partial F/\partial N = \mu, \qquad \partial F/\partial T = -S. \tag{7.2.15}$$

The quantities $V, N, T$ are called the *natural variables* of $F$. They represent a complete set of thermal variables. By forming the derivatives of $F$ with respect to the natural variables we obtain another complete set of thermal variables. This connection is similar to the connection between forces and potentials in a mechanical system. Therefore $F$ as a function of its natural variables is called a *thermodynamic potential.*

By a Legendre transformation we can proceed to other thermodynamical potentials. For example, the Gibbs free energy

$$G = F + \Pi V \tag{7.2.16}$$

is connected with $F$ by a Legendre transformation from the variable $V$ to the variable $\Pi$. The Gibbs fundamental equation takes on the form

$$\mathrm{d}G = V\,\mathrm{d}\Pi + \mu\,\mathrm{d}N - S\,\mathrm{d}T. \tag{7.2.17}$$

Therefore $T, \Pi$ and $N$ are the natural variables of $G$. For $G(T, \Pi, N)$ this means

$$\partial G/\partial \Pi = V, \qquad \partial G/\partial N = \mu, \qquad \partial G/\partial T = -S. \tag{7.2.18}$$

Generally we can say the following. If we express the Gibbs fundamental equation in the form of a total differential of some thermodynamic quantity, we call the variables that occur as independent differentials the *natural variables* of this quantity. The thermodynamic quantity as a function of the just corresponding natural variables is the corresponding thermodynamic potential. The analogy to mechanics is not only due to eq. (7.2.18) and similar relations between forces and potentials, it extends to the fact that a mechanical equilibrium is determined by a minimum of the potential. Thermodynamic potentials possess an extremum as well (see section 6.3). If we ignore the trivial factor $T$, we can say that the different thermodynamic potentials are different kinds of Legendre transforms of the entropy. The entropy $S(M)$ itself is a thermodynamic potential according to eq. (7.2.4).

**The energetic and the entropic scheme** Eqs. (7.2.8) and (7.2.9) are identical in principle. Historically, however, they correspond to different ways of deriving thermodynamics from macroscopic experience. The preference for the form (7.2.9) rather than (7.2.8) corresponds to the distinction of energy as a fundamental quantity, which, indeed, in dynamics is distinguished as the generator of time shift and thus as the most important constant of motion. In general statistics, however, the entropy $S$ is distinguished as it is fully determined by the probability distribution alone, whereas the other extensities are associated with the single microstates. Thus the form (7.2.8) seems to be the more fundamental one. This scheme favours the definition of the coefficients $\beta_\sigma$ as intensities conjugated to the extensities $M^\sigma$ and favours the definition of $\Psi$ as a generalized 'free energy'. Up to the trivial factor $1/T$ it can be the Helmholtz, Gibbs, or another type of free energy.

In statistical mechanics, the discrimination between the two schemes is of no great importance. In phenomenological thermodynamics, however, the two schemes, which are called 'energetic' and 'entropic' schemes respectively by some authors (Ruppeiner 1983; Salamon *et al.* 1984, 1985), are connected with different ways of defining the entropy operationally by experimental processes. For the energetic scheme calorimetric processes are used, for the entropic scheme, on the other hand, processes are used that transform heat into work.

## *7.3 The Gibbs–Duhem equation

The general thermodynamic relations simplify for homogeneous systems. A physical system is called homogeneous if it possesses the same physical properties in all of its spatial parts. This means that the material composition coincides everywhere in the system. Moreover, on the microscopic level the dynamical laws are the same everywhere. We shall assume that the system is in a homogeneous thermal equilibrium state, where the intensities of all partial subsystems coincide (strictly speaking such an equilibrium state only exists if there is no external force such as gravitation and if the surface tension is neglected).

Let us compare the thermal variables of a spatial subsystem with those of the entire system. If $V$ is the volume of the entire system, let $\alpha V$ be the volume of the subsystem, with $0 < \alpha < 1$. Let us assume that the system contains a mixture of particles of different species. We denote the species by the subscript $\tau = 1, 2, \ldots$. The number of particles of species $\tau$ is $N^\tau$. The condition of homogeneity means the following: if $M^\sigma$ are the extensities of the whole system (such as the energy $E$, the volume $V$, and the number of particles $N^\tau$), the corresponding quantities of the subsystem are $\alpha M^\sigma$. This implies for the entropy

$$S(\alpha M) = \alpha S(M), \tag{7.3.1}$$

valid for all $\alpha \in [0, 1]$ (generally any function $f(x_1, \ldots, x_n)$ that fulfils the 'Euler homogeneity relation'

$$f(\alpha x_1, \ldots, \alpha x_R) = \alpha^n f(x_1, \ldots, x_R) \tag{7.3.2}$$

is called a 'homogeneous function of degree $n$'. The entropy is a homogeneous function of degree $n = 1$ of the extensities $M^\sigma$). Differentiation of eq. (7.3.1) with respect to $\alpha$ and putting $\alpha = 1$ afterwards yields

$$S = (\partial S/\partial M^\sigma)M^\sigma. \tag{7.3.3}$$

Using eq. (7.2.7), namely

$$\partial S/\partial M^\sigma = \beta_\sigma, \tag{7.3.4}$$

we obtain

$$S = \beta_\sigma M^\sigma \tag{7.3.5}$$

or explicitly

$$TS = E + \Pi V - \sum_\tau \mu_\tau N^\tau. \tag{7.3.6}$$

According to eqs. (7.2.13) and (7.2.16) we may also write eq. (7.3.6) in the form

$$G = \sum_\tau \mu_\tau N^\tau. \tag{7.3.7}$$

This relation is by convention called the *Gibbs–Duhem equation.*

In the case that there is only one species of particle, we obtain

$$G = \mu N. \tag{7.3.8}$$

This equation provides us with an important interpretation of the chemical potential $\mu$ of a pure substance: it is the Gibbs free energy per particle. Notice that usually, in physical chemistry, the mole number rather than the particle number $N$ is used. Then the 'molar' chemical potential $\mu$ is the Gibbs free energy per mole. It is $L$-times larger than the former, where $L$ is the 'Loschmidt number', the number of particles in one mole.

In general statistics, we may use the name 'Gibbs–Duhem equation' for the general form of eq. (7.3.5). In section 20.4 we shall see that eq. (7.3.8) has an analogue in the theory of dynamical systems.

## 7.4 Susceptibilities and fluctuations

Now we come to the second consequence of the inequality (7.2.3), namely that the second order term in the variations $\delta\beta$ is never positive. Eq. (7.2.5) can also be written in the form

$$\delta M^\sigma\, \delta\beta_\sigma = (\partial M^\sigma/\partial\beta_\tau)\, \delta\beta_\sigma\, \delta\beta_\tau \leqslant 0. \tag{7.4.1}$$

We call the quantities

$$Q^{\sigma\tau} = -\partial M^\sigma/\partial\beta_\tau \tag{7.4.2}$$

'susceptibilities'. They are the elements of the *susceptibility matrix*. We can also write eq. (7.4.1) in the original form of eq. (7.2.5):

$$-(\partial\beta_\sigma/\partial M^\tau)\, \delta M^\sigma\, \delta M^\tau \geqslant 0. \tag{7.4.3}$$

The quantities

$$R_{\sigma\tau} = -\partial\beta_\sigma/\partial M^\tau \tag{7.4.4}$$

are also called susceptibilities. They are the elements of the inverse matrix of $Q^{\sigma\tau}$, since

$$Q^{\sigma\mu} R_{\mu\tau} = \frac{\partial M^\sigma}{\partial\beta_\mu}\frac{\partial\beta_\mu}{\partial M^\tau} = \frac{\partial M^\sigma}{\partial M^\tau} = \delta_{\sigma\tau}. \tag{7.4.5}$$

Differentiating

$$M^\sigma = \sum_i M_i^\sigma \exp(\Psi - \beta_\tau M_i^\tau) \tag{7.4.6}$$

with respect to $\beta_\tau$ we are led to

$$Q^{\sigma\tau} = \langle \tilde{M}^\sigma \tilde{M}^\tau \rangle - \langle \tilde{M}^\sigma \rangle \langle \tilde{M}^\tau \rangle, \tag{7.4.7}$$

where we have used eq. (7.2.7). For the *fluctuations* of $M^\sigma$

$$\Delta M_i^\sigma = M_i^\sigma - \langle \tilde{M}^\sigma \rangle \tag{7.4.8}$$

we can define the *correlation matrix* of the fluctuations

$$K^{\sigma\tau} = \langle \Delta \tilde{M}^\sigma \, \Delta \tilde{M}^\tau \rangle. \tag{7.4.9}$$

This is equal to the susceptibility matrix $Q^{\sigma\tau}$, because

$$\langle (\tilde{M}^\sigma - \langle \tilde{M}^\sigma \rangle)(\tilde{M}^\tau - \langle \tilde{M}^\tau \rangle) \rangle = \langle \tilde{M}^\sigma \tilde{M}^\tau \rangle - \langle \tilde{M}^\sigma \rangle \langle \tilde{M}^\tau \rangle. \tag{7.4.10}$$

We may write eqs. (7.4.1) and (7.4.3) in the form

$$\delta^2 S = \frac{\partial^2 S}{\partial M^\sigma \, \partial M^\tau} \delta M^\sigma \, \delta M^\tau \leqslant 0 \tag{7.4.11}$$

$$\delta^2 \Psi = \frac{\partial^2 \Psi}{\partial \beta_\sigma \, \partial \beta_\tau} \delta \beta_\sigma \, \delta \beta_\tau \leqslant 0. \tag{7.4.12}$$

The statements mean that the entropy $S(M)$ is a concave function of the extensities $M^\sigma$ and that the free energy $\Psi(\beta)$ is a concave function of the intensities $\beta_\sigma$. Independently of that, we have already stated in section 5.2 that the entropy is concave in the probability space as well, that is to say as a function of the probabilities $p_i$.

*Remark*: Notice that the concavity of $\Psi$ with respect to $\beta_\sigma$ does not contradict the principle of minimum free energy described in section 6.3. This principle states that for fixed intensities $\Psi\,[p]$ takes on a minimum in the space of *all possible probability distributions*. Thus it is convex as a function of the probabilities. This, however, has nothing to do with the concavity of the equilibrium free energy $\Psi(\beta)$ in the space of the intensities $\beta_\sigma$.

The equivalence of the correlation and the susceptibility matrix in particular yields a relation for the variance of an extensity $M^\sigma$. The variance is the mean square of the fluctuations, it is related to the diagonal elements of the susceptibility matrix:

$$\langle (\Delta \tilde{M}^\sigma)^2 \rangle = -\partial M^\sigma / \partial \beta_\sigma = Q^{\sigma\sigma}. \tag{7.4.13}$$

As a rule, the right hand side is of the same order of magnitude as $M^\sigma/\beta_\sigma$ (there are exceptions if the system exhibits a phase transition: then the right hand side of eq. (7.4.13) may diverge, see next section). The square root of the variance of a random quantity is a measure of the average size of its fluctuations. The relative variance of $M^\sigma$ is of the order

$$\langle(\Delta\tilde{M}^\sigma)^2\rangle/(M^\sigma)^2 \approx 1/|\beta_\sigma M^\sigma|. \tag{7.4.14}$$

The product $|\beta_\sigma M^\sigma|$ is an essential part of the bit-number necessary to fix a microstate. It is an extremely large number for a macroscopic system. This means that the fluctuations decrease with the size of the system according to eq. (7.4.14) and become extremely small in a macroscopic system. This was used in section 6.2 to identify the mean value of an extensity in one ensemble with the sharp value in another ensemble, for instance the mean energy in the canonical ensemble with the sharp value of the energy in the microcanonical ensemble.

## 7.5 Phase transitions

Some of the most striking and interesting phenomena described in statistical mechanics are the so called *phase transitions.* A physical system is said to exhibit a phase transition if by a very small continuous change of an appropriate thermal variable the behaviour of the system changes abruptly. Examples are solid–liquid and liquid–gas transitions, transitions between paramagnetic and ferromagnetic states, transitions between different crystal structures, transitions between homogeneous and inhomogeneous states of liquid mixtures, transitions between the normal conducting and the superconducting state, and the transition between the normal fluid and the superfluid state of $He^4$. Moreover, one has the so called non-equilibrium phase transitions between different stationary nonequilibrium states. Common to all these phase transitions is the fact that there is a critical point of an appropriate control variable (or more general, a critical line) where certain properties of the system change in a nonsmooth way. For equilibrium systems this change usually manifests itself in a non-analytic behaviour of the corresponding thermodynamical potential. We classify phase transitions according to the degree of this nonanalyticity. If the $k$th derivative of the (generalized) free energy with respect to the relevant thermal variable does not exist at the critical point, whereas the $(k + 1)$th derivative does exist, the phase transition is said to be of *kth order*. We refer the reader to the large amount of literature on phase transitions (Stanley 1971; Griffiths 1972; Ma 1976 and references therein). Phase transitions of first and second order possess analogues in the thermodynamics of chaotic systems as well (see chapter 21).

**Critical exponents** At second order phase transition points, in typical cases susceptibilities (or, in general, appropriate order parameters) either diverge or go to zero, obeying a nontrivial scaling law. For example the specific heat for constant volume

$$c_V = -T\left(\frac{\partial^2 F}{\partial T^2}\right)_V \tag{7.5.1}$$

or the compressibility

$$\kappa_T = -\frac{1}{V}\left(\frac{\partial^2 G}{\partial \Pi^2}\right)_T \tag{7.5.2}$$

scale near the critical temperature $T_c$ where the two phases of liquid and vapour cease to coexist as

$$c_V \sim |\tau|^{-\alpha} \qquad \alpha = 0.11 \tag{7.5.3}$$

$$\kappa_T \sim |\tau|^{-\gamma} \qquad \gamma = 1.25. \tag{7.5.4}$$

Here

$$\tau = (T - T_c)/T_c \tag{7.5.5}$$

denotes the relative deviation of the temperature from the critical point (a dimensionless quantity). According to the considerations presented in the previous sections, a diverging susceptibility implies macroscopic fluctuations at the critical point. For example, the divergence of the compressibility $\kappa_T$ implies that the density fluctuations become macroscopic. Indeed, they give rise to an optical phenomenon, the so called 'critical opalescence'.

In the theory of critical phenomena, a variety of scaling laws similar to eqs. (7.5.3) and (7.5.4) are known, for example, for the liquid–gas density difference, correlation functions, the magnetization of spin systems, and many other quantities (Stanley 1971). In general, $\tau$ need not be a temperature difference but is the relative deviation of an appropriate control parameter from the critical value. Quite generally, in all these cases a characteristic quantity $f$ (the so called order parameter) scales as

$$|f(\tau)| \sim |\tau|^{\lambda} \qquad (\tau \to 0). \tag{7.5.6}$$

The exponent

$$\lambda = \lim_{\tau \to 0} \frac{\ln|f(\tau)|}{\ln|\tau|} \tag{7.5.7}$$

is called a *critical exponent*. A powerful method of calculating numerical values for critical exponents by means of statistical mechanics is the renormalization group approach (see, for example, Ma (1976)). A remarkable fact is the *universality* of the critical exponents: they do not depend on the details of the thermodynamic system, but have the same value for entire classes of systems.

In nonlinear dynamics, the period doubling scenario from ordered to chaotic behaviour (see section 1.3) can also be regarded as a critical phenomenon. The critical exponents are essentially given by the Feigenbaum constants. We also have universality, and a renormalization group treatment is possible (see section 21.7).

# 8
# Spin systems

We present a short introduction to various types of spin models commonly used in statistical mechanics. Different types of interactions are discussed, and we report on some of the known results on the possibility of phase transitions for spin systems with a certain interaction. The equivalence between symbolic stochastic processes and one-dimensional spin systems is elucidated. Finally we present a short introduction to the transfer matrix method.

## 8.1 Various types of spin models

A ferromagnet can be regarded as a system of a large number of elementary magnets that are placed at the sites of a crystal lattice. To model and understand the magnetic properties of solids, various types of lattice spin models have been introduced in statistical mechanics. Such a model of a spin system is defined by the lattice type, the possible values of the spins at each lattice site, and the interaction between the spins.

In this context the name '*spin*' is used in quite a general sense: the spin is a random variable with certain possible values. These may be either discrete or continuous; moreover, in this generalized sense, the spin can be a scalar, vector, or tensor. Also the lattice, originally introduced to model the crystal structure of the solid, sometimes does not possess a physical realization in nature.

Quite common choices of lattice structures are cubic, triangular, or hexagonal lattices in dimensions $d = 1, 2, 3, 4, \ldots$. But one also studies so called hierarchical lattices, which have a self-similar structure. Given a lattice, let us label the lattice sites in an appropriate way by an index $n = 1, \ldots, N$, where $N$ is the total number of lattice sites. In the thermodynamic limit, the number $N$ goes to infinity. Actually, in $d$

dimensions the thermodynamic limit usually means extending the lattice to infinity for *all* directions of the $d$-dimensional space. After having chosen the lattice type, the next step to define a spin model is to determine the possible values of the spin $s_n$ at the lattice sites. The most popular choices are the following:

**Ising model** Only the $z$-component of the physical spin is taken into account, the other components are neglected. $s_n$ can take on just two values $s_n = +1$ or $s_n = -1$.

***XY* model** Each spin is a complex number of absolute value 1:

$$s_n = \exp(\mathrm{i}\phi_n)$$

**Heisenberg model** In the original Heisenberg model, each spin is described by three quantum mechanical Pauli matrices. In a classical model, each spin is a unit vector pointing in an arbitrary direction of the three-dimensional space.

**Potts model of *R*th order** Each spin is directed in the $z$-direction and can take on $R$ different discrete values $s_n$ labelled $1, 2, \ldots, R$ (the labelling of the states is, of course, arbitrary). The Potts model of second order is also called 'lattice gas system' and is equivalent to the Ising spin system.

The final step in defining the lattice spin model is to determine the *interaction* between the spins. This is implicitly done by defining a Hamilton function $H_N(s_1, \ldots, s_N)$ which attributes to each spin configuration $(s_1, \ldots, s_N)$ a certain energy $H_N$. The function $H_N$ determines the probability $p_N(s_1, \ldots, s_N)$ to observe the microstate $(s_1, \ldots, s_N)$ in a canonical ensemble. Namely, for a system with inverse temperature $\beta$ the maximum entropy principle (see section 6.1) yields

$$p_N(s_1, \ldots, s_N) = \frac{1}{Z_N(\beta)} \exp[-\beta H_N(s_1, \ldots, s_N)]. \tag{8.1.1}$$

Here the partition function $Z_N$ is again defined by the sum over all possible microstates:

$$Z_N(\beta) = \sum_{s_1, \ldots, s_N} \exp[-\beta H_N(s_1, \ldots, s_N)]. \tag{8.1.2}$$

It may happen that certain spin configurations never occur; they are forbidden due to certain constraints on the system. For example, consider a lattice gas model with $s_n = 0$ or $s_n = 1$. Such a system can model a liquid consisting of two different species of atoms, represented by the two different symbols. Indeed, it is often observed in physical systems that certain atomic configurations never occur. Formally these configurations have an infinite energy. It is then convenient to restrict the summation in the partition function (8.1.2) to the allowed spin configuration.

It is intuitively clear that in a physical system spins that are sited at neighbouring lattice sites will strongly influence each other, whereas spins far away from each other will hardly have any influence on each other. This means that in many cases the energy $H_N(s_1, \ldots, s_N)$ does not explicitly depend on all spins but only on the relative states of nearest-neighbour spins. As the most common example, let us consider the Ising system with nearest-neighbour interaction on a $d$-dimensional cubic lattice in a constant magnetic field $\vec{B}$. To each microstate $(s_1, \ldots, s_N) = \{s\}$ of the spin system we attribute the energy

$$H_N\{s\} = -J \sum_{[ij]} s_i s_j - B \sum_{i=1}^{N} s_i. \tag{8.1.3}$$

Here $[ij]$ denotes summation over nearest neighbours on the lattice, and $J$ is a coupling constant. This choice of $H$ is physically motivated by the fact that the interaction energy of two quantum mechanical momenta $\vec{L}$, $\vec{S}$ is proportional to the scalar product $(\vec{L} \cdot \vec{S})$. When $\vec{L}$ and $\vec{S}$ are spins with only two possible values of the $z$-component $L_z = \pm\frac{1}{2}$, $S_z = \pm\frac{1}{2}$, the first term on the right hand side of eq. (8.1.3) corresponds to the ansatz that the interaction energy is proportional to $L_z S_z$. This ansatz is a simplification and characterizes the Ising model. The interaction of a quantum mechanical angular momentum $\vec{S}$ with a constant magnetic field $\vec{B}$ in the $z$-direction is proportional to $(\vec{B} \cdot \vec{S}) = BS_z$, which leads to the second term in eq. (8.1.3).

Solving a spin model of statistical mechanics means finding an expression for the free energy $f$ per spin in the thermodynamic limit of infinitely many spins:

$$f(\beta) = -\lim_{N \to \infty} \frac{1}{\beta N} \ln Z_N(\beta). \tag{8.1.4}$$

It is clear that generally the free energy does not only depend on the inverse temperature $\beta$ but also on further coupling constants that enter via the Hamiltonian $H_N$.

The one-dimensional Ising model with nearest-neighbour interaction was solved by Ising in 1925 (Ising 1925) and the two-dimensional model by Onsager in 1944 (Onsager 1944). An analytic solution of the model in $d \geqslant 3$ is still lacking. But, of course, all Ising models have been carefully studied in numerical experiments.

More general interactions than nearest-neighbour interactions can be defined as follows. Consider an arbitrary subset $A$ of the lattice containing $N_A$ sites. These lattice sites of $A$ do not have to be neighbouring ones, they may be distributed in an arbitrary way on the lattice. We define a more general spin variable $S_A$ of the subset $A$ by the product of the spin variables in this subset:

$$S_A = \prod_{n \in A} s_n. \tag{8.1.5}$$

To each $A$ we associate a coupling constant $J_A$, taking on values in the reals. $J_A$ determines the strength of the interaction of the spins contained in $A$. Now the Hamiltonian is defined as

$$H_N = -\sum_A J_A S_A. \tag{8.1.6}$$

The summation runs over all possible subsets $A$ of the lattice. In practice, of course, many of the coupling constants $J_A$ are chosen to be zero. For example, one usually restricts oneself to pair interactions, i.e., to subsets $A$ containing only two lattice sites. These, however, may be far apart from each other, i.e., the interaction is still much more general than the nearest-neighbour interaction. The coupling constants $J_A$ are then indexed by two indices $i, j$ representing the two lattice sites belonging to $A$.

Of course, a well defined thermodynamic limit of the free energy per spin will only exist if the coupling constants satisfy certain conditions. For Ising type models it is sufficient to postulate that for arbitrary lattice sites $i$

$$\sum_j |J_{ij}| < \infty. \tag{8.1.7}$$

This means essentially that for a $d$-dimensional lattice the coupling strength $|J_{ij}|$ decreases slightly more strongly than $|\vec{r}_i - \vec{r}_j|^{-d}$, where $\vec{r}_i, \vec{r}_j$ are the positions of the lattice sites $i, j$. For arbitrary interactions the condition (8.1.7) is generalized to

$$\sum_A{}' |J_A|/N_A < \infty. \tag{8.1.8}$$

Here the sum runs over all subsets $A$ that contain the spin $i$, and $N_A$ is the number of lattice sites contained in $A$ (see, for example, Griffiths (1972)).

## 8.2 Phase transitions of spin systems

Many spin models of statistical mechanics exhibit a phase transition behaviour. In dimensions $d \geqslant 2$ this is a generic phenomenon even if there is just a nearest-neighbour interaction. For example, the Ising model with nearest-neighbour interaction exhibits a phase transition for all dimensions $d \geqslant 2$. For the thermodynamic analysis of chaotic systems the case of one-dimensional rather than of higher-dimensional lattices is chiefly of interest, because here we have an equivalence with symbolic stochastic processes (see next section). The phase transition behaviour of one-dimensional lattice spin systems is somewhat exceptional compared to the higher-dimensional case. For example, the one-dimensional Ising model with nearest-neighbour interaction does not exhibit a phase transition, whereas models with dimension $d \geqslant 2$ do. Quite generally one can prove that the Ising model in one dimension can only exhibit a phase transition if there is an appropriate long-range interaction between the spins. Consider a Hamiltonian of the pair interaction form

$$H = \sum_n \sum_{m \geqslant 1} J(m) s_n s_{n+m} \qquad J(m) \geqslant 0, \tag{8.2.1}$$

for a nearest-neighbour interaction

$$J(m) = \begin{cases} 1 & \text{for } m = 1 \\ 0 & \text{otherwise,} \end{cases} \tag{8.2.2}$$

for an exponentially decreasing interaction

$$J(m) \sim \exp(-\gamma m) \qquad \gamma > 0, \tag{8.2.3}$$

and for a polynomially decreasing interaction

$$J(m) \sim 1/m^{\alpha} \qquad \alpha > 0. \tag{8.2.4}$$

In order for the thermodynamic limit to exist, we must have $\alpha > 1$ (see the consistency condition (8.1.7)). Ruelle has proved that an Ising system with $\alpha > 2$ does not exhibit a (first order) phase transition (Ruelle 1968), whereas Dyson showed that for $1 < \alpha < 2$ there is a first order phase transition (Dyson 1969) (no statement is given for the limiting case $\alpha = 2$). We notice that for one-dimensional lattices an exponentially decreasing interaction (and, of course, a nearest-neighbour interaction) is too weak to lead to a phase transition behaviour. In the one-dimensional case a strong interaction of spins is necessary in order to observe nonanalytic behaviour of the free energy.

Reviews of rigorous results on spin models and their phase transition behaviour can, for example, be found in Ruelle (1969), Griffiths (1972), Mayer (1980) and Baxter (1982). Phase transitions play an important role in almost all branches of physics – from solid state physics to elementary particle physics (lattice field theories) and astrophysics (cosmology). It is a relatively recent observation that phase transitions also occur quite frequently in the thermodynamic description of chaotic systems. We shall come back to this point in chapter 21.

## 8.3 Equivalence between one-dimensional spin systems and symbolic stochastic processes

Let us consider a one-dimensional spin system with the Hamilton function $H_N(s_1, \ldots, s_N)$. We assume that the spins can take on $R$ different discrete states labelled by the subscript $1, \ldots, R$, i.e., we consider a *Potts model.* The total number $N$ of spins is assumed to be large, but fixed. As already mentioned in section 8.1, the probability of observing the spin configuration $(s_1, \ldots, s_N)$ is given by

$$p_N(s_1, \ldots, s_N) = \frac{1}{Z_N(\beta)} \exp[-\beta H_N(s_1, \ldots, s_N)]. \tag{8.3.1}$$

Let us now choose an arbitrary subset $A = (n_1, \ldots, n_K)$ of $K$ lattice sites (not necessarily neighbouring ones). The probability $P_{\{A\}}$ of observing the spins $s_{n_1}, \ldots, s_{n_K}$ on this subset $A$ of lattice sites is

$$P_{\{A\}}(s_{n_1}, \ldots, s_{n_K}) = \sum_{s^*} p_N(s_1, \ldots, s_N). \tag{8.3.2}$$

On the right hand side the symbol $s^*$ indicates that we sum over all spins $s_1, \ldots, s_N$ with the exception of those at the sites $n_1, \ldots, n_K$. In particular, we can choose for $A = (n_1, \ldots, n_K)$ the first $K$ sites $1, 2, \ldots, K$ of the lattice, obtaining

$$\begin{aligned} p_K(s_1, \ldots, s_K) &= \sum_{s_{K+1}, \ldots, s_N} p_N(s_1, \ldots, s_N) \\ &= \frac{1}{Z_N(\beta)} \sum_{s_{K+1}, \ldots, s_N} \exp[-\beta H_N(s_1, \ldots, s_N)] \\ &= \frac{1}{Z_K(\beta)} \exp[-\beta H_K(s_1, \ldots, s_K)] \end{aligned} \tag{8.3.3}$$

(for simplicity we neglect all boundary effects). We can now interpret the hierarchy of joint probabilities $p_K(s_1, \ldots, s_K)$ as the probabilities of a symbolic stochastic process. The discrete time variable of the process corresponds to the lattice coordinate $K$, the time evolution of the stochastic process corresponds to translation on the one-dimensional lattice. A stationary stochastic process is defined by the property

$$p_K(s_1, \ldots, s_K) = p_K(s_{1+\tau}, \ldots, s_{K+\tau}) \tag{8.3.4}$$

for arbitrary $\tau \in \mathbb{N}$. For the spin model this means that the Hamiltonian $H_K(s_1, \ldots, s_K)$ is invariant under translation on the lattice:

$$H_K(s_1, \ldots, s_K) = H_K(s_{1+\tau}, \ldots, s_{K+\tau}). \tag{8.3.5}$$

We may also introduce conditional probabilities $p_{1|K}(s_{K+1} \,|\, s_1, \ldots, s_K)$ defined by

$$p_{K+1}(s_1, \ldots, s_{K+1}) = p_{1|K}(s_{K+1} \,|\, s_1, \ldots, s_K) p_K(s_1, \ldots, s_K). \tag{8.3.6}$$

For the spin model the difference

$$H_{K+1}(s_1, \ldots, s_{K+1}) - H_K(s_1, \ldots, s_K) = W_{1|K}(s_{K+1} \,|\, s_1, \ldots, s_K) \tag{8.3.7}$$

of the energies of the spin configurations $(s_1, \ldots, s_{K+1})$ and $(s_1, \ldots, s_K)$ is called the *interaction energy* of the spin $s_{K+1}$ with the configuration $(s_1, \ldots, s_K)$. As

$$\begin{aligned} p_{1|K}(s_{K+1} \,|\, s_1, \ldots, s_K) &= \frac{p_{K+1}(s_1, \ldots, s_{K+1})}{p_K(s_1, \ldots, s_K)} \\ &= \frac{Z_K(\beta)}{Z_{K+1}(\beta)} \exp\{-\beta[H_{K+1}(s_1, \ldots, s_{K+1}) - H_K(s_1, \ldots, s_K)]\} \end{aligned} \tag{8.3.8}$$

the conditional probability $p_{1|K}$ is proportional to the exponential of the interaction energy times $-\beta$:

$$p_{1|K}(s_{K+1} \,|\, s_1, \ldots, s_K) \sim \exp[-\beta W_{1|K}(s_{K+1} \,|\, s_1, \ldots, s_K)]. \tag{8.3.9}$$

Thus conditional probabilities of the symbolic stochastic process correspond to interaction energies of the spin system and vice versa.

In general, all properties of the spin system can be 'translated' into properties of the symbolic stochastic process and vice versa. This, indeed, is the root for the applicability of methods of statistical mechanics for the dynamical analysis of chaotic systems.

## 8.4 The transfer matrix method

The transfer matrix method is an important tool for calculating the free energy of a spin system in the thermodynamic limit. We shall illustrate this method by means of a very simple example, namely the one-dimensional Ising model with nearest-neighbour interaction, and shall afterwards point out the generalization to more general spin systems. The partition function for a system in a magnetic field is

$$Z_N = \sum_{s_1,\ldots,s_N} \exp\left(\beta J \sum_{j=1}^{N} s_j s_{j+1} + \beta B \sum_{j=1}^{N} s_j\right). \tag{8.4.1}$$

It is useful to consider periodic boundary conditions, i.e., the spin $s_{N+1}$ is identified with $s_1$. Now the key observation is that the exponential in eq. (8.4.1) can be written as a product of factors $v$ that depend on nearest-neighbouring spins only: defining

$$v(s, s') = \exp[\beta J s s' + \tfrac{1}{2}\beta B(s + s')], \tag{8.4.2}$$

we can write

$$Z_N = \sum_{s_1,\ldots,s_N} v(s_1, s_2)v(s_2, s_3)\cdots v(s_{N-1}, s_N)v(s_N, s_1) \tag{8.4.3}$$

(the choice of $v$ is not unique, but the above choice has the advantage that $v(s, s')$ is symmetric under the exchange of $s$ and $s'$). The tuple $(s, s')$ can take on four different values $(+1, +1)$, $(+1, -1)$, $(-1, +1)$, $(-1, -1)$. Hence we may define a matrix $V$, the so called transfer matrix, that contains the four possible values of $v(s, s')$ as matrix elements:

$$V = \begin{pmatrix} v(+1, +1) & v(+1, -1) \\ v(-1, +1) & v(-1, -1) \end{pmatrix} = \begin{pmatrix} \exp[\beta(J + B)] & \exp(-\beta J) \\ \exp(-\beta J) & \exp[\beta(J - B)] \end{pmatrix}. \tag{8.4.4}$$

Now we observe that the summation over $s_2, \ldots, s_N$ in eq. (8.4.3) can be interpreted as corresponding to a matrix multiplication of $N$ matrices $V$. Moreover, the summation over $s_1$ can be interpreted as taking the trace of the product matrix $V^N$. Hence, we can write

$$Z_N = \text{trace } V^N. \tag{8.4.5}$$

As $V$ is a symmetric matrix, we can find two orthogonal and linear independent eigenvectors $x_1, x_2$ and two corresponding eigenvalues $\lambda_1, \lambda_2$ such that

$$Vx_j = \lambda_j x_j. \tag{8.4.6}$$

It is well known that the trace of a matrix is invariant under equivalence transformations. Hence

$$Z_N = \text{trace}\begin{pmatrix} \lambda_1 & 0 \\ 0 & \lambda_2 \end{pmatrix}^N = \lambda_1{}^N + \lambda_2{}^N. \tag{8.4.7}$$

Let $\lambda_1 > \lambda_2$. Then we obtain for the free energy $f$ per spin in the thermodynamic limit $N \to \infty$

$$\begin{aligned} f(\beta) &= -\lim_{N\to\infty} \frac{1}{\beta N} \ln Z_N \\ &= -\lim_{N\to\infty} \frac{1}{\beta N} \ln\left\{\lambda_1{}^N\left[1 + \underbrace{\left(\frac{\lambda_2}{\lambda_1}\right)^N}_{\to 0}\right]\right\} \\ &= -\frac{\ln \lambda_1}{\beta}, \end{aligned} \tag{8.4.8}$$

i.e., in the thermodynamic limit the free energy is determined by the largest eigenvalue of the transfer matrix. This is quite a general result of statistical mechanics, valid for much more general classes of spin systems than the one-dimensional Ising system with nearest-neighbour interaction.

In our case, the eigenvalues of the matrix $V$ given by eq. (8.4.4) can easily be determined:

$$\lambda_{1/2} = \exp(\beta J) \cosh \beta B \pm \exp(\beta J)(\cosh^2 \beta B - 2\exp(-2\beta J)\sinh 2\beta J)^{1/2}. \tag{8.4.9}$$

The largest eigenvalue corresponds to the plus sign and yields the free energy $f$ of our spin system in the thermodynamic limit via eq. (8.4.8). Notice that $f$ is a smooth analytical function; the one-dimensional Ising model with nearest-neighbour interaction does not exhibit a phase transition.

The transfer matrix method can be easily generalized to more general one-dimensional spin systems with finite range interaction. An interaction of finite range $K$ means

$$J_{ij} = 0 \qquad \text{for } |i - j| > K \tag{8.4.10}$$

(see section 8.1 for the definition of the coupling constant $J_{ij}$). If the spins take on discrete values, the transfer matrix for a spin system with finite range interaction is a finite-dimensional matrix. In the case of continuously varying spins, as well as for exponentially decreasing interactions of discrete spins, the transfer matrix is no longer finite but has to be replaced by an appropriate operator (see, for example, Mayer

(1980)). Also for higher-dimensional lattices the dimension of the transfer matrix grows rapidly, which, of course, makes it much more difficult to determine the largest eigenvalue. For example, for an Ising system with nearest-neighbour interaction on a *finite* two-dimensional lattice of size $L^2$ the transfer matrix is a $2^L \times 2^L$ matrix, i.e., the dimension of the matrix goes to infinity in the thermodynamic limit (see, for example, Baxter (1982)).

Part III · Thermostatistics of multifractals

# 9
# Escort distributions

Suppose we have an arbitrary, possibly fractal, probability distribution. For example, this might be the natural invariant density of a chaotic map. We wish to analyse the most important properties of this complicated distribution in a quantitative way. A fundamental idea, which has turned out to be very useful in nonlinear dynamics, is the following. To a given probability distribution a set of further probability distributions is attributed, which, in a way, have the ability to scan the structure of the original distribution. These distributions have the same form as thermodynamic equilibrium distributions and yield the key for various analogies between chaos theory and thermodynamics. We shall call them 'escort distributions'. The considerations of this chapter are valid for arbitrary probability distributions, no matter how they are generated.

## 9.1 Temperature in chaos theory

Whereas in statistical thermodynamics probability distributions are sought on the basis of incomplete knowledge (usually only a few macroscopic variables are given, which are interpreted as statistical mean values), in chaos theory the central question can sometimes be reversed in a sense: here the distributions $p$ may be given in the form of observed relative frequencies in a computer experiment. Then a characterization of the system is sought using relatively few global quantities that describe the relevant features. In thermodynamics we were led to the generalized canonical distribution by the unbiased guess. In nonlinear dynamics, we shall also be formally led to canonical distributions, but in a different way, which we shall discuss in this chapter and many of those which follow.

Let us consider general statistics and let

$$p_i = \exp(-b_i) \tag{9.1.1}$$

be the observed relative frequencies of events $i$ of a sample set, which we also call microstates. In nonlinear dynamics, such an event may be the occurrence of an iterate in the $i$th cell of a grid, as already described in section 3.1. Or in another context, an entire symbol sequence may be considered as an event $i$. Provided $p_i \neq 0$, we can always form the normalized powers $(p_i)^\beta$, that means the probability distributions

$$P_i = (p_i)^\beta \Big/ \sum_{j=1}^{r} (p_j)^\beta, \tag{9.1.2}$$

Here $\beta$ is an arbitrary real parameter. As it may take on negative values, we restrict the summation to events $j$ with nonzero probability. The number $r$ of events with nonzero probability is, in general, different from the total number $R$ of events of the entire sample set. We shall call $P$ the *escort distribution* of $p$ of order $\beta$. A change of the order $\beta$ will change the relative weights of how the events $i$ enter into the escort distribution. It is obvious that quite generally global quantities, such as the different types of information or mean values, formed with several of these escort distributions of different order, will give more revealing information than those formed with $p$ only. Indeed, changing $\beta$ is a tool for scanning the structure of the distribution $p$: the smaller the parameter $\beta$, the more heavily the small probabilities are weighted. On the other hand, for larger $\beta$ the large probabilities are weighted more heavily. Indeed, for $\beta \to +\infty$ the largest $p_i$ dominates the sum, whereas for $\beta \to -\infty$ the smallest (nonzero) $p_i$ yields the dominating contribution. To illustrate the analogy with statistical mechanics, we write the escort distributions as

$$P_i = \exp(\Psi - \beta b_i), \tag{9.1.3}$$

where $\Psi(\beta) = -\ln Z(\beta)$, and the partition function $Z(\beta)$ is defined by

$$Z(\beta) = \sum_{i=1}^{r} \exp(-\beta b_i) = \sum_{i=1}^{r} (p_i)^\beta. \tag{9.1.4}$$

This shows that an escort distribution has the form of a canonical distribution. The positive bit-number $b_i$ of $p_i$ plays the role of the energy $E_i$, and $1/\beta$ plays the role of the temperature $T$ in the thermodynamic analogy. The scanning of $p$ by a change of $\beta$ corresponds to a change of the temperature. For the complicated distributions appearing in nonlinear dynamics we shall consider negative values of $\beta$ as well. We shall see that in this context negative temperatures are indeed a useful tool to get more information.

Usually the free energy and other quantities diverge for $r$ going to infinity. Then the divergence rate is of interest. It will be described by the so called generalized *dimensions* of different type. This will be of central interest in the following chapters. In this chapter, however, we still restrict ourselves to the case of finite $r$.

We can see that the $\Psi(\beta)$ of an escort distribution $P$, up to the factor $\beta^{-1}$, is formally a Helmholtz free energy

$$F(\beta) = -\frac{1}{\beta} \ln Z(\beta) = \frac{1}{\beta} \Psi(\beta). \tag{9.1.5}$$

It is directly connected with the Rényi information of order $\beta$ of the original distribution by

$$I_\beta(p) = \frac{1}{\beta - 1} \ln \sum_{i=1}^{r} (p_i)^\beta = -\frac{1}{\beta - 1} \Psi(\beta). \tag{9.1.6}$$

For $\beta = 1$ we have already derived in section 5.1 that in this special case the Rényi information $I_\beta(p)$ reduces to the Shannon information of the original distribution $p$. For $\beta = 0$ the partition function $Z(0)$ is equal to the number $r$ of events $i$ with nonzero probability. Already these two special values of $\beta$ correspond to very different features of the distribution $p$. This example once more elucidates that the use of escort distributions with different values of $\beta$ enables us to scan the original distribution $p$ with respect to various properties.

In nonlinear dynamics we are usually concerned with probability distributions $p_i$ that depend on $i$ in a highly irregular way. Then it is a great advantage to find a description by a global quantity such as $I_\beta(p)$ that depends on the scanning parameter $\beta$ in a smooth way, at least for most values of $\beta$. Whereas in traditional thermodynamics temperature is always positive (with the exception of nonequilibrium systems with energy bounded from above, such as two-level systems), the escort distributions discussed here can also be constructed for negative $\beta$, since we agree upon restricting the sum to those probabilities $p_i$ that do not vanish.

## 9.2 Two or more intensities

Sometimes the probabilistic description of an event $i$ suggests a certain factorization. To give an example, the event $i$ may consist of a joint event $(j, k)$. Then the following factorization is always possible:

$$p_{jk} = Q(j \mid k) p_k. \tag{9.2.1}$$

Here $p_{jk}$ denotes the probability of the joint event $(j, k)$, $p_k$ is the probability of the event $k$, and $Q(j \mid k)$ is the conditional probability of $j$ under the condition that $k$ has already occurred. This is only one example of a possible factorization. If, in general, we have a factorization

$$p_i = u_i' u_i'', \tag{9.2.2}$$

we may again introduce the corresponding bit-numbers

$$b_i' = -\ln u_i', \qquad b_i'' = -\ln u_i''. \tag{9.2.3}$$

The escort distribution of $p$ then takes the form of a generalized canonical distribution with two random variables $b', b''$, corresponding to two intensities $\beta', \beta''$

$$P_i = \exp(\Psi - \beta' b_i' - \beta'' b_i''). \tag{9.2.4}$$

This can be considered as the analogue of, say, the grand canonical ensemble with chemical potential $-\beta''/\beta'$, or of the pressure ensemble with pressure $\beta''/\beta'$. It is no problem to generalize this procedure to distributions that factorize in three or more factors, corresponding to three or more intensities $\beta_\sigma$.

In general we can say that the escort distributions possess all the formal properties of the generalized canonical distributions. This allows us to apply to them all the general thermodynamic techniques described in the preceding chapters. Interesting behaviour is expected if the probability distribution under study has a sufficiently complex structure. In this case we expect a nontrivial dependence of the free energy on the inverse temperature $\beta$.

## *9.3 Escort distributions for general test functions

This section will deal with a generalization of the concept of escort distributions. Suppose that again a certain probability distribution $p$ is given. We can then choose some positive test function $\phi$ defined for all $x \in [0, 1]$ and form a new set of normalized probabilities as follows:

$$P_i = \frac{\phi(p_i)}{\sum_{j=1}^{r} \phi(p_j)}. \tag{9.3.1}$$

In the previous section we dealt with special test functions of the form

$$\phi(x) = x^\beta, \tag{9.3.2}$$

but it is clear that generally the function $\phi$ can be chosen to be any positive function in order to yield a probability distribution $P$. However, to scan the original distribution $p$, certain classes of functions are of greater interest than others. For example, it is reasonable to restrict the choice to convex (or concave) functions $\phi$, because then the quantity

$$I_{\phi(p)} = \sum_{j=1}^{r} \phi(p_j) \tag{9.3.3}$$

satisfies at least the Khinchin axioms I–III, i.e., the most elementary postulates for an information measure (Schlögl 1980). The consideration of general information-like measures of type (9.3.3) goes back to early work by Csiszàr (1963, 1967). Their importance for statistical mechanics was stressed by Ruch (1975).

An interesting question concerning generalized escort distributions is the following: what happens if we form the escort distribution of an escort distribution? Suppose we have two different test functions $\phi_1$, $\phi_2$ and the escort distributions

$$P_i = \frac{\phi_1(p_i)}{\sum_j \phi_1(p_j)}, \tag{9.3.4}$$

$$Q_i = \frac{\phi_2(p_i)}{\sum_k \phi_2(p_k)}. \tag{9.3.5}$$

We denote the escort distribution $P$ of the escort distribution $Q$ by $R$:

$$R_i = \frac{\phi_1(Q_i)}{\sum_j \phi_1(Q_j)}. \tag{9.3.6}$$

This expression can be further simplified if $\phi_1$ is a homogeneous function of degree $\beta$ (see section 7.3), i.e., if it satisfies

$$\phi_1(xy) = x^\beta \phi_1(y). \tag{9.3.7}$$

We then obtain

$$\phi_1(Q_i) = \phi_1\left\{\phi_2(p_i)\left[\sum_k \phi_2(p_k)\right]^{-1}\right\} = \left[\sum_k \phi_2(p_k)\right]^{-\beta} \phi_1(\phi_2(p_i)) \tag{9.3.8}$$

and

$$R_i = \frac{\phi_1 \circ \phi_2(p_i)}{\sum_j \phi_1 \circ \phi_2(p_j)}, \tag{9.3.9}$$

where the symbol $\circ$ denotes the composition of two functions, as already explained in section 2.3. In particular, for the standard choice of the previous sections

$$\phi_1(x) = x^{\beta_1}, \qquad \phi_2(x) = x^{\beta_2} \tag{9.3.10}$$

we obtain the trivial result that the escort distribution of order $\beta_1$ of an escort distribution of order $\beta_2$ is an escort distribution of order $\beta_1\beta_2$. This means that by forming escort distributions of escort distributions we do not get more information on the system. Mathematically, we may say: the escort distributions of order $\beta$ form a *group* with respect to the operation of taking the escort distribution of an escort distribution. The neutral element corresponds to $\beta = 1$, the inverse element of an escort distribution of order $\beta$ is an escort distribution of order $1/\beta$. The group is even commutative. This very simple group structure, however, is only observed for the test function $\phi(x) = x^\beta$. For nonhomogeneous test functions the escort distribution of an escort distribution cannot be expressed by the simple formula (9.3.9).

# 10
# Fractals

Fractals are complex geometrical objects that possess nontrivial structure on arbitrary scales. In this chapter we first describe a few examples of fractals with a simple recurrent structure. Starting from these simple examples we explain the concept of a 'fractal dimension' and 'Hausdorff dimension'. Finally, we consider some more complicated examples of fractals that are of utmost interest in nonlinear dynamics, yielding a glimpse of the beauty inherent in 'self-similar' structures: these are the Mandelbrot set, Julia sets, and fractals generated by iterated function systems.

## 10.1 Simple examples of fractals

**The Koch curve** A standard example of a fractal is the so called 'Koch curve'. It is constructed as follows. We start with an equilateral triangle with sides of unit length and divide each side into three equal parts. Then, as illustrated in fig. 10.1, we put onto the middle part of each side a smaller equilateral triangle with a third of the side length. This step is then repeated for each of the new sides that were generated in the preceding step. The figure that arises after an infinite number of steps is the famous 'Koch island'. Its border is called the 'Koch curve'. It does not possess a finite length nor a tangent at any point. In contrast to the smooth lines and curves of Euclidean geometry such a geometric creation is called a 'fractal'.

Let us use the following procedure to measure the length of the Koch curve or of an irregularly shaped coastline of an island. We choose a yardstick of a certain length $\varepsilon_0$ and put one end onto a point of the coastline. Then we swivel the stick until another point of the curve touches the other end. This means we interpolate the coastline in a linear way on our yardstick scale. Starting from the new endpoint of the yardstick we repeat the procedure: we always start from the newly found point on the coast. Returning to the initial point, we measure a total length of the coastline that is substantially dependent on the yardstick. The smaller the yardstick, the longer the measured length of the coastline. For the Koch island we may choose at the beginning a yardstick length $\varepsilon_0$. After doing the measurement, we repeat the measurement with a new yardstick one third as long. In an unlimited iteration process we successively shorten the yardstick to

$$\varepsilon_N = \tfrac{1}{3}\varepsilon_{N-1} = (\tfrac{1}{3})^N \varepsilon_0 \tag{10.1.1}$$

in the $N$th step. The yardstick is laid on $G_N = 3 \times 4^{N-1}$ times. Hence the measured length is

$$L_N = G_N \varepsilon_N = (\tfrac{4}{3})^{N-1} \varepsilon_0. \tag{10.1.2}$$

For increasing $N$ this expression grows without limit, in contrast to a smooth line of Euclidean geometry.

Fig. 10.1 The Koch curve: (*a*) method of construction and (*b*) the outcome of one of the three original triangle sides after a large number of steps.

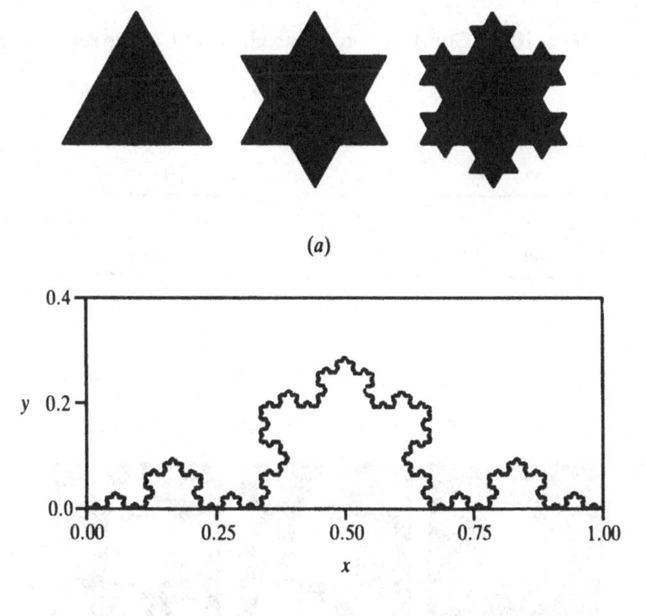

**Cantor sets** As another simple example of a fractal let us discuss the so called *classical Cantor set*. We start with the unit interval, divide it into three subintervals of equal size $\varepsilon_1 = \frac{1}{3}$, and eliminate the central subinterval. The remaining two intervals are again divided into three subintervals of size $\varepsilon_2 = \frac{1}{9}$, and the central subintervals are removed again (see fig. 10.2). If we repeat this procedure infinitely many times, we obtain the classical Cantor set. It is the set of the real numbers in the interval [0, 1] that in ternary notation can be represented as

$$x = \sum_{k=1}^{\infty} s_k 3^{-k} \tag{10.1.3}$$

where the $s_k$ take on the values 0 and 2.

Similar procedures can be defined in higher dimensions as well.

**Sierpinski carpet** (Sierpinski 1916) Let us divide a two-dimensional square of side length 1 into nine smaller squares, the side length of which is $\frac{1}{3}$ of the original one. Then we eliminate the central subsquare. The same procedure is then applied to the remaining squares in the next step (see fig. 10.3). The infinite repetition of this iteration process yields the so called 'Sierpinski carpet'. The total remaining area approaches zero for $N \to \infty$. Later we shall examine *how* it approaches zero. The classical

Fig. 10.2 Construction of the classical Cantor set.

1

1/3 1/3

1/9 1/9 1/9 1/9

Fig. 10.3 Construction of the Sierpinski carpet.

one-dimensional Cantor set is the set of points lying on the central straight line through the Sierpinski carpet, parallel to one side.

**Sierpinski gasket** Similarly to the Sierpinski carpet, we may also divide an equilateral triangle into four subtriangles, eliminating the central one. Repeating this process *ad infinitum* we obtain the so called 'Sierpinski gasket' (fig. 10.4).

**Sierpinski sponge** The three-dimensional extension of the Sierpinski carpet leads us to the 'Sierpinski sponge' (also called the 'Monger sponge'). It is obtained by an iteration process where a cube of volume 1 is divided into $3^3 = 27$ subcubes with the side length $\frac{1}{3}$. Then seven central subcubes are omitted, as shown in fig. 10.5. The number of the remaining subcubes is $27 - 7 = 20$. Repeating this process, the remaining volume of the sponge approaches zero. This happens in a way that we shall discuss in the next section.

## 10.2 The fractal dimension

It is intuitively clear that simple geometrical objects possess an integer dimension. For example, a thread has dimension 1, a piece of paper dimension 2, and a brick dimension 3. We call this dimension $D(0)$ (the argument 0 in the notation is only used to distinguish this dimension from other kinds of dimensions, which will be discussed later). These three examples are objects of traditional geometry. But what is the dimension of a fractal? We may relate this question to the growth rate of the number

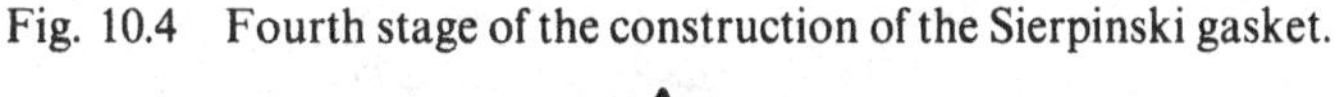

Fig. 10.4 Fourth stage of the construction of the Sierpinski gasket.

$r(\varepsilon)$ of small $d$-dimensional cells of volume $\varepsilon^d$ that we need to cover the object under consideration. Here $d$ is an arbitrary integer, called the *embedding dimension*, and is chosen large enough such that the dimension $D(0)$ of the object under consideration satisfies $D(0) \leqslant d$. For example for a thread, a piece of paper, or a brick we may choose $d = 3$, i.e., little cells of volume $\varepsilon^3$. If we cover these objects completely, we need for small $\varepsilon$

$$r(\varepsilon) \sim \varepsilon^{-D(0)} \tag{10.2.1}$$

cells, where $D(0)$ takes on the values 1, 2, 3, respectively (the symbol $\sim$ expresses an asymptotic proportionality). Due to eq. (10.2.1), we may identify the growth rate of $r(\varepsilon)$ for $\varepsilon \to 0$ as the dimension $D(0)$ of the object under consideration. Now, to determine the dimension of a fractal, we can just apply the same method. The only new thing is that we may now also obtain noninteger dimensions $D(0)$. The *fractal dimension*, also called the *box dimension* or *capacity*, is defined as

$$D(0) = -\lim_{\varepsilon \to 0} \frac{\ln r(\varepsilon)}{\ln \varepsilon}. \tag{10.2.2}$$

Often, in nonlinear dynamics a fractal is given numerically by a large number of iterates in the phase space. Having no knowledge about its geometrical structure, the fractal dimension has to be determined by a computer experiment. Here one simply takes the $d$-dimensional phase space as an embedding space and counts the number of boxes necessary to cover the fractal for $\varepsilon$ small enough (the 'box counting algorithm').

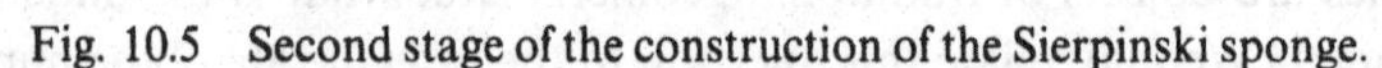

Fig. 10.5 Second stage of the construction of the Sierpinski sponge.

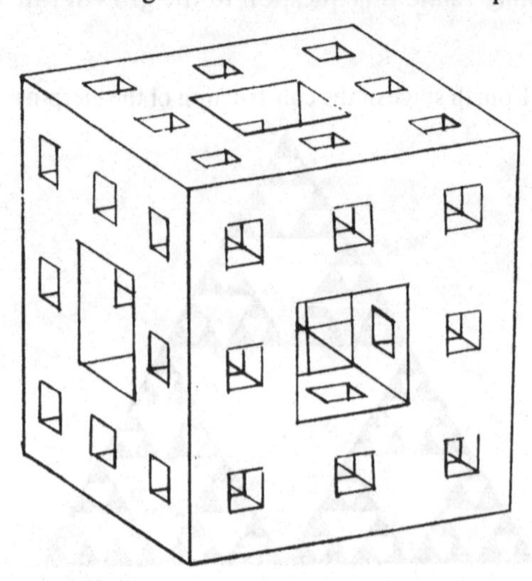

However, for simple fractals with an exact self-similar structure, it is sometimes possible to determine the fractal dimension by a simple theoretical argument. Examples where this is possible are those fractals considered in section 10.1. Here we know the construction recipe, and this very easily allows us to determine the fractal dimension. In the case of the Koch curve we may use two-dimensional spheres of diameter $\varepsilon$ as cells. In the case of the classical Cantor set the boxes can be chosen to be intervals of length $\varepsilon$. In the case of the Sierpinski carpet we may use two-dimensional squares of area $\varepsilon^2$, and for the Sierpinski gasket little triangles of area $\varepsilon^2/2$. In the case of the Sierpinski sponge, the cells may be chosen to be cubic boxes of volume $\varepsilon^3$. The only important point is that the embedding dimension $d$ is large enough. The shape of the cells is irrelevant. We start with an initial diameter $\varepsilon_0 = 1$. Then we shorten $\varepsilon$ step by step by a factor $a$. For the Sierpinski gasket we choose $a = \frac{1}{2}$, for all other examples $a = \frac{1}{3}$. After the $N$th step the cell diameter is

$$\varepsilon = a^N \varepsilon_0. \tag{10.2.3}$$

In more general cases, $a$ may be an arbitrary positive number smaller than 1. For the above examples the smallest number $r(\varepsilon)$ of cells needed to cover the entire fractal increases by a certain factor $G$ in each step. After the $N$th step

$$r(\varepsilon) = G^N r(\varepsilon_0). \tag{10.2.4}$$

Eliminating $N$ from eq. (10.2.3) this yields for $N \to \infty$

$$N \sim \ln \varepsilon / \ln a, \tag{10.2.5}$$

$$r(\varepsilon) \sim \exp[(\ln G)(\ln \varepsilon)/\ln a]. \tag{10.2.6}$$

This means the fractal dimension is

$$D(0) = -\lim_{\varepsilon \to 0} \frac{\ln r(\varepsilon)}{\ln \varepsilon} = -\frac{\ln G}{\ln a}. \tag{10.2.7}$$

For the Koch curve we obtain

$$G = 4, \ D(0) = \frac{\ln 4}{\ln 3} = 1.2619\ldots. \tag{10.2.8}$$

Hence it is an object 'inbetween' a straight line and a plane, yet still 'nearer' to a straight line. For the classical Cantor set, the Sierpinski carpet, and the Sierpinski sponge we obtain

$$G = 2, \quad D(0) = \frac{\ln 2}{\ln 3} = 0.6309\ldots, \tag{10.2.9}$$

$$G = 8, \quad D(0) = \frac{\ln 8}{\ln 3} = 1.8928\ldots, \tag{10.2.10}$$

$$G = 20, D(0) = \frac{\ln 20}{\ln 3} = 2.7268\ldots, \tag{10.2.11}$$

respectively. For the Sierpinski gasket we have $a = \frac{1}{2}$. Thus we obtain

$$G = 3, D(0) = \frac{\ln 3}{\ln 2} = 1.5850\ldots. \tag{10.2.12}$$

All the examples of fractals that we have discussed above were constructed by a simple mathematical prescription. The process of decreasing $\varepsilon$ can be continued without any restriction. This is also true for the fractal attractors of mappings. We speak, however, of fractals in nature as well. Examples are coastlines, surfaces of mountains, structures of trees, distributions of galaxies, and many other objects in nature. Here the characteristic features of fractals are realized only on certain limited length scales. This means that, in practice, we already speak of fractals if the typical fractal structure is realized in a restricted region of length scales only, let us say in a region of some powers of 10.

A generalization of the definition (10.2.2) leads us to more general 'dimensions'. If a quantity $M(\varepsilon)$ attributed to a fractal scales for $\varepsilon \to 0$ as

$$M(\varepsilon) \sim \varepsilon^{-D}, \tag{10.2.13}$$

we may consider $D$ to be a dimension of the fractal with respect to the property $M$. We shall discuss examples of such generalized dimensions in later chapters. In the next section, however, we shall discuss an alternative definition that replaces the fractal dimension $D(0)$ by a mathematically more rigorous concept, the *Hausdorff dimension.*

## 10.3 The Hausdorff dimension

So far we have only considered the covering of a fractal object $A$ by cells of equal size $\varepsilon^d$. Moreover, we have assumed that the dimension obtained in the limit $\varepsilon \to 0$ does not depend on the shape of the cells, i.e., whether we choose cubes, triangles, or spheres. In the mathematical literature, the corresponding fractal dimension, obtained with cells of just one shape, is called the *capacity*. From a rigorous mathematical standpoint, the limit in eq. (10.2.2) may not always exist for complicated fractal objects. A well defined quantity, however, is the *Hausdorff dimension* $D_H$. The definition of this dimension uses cells of variable size. The fractal set $A$ is covered by cells $\sigma_k$ of variable diameter $\varepsilon_k$, where all $\varepsilon_k$ satisfy $\varepsilon_k < \varepsilon$. One then defines for a positive parameter $\beta$

$$m(\beta, \varepsilon) = \inf_{\{\sigma_k\}} \sum_k (\varepsilon_k)^\beta. \tag{10.3.1}$$

The infimum is taken over all possible coverings of $A$. In the limit $\varepsilon \to 0$, the quantity $m(\beta, \varepsilon)$ will go to zero for $\beta > \beta_0$, and will diverge for $\beta < \beta_0$. The point $\beta_0$, where $m(\beta, \varepsilon)$ neither diverges nor goes to zero, is defined to be the Hausdorff dimension $D_H$ of $A$:

$$D_H = \beta_0. \tag{10.3.2}$$

The advantage of this definition is that $D_H$ is a well defined mathematical quantity. The disadvantage is that the infimum over all partitions is quite a cumbersome condition for practical purposes. In a computer experiment, it is much easier to determine the capacity $D(0)$. Fortunately, it has turned out that in practically all interesting cases, up to very pathological examples, Hausdorff dimension and capacity coincide. Therefore, we shall often follow the commonly used language of physicists, and will not distinguish between the two quantities.

To illustrate the definition (10.3.1), (10.3.2) of the Hausdorff dimension, let us determine it once more for a trivial example, namely the classical Cantor set, introduced in section 10.1. We cover this set by little intervals of size $\varepsilon_k = (\frac{1}{3})^N$ obtained at the $N$th level of construction. It is clear that this covering is the optimum choice, i.e., the infimum in eq. (10.3.1) is taken on for this choice. We obtain

$$m(\beta, \varepsilon) = 2^N \times 3^{-N\beta} = \exp[N(\ln 2 - \beta \ln 3)]. \tag{10.3.3}$$

This quantity indeed stays finite for $N \to \infty$ if we choose

$$\beta = \beta_0 = \ln 2/\ln 3 = D_H = D(0). \tag{10.3.4}$$

A less trivial example is provided by the *two-scale Cantor set*. It is a generalization of the classical Cantor set and is constructed as follows. Starting from the unit interval, this interval is replaced by two smaller intervals of length $a_1$ and $a_2$, where $a_1 \neq a_2$ and $a_1 + a_2 < 1$. Each of these two intervals is then replaced by two new intervals whose lengths again decrease by a factor $a_1$, respectively $a_2$ (see fig. 10.6). Repeating this construction *ad infinitum*, we obtain a two-scale Cantor set. At the $N$th level of construction there are $2^N$ intervals of different lengths $l_i^{(N)}$. The number $n_k^{(N)}$ of intervals of length $a_1^k a_2^{N-k}$ is

$$n_k^{(N)} = \binom{N}{k} = \frac{N!}{k!\,(N-k)!}. \tag{10.3.5}$$

Hence, the sum over all length scales raised to the power $\beta$ is

$$\sum_i (l_i^{(N)})^\beta = \sum_{k=0}^{N} n_k^{(N)} a_1^{\beta k} a_2^{\beta(N-k)} = (a_1^\beta + a_2^\beta)^N. \tag{10.3.6}$$

Let us cover the Cantor set with little pieces of size $\varepsilon_i < \varepsilon$. It is clear that the optimum choice is

$$\varepsilon_i = l_i^{(N)} \tag{10.3.7}$$

($N$ sufficiently large) because then we have really chosen a covering of the two-scale Cantor set by a set of smallest possible pieces. Hence, the infimum on the right hand side of eq. (10.3.1) is taken on for the choice (10.3.7). We obtain

$$m(\beta, \varepsilon) = \sum_i (l_i^{(N)})^\beta = (a_1^\beta + a_2^\beta)^N. \tag{10.3.8}$$

This quantity neither diverges nor goes to zero for the special value $\beta = \beta_0$ determined by the condition

$$a_1^{\beta_0} + a_2^{\beta_0} = 1. \tag{10.3.9}$$

This equation implicitly determines the Hausdorff dimension $D_H = \beta_0$ of the two-scale Cantor set.

## 10.4 Mandelbrot and Julia sets

Let us now mention some more complicated examples of fractals. The notion 'complicated' relates to the structure of the fractal, not to the way

Fig. 10.6 Construction of a two-scale Cantor set ($a_1 = 0.25$, $a_2 = 0.4$).

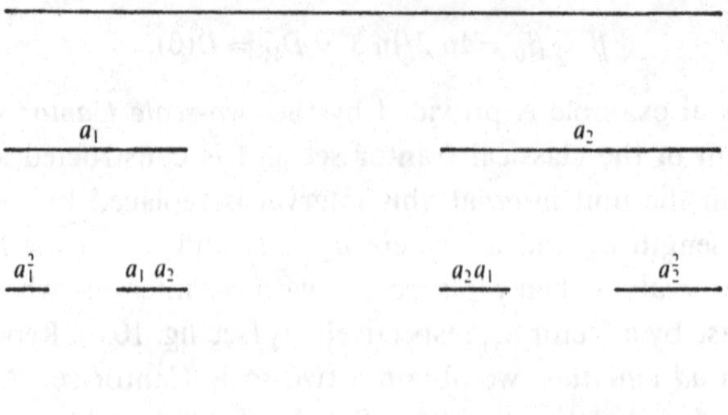

$a_1$ $a_2$

$a_1^2$ $a_1 a_2$ $a_2 a_1$ $a_2^2$

it is generated. Indeed, some of the most beautiful fractals are generated by very simple rules. As an example let us consider the famous 'Mandelbrot set' (Mandelbrot 1980, 1982). It is intimately connected to the complex generalization of the logistic map, given, for example, in the form

$$x_{n+1} = x_n^2 + C, \tag{10.4.1}$$

where both $x_n$ and the control parameter $C$ take on complex values.

Let us choose the initial value $x_0 = 0$ and ask the following question: for which values of $C$ does the orbit not escape to infinity? The border line of these parameter values in the complex plane is shown in fig. 10.7. For parameter values inside this boundary, the trajectory does not escape. A beautiful fractal object is created, which is called the 'Mandelbrot set'. In the parameter space it corresponds to the set of parameters that generate bounded motion. There is self-similarity on each scale. Along the real axis the rescaling factor of the fractal 'bubbles' is given by the Feigenbaum constant $\delta$, which we got to know in section 1.3. Along other directions in the complex plane, there are other scaling constants that describe the self-similarity. Indeed, there exists an infinite set of complex generalizations of the Feigenbaum constant with universal character (Cvitanović and Myrheim 1989). The complexity inherent in the

Fig. 10.7 The Mandelbrot set.

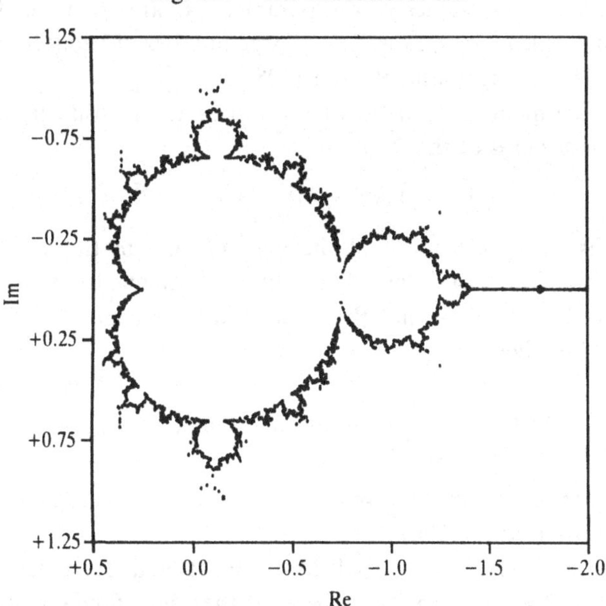

Mandelbrot set is remarkable – and quite surprising, if one looks at the 'simple' recurrence relation (10.4.1) that is responsible for it.

Another interesting fractal object connected with the complex version of the logistic map is its *Julia set* (Julia 1918). In contrast to the Mandelbrot set, which is a fractal in the *parameter space*, the Julia set is a fractal set in the *phase space*. A simple method of generating it is the following. For a fixed parameter value $C$ the map (10.4.1) is inverted:

$$y_{n+1} = \pm(y_n - C)^{1/2}. \tag{10.4.2}$$

Notice that for each $x_{n+1}$ there are *two* possible complex preimages $x_n$. That is why the map (10.4.2) is not unique but actually consists of two maps, corresponding to the plus and minus signs. We now iterate eq. (10.4.2) for the initial value $y_0 = 0$. The starting value $y_0$ is mapped onto two values $y_1$, namely

$$y_1 = \pm C^{1/2}. \tag{10.4.3}$$

Iterating again, we obtain four complex values

$$y_2 = \pm(\pm C^{1/2} - C)^{1/2}, \tag{10.4.4}$$

eight values

$$y_3 = \pm[\pm(\pm C^{1/2} - C)^{1/2} - C]^{1/2}, \tag{10.4.5}$$

and so on. The limit set $\{y_n\}$ for $n \to \infty$ is called the 'Julia set'. A few examples for various values of the parameter $C$ are shown in fig. 10.8. More details of Julia sets, as well as a large number of beautiful pictures, can be found in Peitgen and Richter (1986).

Here we just mention an interesting result on the Hausdorff dimension of Julia sets of maps of the form

$$x_{n+1} = |x_n|^z + C. \tag{10.4.6}$$

This map is a generalization of the map (10.4.1) in the sense that it possesses a maximum of order $z$. Using thermodynamic formalism, Ruelle (1982) was able to prove that the Hausdorff dimension of the Julia set for small $|C|$ is given by

$$D(0) = 1 + \frac{|C|^2}{4 \ln z} + \text{higher order terms in } C. \tag{10.4.7}$$

For the higher order terms, see Widom, Bensimom, Kadanoff and Shenker (1983) and Michalski (1990).

Although we have introduced Mandelbrot and Julia sets for the example of the logistic map, it is quite clear that these fundamental fractal

Fig. 10.8 The Julia set of the logistic map for (*a*) $C = 0.6$, (*b*) $-0.2$.

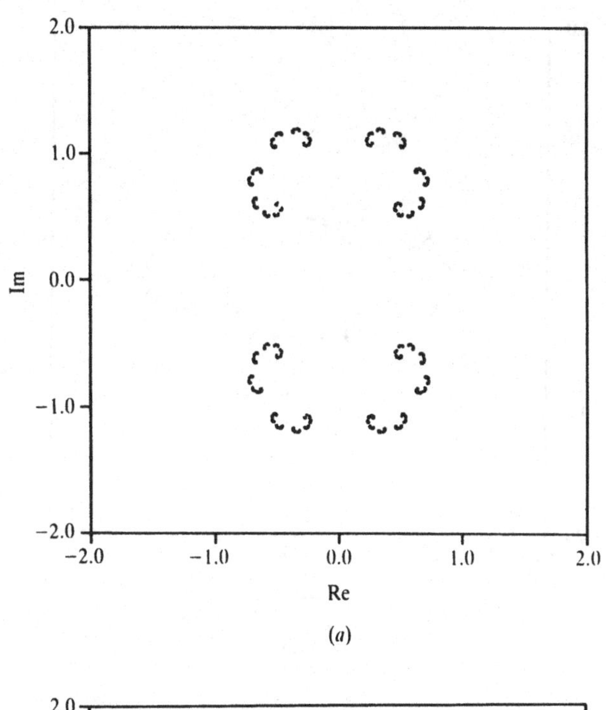


(*a*)

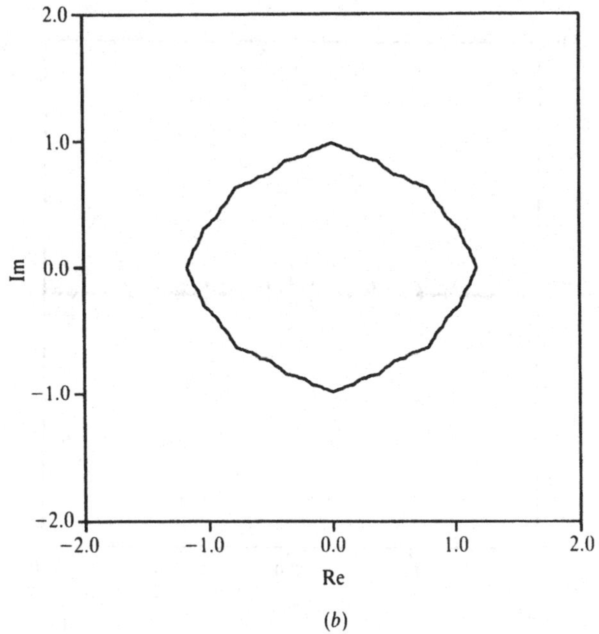


(*b*)

*(continued)*

Fig. 10.8 (*continued*) (*c*) $C = -1.0$, (*d*) $-1.8$.

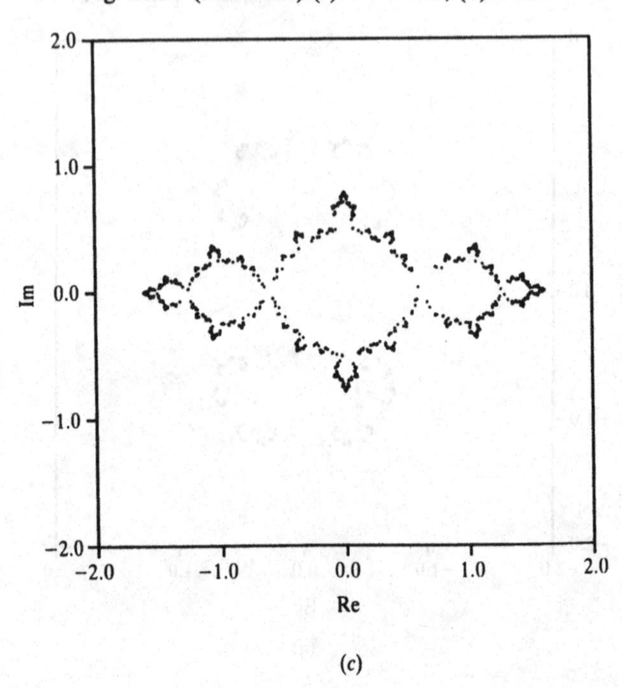


(*c*)

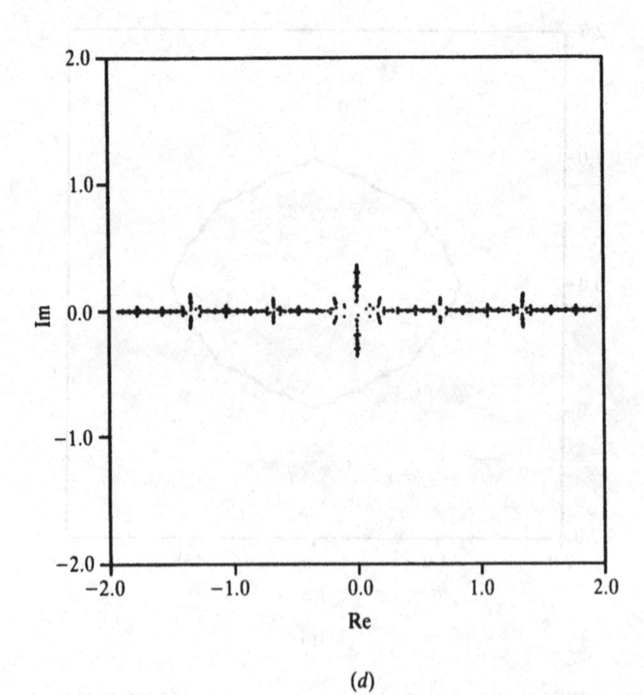


(*d*)

Fig. 10.8 (*continued*) As well as for (*e*) $C = 0.6 - 0.5i$, (*f*) $-0.2 - 0.5i$.

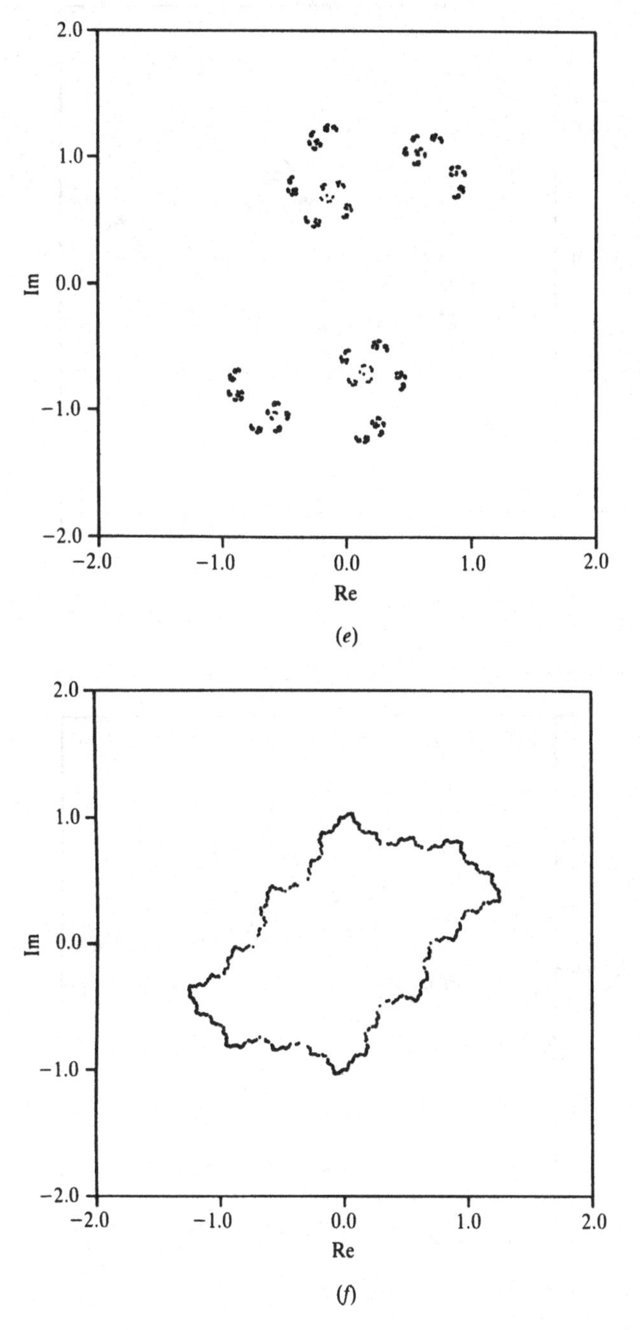


(*continued*)

Fig. 10.8 (*continued*) (*g*) $C = -1.0 - 0.5i$, (*h*) $-1.8 - 0.5i$.

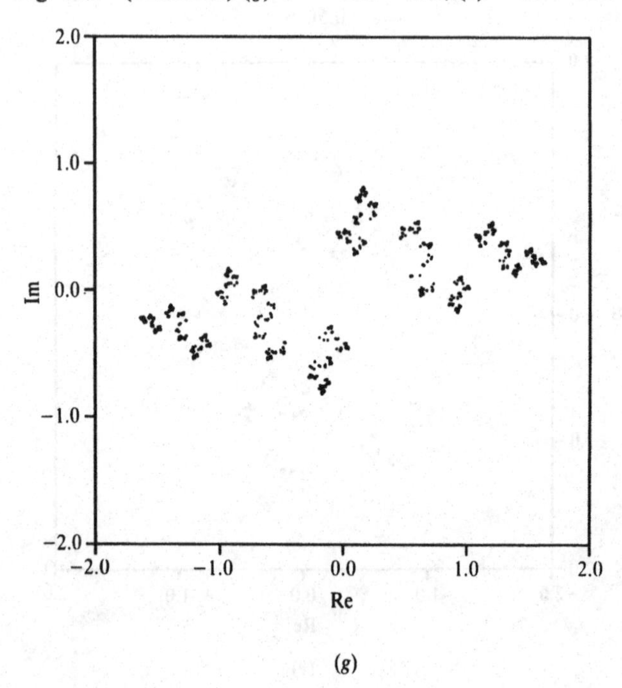


(*g*)

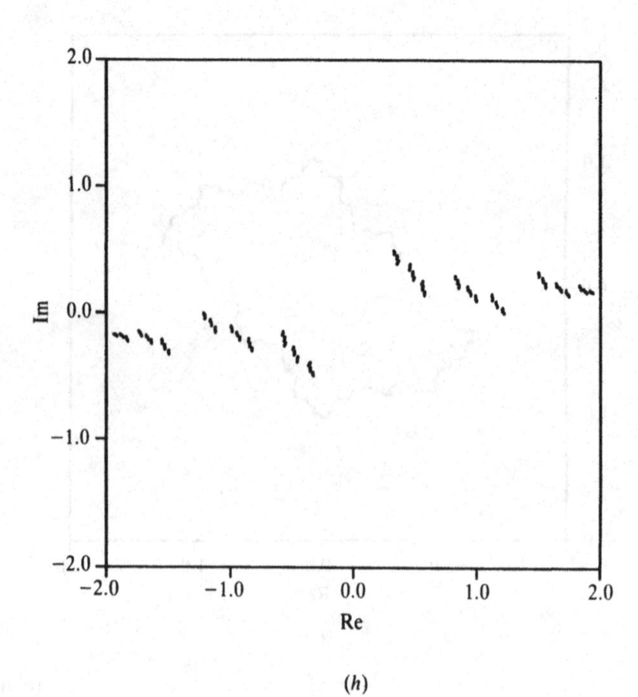


(*h*)

sets can be defined for any complex mapping by a straightforward generalization of the above concepts.

## *10.5 Iterated function systems

The recurrence relation (10.4.2) represents something new in the sense that instead of one unique mapping we consider a branching process generated by an entire set of mappings (in our example two mappings, distinguished by the plus and minus signs). This is the basic idea of the so called *iterated function systems* (IFS), which quite recently have become very popular (see, for example, Barnsley (1988) and Jürgens, Peitgen and Saupe (1989)). A Julia set can be regarded as the *attractor* of such an iterated function system, indeed of an IFS consisting of nonlinear maps. But even IFSs defined for linear maps can produce beautiful fractal patterns. The most popular one is probably *Barnsley's fern* shown in fig. 10.9. The iterated function system consists of four linear two-dimensional maps $T_i$ of the form

$$T_i(x, y) = (a_{11}x + a_{12}y + b_1, a_{21}x + a_{22}y + b_2), \qquad (10.5.1)$$

where the values of the constants are listed in table 10.1($a$). Starting with an arbitrary two-dimensional object in the $(x, y)$-plane, for example, a

Fig. 10.9 Barnsley's fern.

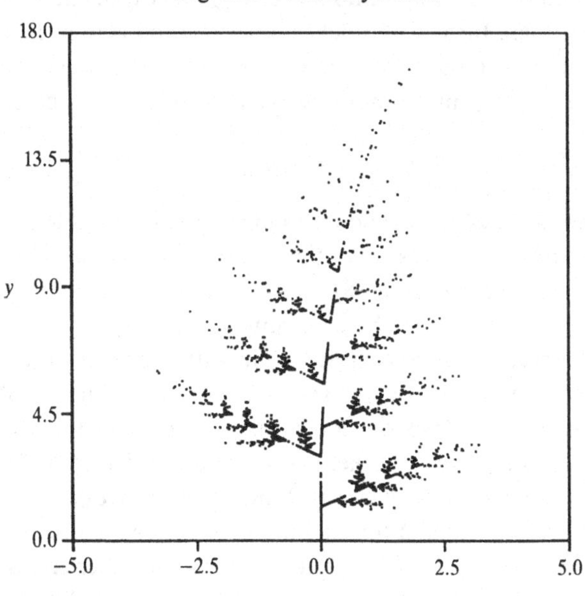

Table 10.1. *Parameter values of the iterated function systems that generate (a) Barnsley's fern, (b) Barnsley's leaf, and (c) the Koch curve.*

| | | $a_{11}$ | $a_{12}$ | $a_{21}$ | $a_{22}$ | $b_1$ | $b_2$ |
|---|---|---|---|---|---|---|---|
| (*a*) | $T_0$ | 0 | 0 | 0 | 0.17 | 0 | 0 |
| | $T_1$ | 0.8496 | 0.0255 | −0.0255 | 0.8496 | 0 | 3 |
| | $T_2$ | 0.1554 | 0.235 | −0.1958 | 0.1865 | 0 | 1.2 |
| | $T_3$ | 0.1554 | −0.235 | 0.1958 | 0.1865 | 0 | 3 |
| (*b*) | $T_0$ | 0.650 | −0.013 | 0.013 | 0.650 | 0.175 | 0 |
| | $T_1$ | 0.650 | −0.026 | 0.026 | 0.650 | 0.165 | 0.325 |
| | $T_2$ | 0.318 | −0.318 | 0.318 | 0.318 | 0.2 | 0 |
| | $T_3$ | −0.318 | 0.318 | 0.318 | 0.318 | 0.8 | 0 |
| (c) | $T_0$ | 1/3 | 0 | 0 | 1/3 | 0 | 0 |
| | $T_1$ | 1/3 | 0 | 0 | 1/3 | 2/3 | 0 |
| | $T_2$ | 0.1667 | −0.2887 | 0.2887 | 0.1667 | 1/3 | 0 |
| | $T_3$ | −0.1667 | 0.2887 | 0.2887 | 0.1667 | 2/3 | 0 |

square, all four maps act on this square, leading to four squeezed squares. Now each $T_i$ again acts on all of the squeezed squares, leading to 16 objects, and so on. In the limit $n \to \infty$ we obtain the fractal pattern of fig. 10.9, which has remarkable similarity to a fern leaf.

In practice, the above method takes much too much computing time, because the number of computational steps grows exponentially (as $4^n$ in the above example). Instead, a much simpler method is to iterate just one trajectory, but choose the map $T_i$ out of the set of the possible maps (in our case $T_0, T_1, T_2, T_3$) in a random way. This means, we iterate

$$x_{n+1} = T_i(x_n) \tag{10.5.2}$$

and choose the index $i$ by a random number generator. It can be shown that this method produces just the same attractor as the original branching process (Barnsley 1988).

If we change the values of the constants $a_{11}, a_{12}, a_{21}, a_{22}, b_1, b_2$, other interesting structures can be produced. The attractor shown in fig. 10.10 reminds us very much of a maple leaf. The corresponding constants are listed in table 10.1(*b*). Also the Koch curve, shown in fig. 10.1(*b*), has actually been plotted with the help of an IFS consisting of four linear maps, with an appropriate set of constants (table 10.1(*c*)).

So far we have considered iterated function systems consisting of four different linear maps $T_0, T_1, T_2, T_3$, but even with two different linear maps $T_0$ and $T_1$ interesting complex patterns can be produced. A few examples

Fig. 10.10 Barnsley's leaf.

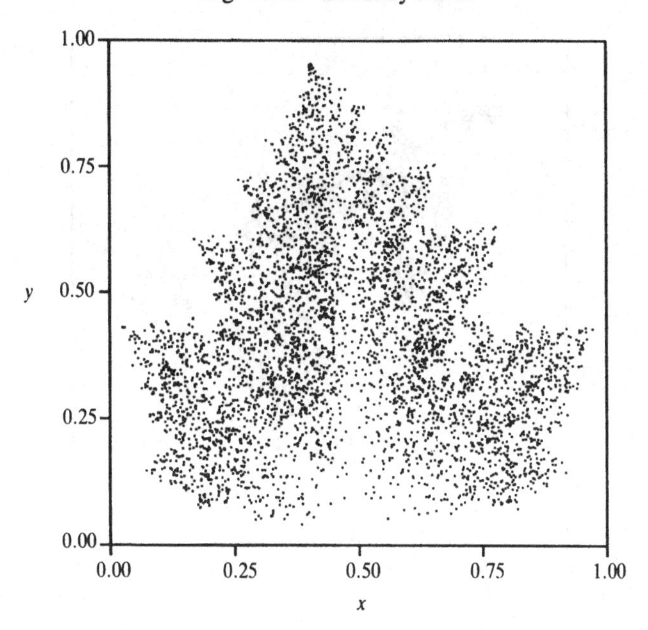


Fig. 10.11 Various attractors of iterated function systems consisting of two linear maps.

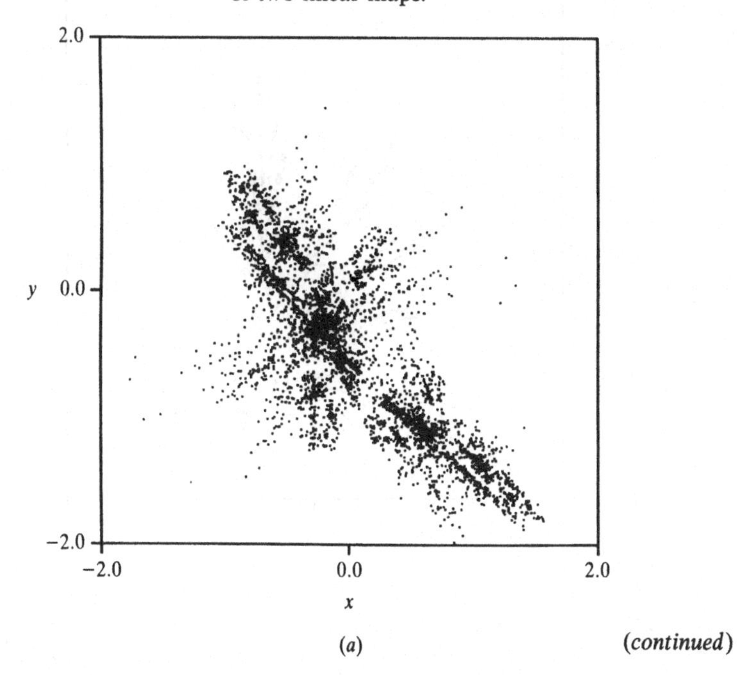


(*a*)

(*continued*)

Fig. 10.11 (*continued*).

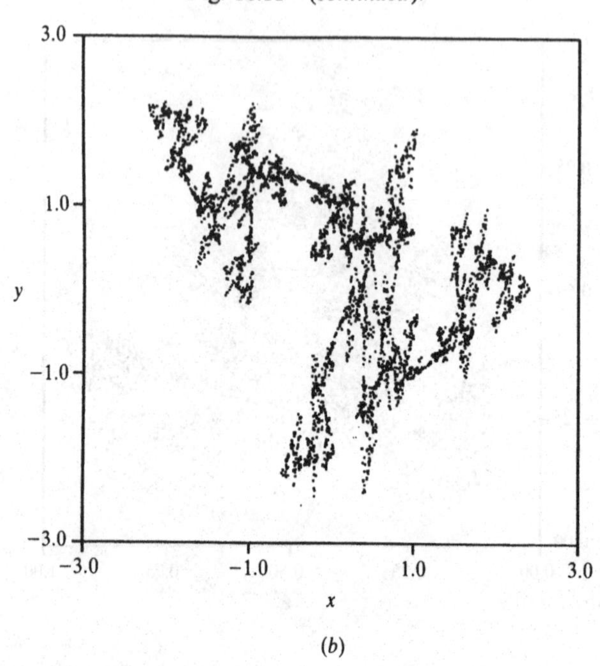


(*b*)

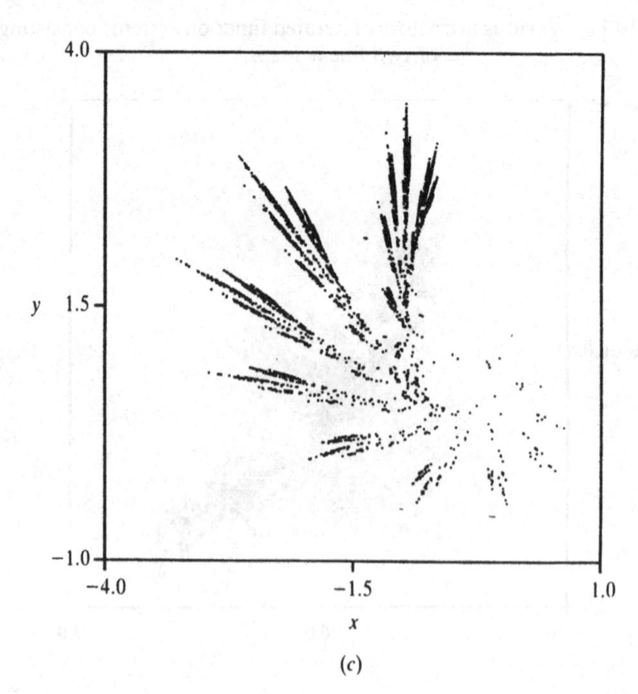


(*c*)

**(*continued*)**

Table 10.2. *Parameter values of the IFSs that generate the attractors of fig.* 10.11(*a*)–(*d*).

| | | $a_{11}$ | $a_{12}$ | $a_{21}$ | $a_{22}$ | $b_1$ | $b_2$ |
|---|---|---|---|---|---|---|---|
| (*a*) | $T_0$ | −0.506 | 0.475 | 0.921 | −0.002 | 0.629 | −0.905 |
| | $T_1$ | −0.867 | −0.492 | 0.008 | −0.797 | −0.534 | −0.502 |
| (*b*) | $T_0$ | −0.609 | 0.088 | −0.821 | −0.736 | 0.858 | −0.189 |
| | $T_1$ | −0.588 | 0.243 | 0.497 | 0.380 | −0.821 | 0.970 |
| (*c*) | $T_0$ | 0.147 | 0.115 | −0.410 | 0.526 | −0.919 | 0.872 |
| | $T_1$ | 0.570 | −0.680 | 0.340 | 0.885 | 0.248 | 0.233 |
| (*d*) | $T_0$ | 0.348 | 0.807 | −0.577 | 0.960 | 0.187 | 0.381 |
| | $T_1$ | −0.132 | 0.079 | 0.541 | 0.767 | 0.163 | −0.554 |

are shown in fig. 10.11(*a*)–(*d*), the corresponding data for the maps are listed in table 10.2(*a*)–(*d*).

Of course, IFSs can be defined for all kinds of maps $T_i$. Here we have just described a few simple examples. More on the mathematical theory of IFSs can be found in Barnsley's book (Barnsley 1988).

Fig. 10.11 (*continued*).

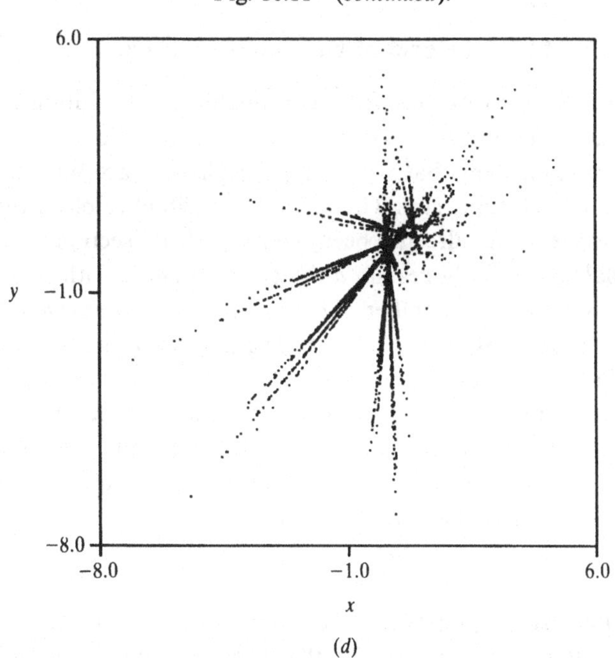


(*d*)

# 11
# Multifractals

A multifractal (Benzi, Paladin, Parisi and Vulpiani 1984; Frisch and Parisi 1985; Halsey *et al.* 1986; Feigenbaum 1987) is a fractal for which a probability measure on the fractal support is given. For example, if the fractal is the attractor of a map in a numerical experiment, the probability measure is given by the relative frequencies of the iterates, which are interpreted as probabilities, as already explained in chapter 2. It is then useful to introduce more general dimensions $D(\beta)$ that also contain information about the probability distribution on the fractal.

## 11.1 The grid of boxes of equal size

Suppose an arbitrary (possibly fractal) probability distribution is given (for example, the natural invariant measure of an ergodic map). For the following it is important that we divide the phase space into boxes of *equal size*, not of variable size. The different possibilities of partitioning the phase space have already been mentioned in section 3.1. In a $d$-dimensional space the boxes are $d$-dimensional cubes with side length $\varepsilon$. Again let us denote the number of boxes with *nonzero probability* by $r$. We label these boxes by $i = 1, 2, \ldots, r$. The number $r$ should be distinguished from the total number of boxes $R \sim \varepsilon^{-d}$. For a given value $\varepsilon$ the probability attributed to a box $i$ centred at some point $x$ will be called $p_i$. As $p_i$ is attributed to a single box, it is a 'local' quantity. A related local quantity is the *crowding index* $\alpha_i(\varepsilon)$ of the box $i$ centred at the point $x$. For finite values $\varepsilon$ it is defined as the ratio

$$\alpha_i(\varepsilon) = \ln p_i / \ln \varepsilon. \tag{11.1.1}$$

We shall often use the equivalent notation $\alpha_i(\varepsilon) \equiv \alpha(\varepsilon, x)$. Let us now consider a sequence of nested boxes, all centred at the point $x$, with sizes

$\varepsilon_j \to 0$. If the limit

$$\alpha(x) = \lim_{\varepsilon \to 0} \alpha(\varepsilon, x) \tag{11.1.2}$$

exists, then $\alpha(x)$ is called a 'local scaling exponent'. Some authors also use the name 'pointwise' or 'local dimension'. The fractal is called 'nonuniform' or 'multifractal' if $\alpha(x)$ is actually dependent on $x$. Generically, fractal attractors of nonlinear mappings are of this type. That is to say, we have an entire spectrum of different crowding indices, respectively of singularities of the invariant density.

Let us return to finite box sizes $\varepsilon$. We may write for the probability

$$p_i = \exp(-b_i). \tag{11.1.3}$$

The bit-number $b_i$ occurring here is connected with the crowding index $\alpha_i(\varepsilon)$ by

$$b_i = -\alpha_i(\varepsilon) \ln \varepsilon. \tag{11.1.4}$$

All the formulas of chapter 9 are again valid. But now they relate to boxes of equal size. The escort distributions of $p$ are

$$P_i = \exp(\Psi - \beta b_i), \tag{11.1.5}$$

with

$$\Psi(\beta) = -\ln \sum_{i=1}^{r} \exp(-\beta b_i) = -(\beta - 1) I_\beta(p), \tag{11.1.6}$$

corresponding to eq. (9.1.6). The partition function is

$$Z(\beta) = \sum_{i=1}^{r} p_i^\beta = \sum_{i=1}^{r} \exp(-\beta b_i) = \exp[-\Psi(\beta)], \tag{11.1.7}$$

and the Rényi informations are related to $Z(\beta)$ by

$$I_\beta(p) = \frac{1}{\beta - 1} \ln Z(\beta). \tag{11.1.8}$$

## 11.2 The Rényi dimensions

So far we have kept the number $R$ of boxes (respectively, the number of events $i$) finite. Now, however, we are interested in the limit behaviour of box size $\varepsilon \to 0$, which implies $R \sim \varepsilon^{-d} \to \infty$. In the limit $\varepsilon \to 0$ the Rényi informations given by eq. (11.1.8) diverge. However, quantities that usually remain finite are the so called *Rényi dimensions* (Rényi 1970;

Mandelbrot 1974; Hentschel and Procaccia 1983; Grassberger 1983)

$$D(\beta) = \lim_{\varepsilon \to 0} \frac{I_\beta(p)}{\ln \varepsilon} = \lim_{\varepsilon \to 0} \frac{1}{\ln \varepsilon} \frac{1}{\beta - 1} \ln \sum_{i=1}^{r} p_i^\beta . \tag{11.2.1}$$

This means, for $\varepsilon \to 0$ the partition function (11.1.7) is expected to scale as

$$Z(\beta) \sim \varepsilon^{(\beta-1)D(\beta)} . \tag{11.2.2}$$

It is important that the probabilities $p_i$ relate to a grid of boxes of equal size. An alternative definition of the Rényi dimensions with cells of variable size will be given in section 11.4. Depending on the order $\beta$, the Rényi dimensions describe different interesting properties of the system. Corresponding to this, some Rényi dimensions have special names.

For $\beta = 0$ the negative Rényi information becomes equal to the logarithm of the number $r(\varepsilon)$ of nonempty boxes, which is the smallest number of boxes needed to cover the entire fractal. Hence, $D(0)$ is equal to the *capacity* or *box dimension* defined by eq. (10.2.2). As explained in the preceding chapter, in most practical cases it is equal to the *Hausdorff dimension.*

For $\beta \to 1$, in accordance with eq. (5.1.5), the Rényi information approaches the Shannon information. Therefore $D(1)$ is called the *information dimension.* It describes how the Shannon information grows with the refinement of the grid scale

$$D(1) = \lim_{\varepsilon \to 0} \frac{1}{\ln \varepsilon} \sum_{i=1}^{r} p_i \ln p_i . \tag{11.2.3}$$

$D(2)$ is called the *correlation dimension.* It can easily be extracted from experimentally observed time series via the Grassberger–Procaccia algorithm (Grassberger and Procaccia 1983a,b). Moreover, it is of relevance to estimate the effects of round-off errors that occur if a digital computer is used for the iteration of a chaotic map. A computer, due to its finite precision $\Delta$, can only distinguish a finite number of phase space cells on a finite phase space. Hence, every rounded trajectory must asymptotically become periodic, no matter what behaviour the exact map exhibits. It has been discovered in Yorke, Grebogi and Ott (1988) and proved in Beck (1989) that for rounded chaotic maps in typical cases the average period length $\langle L \rangle$ scales with the machine precision $\Delta$ as

$$\langle L \rangle \sim \Delta^{-D(2)/2}, \tag{11.2.4}$$

provided certain assumptions (Beck 1991d) are satisfied.

Of further special importance are the *limit dimensions* $D(\pm\infty)$. The dimension $(D+\infty)$ describes the scaling behaviour of the region in the phase space where the probability is most concentrated. $D(-\infty)$, on the other hand, describes the behaviour of the region where the probability is most rarefied.

The general considerations of section 5.3 imply that the Rényi dimensions fulfil certain general inequalities for arbitrary probability measures (Beck 1990a). The most important ones are

$$D(\beta) \geqslant 0 \tag{11.2.5}$$

$$D(\beta') \leqslant D(\beta) \qquad \text{for } \beta' > \beta \tag{11.2.6}$$

$$(\beta' - 1)D(\beta') \geqslant (\beta - 1)D(\beta) \qquad \text{for } \beta' > \beta \tag{11.2.7}$$

$$\frac{\beta' - 1}{\beta'} D(\beta') \geqslant \frac{\beta - 1}{\beta} D(\beta) \qquad \text{for } \beta' > \beta,\ \beta'\beta > 0. \tag{11.2.8}$$

A useful special case of eq. (11.2.8) is obtained for $\beta' \to +\infty$, respectively $\beta \to -\infty$:

$$D(\beta) \leqslant \frac{\beta}{\beta - 1} D(\infty) \qquad \text{for } \beta > 1 \tag{11.2.9}$$

$$D(\beta) \geqslant \frac{\beta}{\beta - 1} D(-\infty) \qquad \text{for } \beta < 0. \tag{11.2.10}$$

A great deal of effort in nonlinear research has been devoted to the extraction of the Rényi dimensions and further characteristic quantities from a time series analysis of experimental data. We shall not treat this subject here but refer the reader to some of the large amount of literature on this subject (Eckmann and Ruelle 1985; Pawelzik and Schuster 1987; Drazin and King 1992).

## 11.3 Thermodynamic relations in the limit of box size going to zero

For simplicity, from now on we shall frequently use the abbreviation

$$V = -\ln \varepsilon. \tag{11.3.1}$$

As we are interested in the limit of small $\varepsilon$, we can always suppose $\varepsilon < 1$. Then $V$ is positive and goes to infinity if $\varepsilon$ goes to zero. $V$ enters into the 'canonical' distribution $P$ as a fixed parameter, like the volume in the canonical distribution of thermodynamics. Therefore we have chosen the letter $V$. If $V$, according to thermodynamic analogue, is called *volume*, it is not to be mistaken for the phase space volume of a box, which is $\varepsilon^d$ in a $d$-dimensional phase space.

We assume that the limit

$$\lim_{\varepsilon \to 0} \alpha(\varepsilon, x) = \alpha(x) \tag{11.3.2}$$

in eq. (11.1.2) exists. In statistical mechanics the limit of volume $V$ going to infinity such that the intensities $\beta_\sigma$ remain constant is called the *thermodynamic limit*. We call a system *dynamically homogeneous* if the dynamics of all its partial systems of equal size are the same. For such a system, we require that the extensities $M^\sigma$, and the entropy $S$ have homogeneous spatial densities (the possible value zero included). This means that these quantities asymptotically become proportional to $V$ for large $V$. In our case we shall only consider probability distributions for which the analogous thermodynamic limit of $V$ going to infinity keeping $\beta$ fixed leads to finite values $S/V$, $\Psi/V$, etc.

In the limit $\varepsilon \to 0$ we may replace in eq. (11.1.6) the sum over $i$ by an integral over the possible values $\alpha$

$$\Psi = -\ln \int_{\alpha_{\min}}^{\alpha_{\max}} d\alpha\, \gamma(\alpha) \exp(-\beta\alpha V). \tag{11.3.3}$$

Here $\gamma(\alpha)\, d\alpha$ denotes the number of boxes with crowding index $\alpha_i$ in the range between $\alpha$ and $\alpha + d\alpha$. The function $\gamma(\alpha)$ is the analogue of a 'state density' in statistical mechanics, where $\alpha V$ corresponds to an energy. In general, for a complicated chaotic attractor the function $\alpha(x)$ will depend on $x$ in a very irregular way. But a reordering of the boxes with respect to the values of $\alpha$ makes $\gamma(\alpha)$ comparatively smooth.

In the limit $\varepsilon \to 0$, equivalent to $V \to \infty$, we expect the asymptotic scaling behaviour

$$\gamma(\alpha) \sim \varepsilon^{-f(\alpha)}, \tag{11.3.4}$$

where $f(\alpha)$ is some bounded function independent of $\varepsilon$, often called the *spectrum of singularities* (Halsey *et al.* 1986). We obtain

$$\Psi \sim -\ln \int_{\alpha_{\min}}^{\alpha_{\max}} d\alpha \exp\{[f(\alpha) - \beta\alpha]V\}. \tag{11.3.5}$$

For $\varepsilon \to 0$, only the maximum value of the integrand contributes. Assuming that there is just one maximum for a well defined value $\alpha$, we can replace the integral by the integrand obtaining

$$\Psi \sim [\beta\alpha - f(\alpha)]V. \tag{11.3.6}$$

This argument is quite commonly used in statistical physics and is usually called the *saddle point method.* The argumentation is as follows. Suppose we want to evaluate an integral of the form

$$I = \int dx \exp[F(x)V] \tag{11.3.7}$$

for large $V$. We assume that $F$ is a smooth differentiable function with a single maximum at $x = x_0$, i.e.,

$$F'(x_0) = 0, \tag{11.3.8}$$

$$F''(x_0) < 0. \tag{11.3.9}$$

We may then expand $F$ in a Taylor series around $x_0$ and approximate the integral $I$ as

$$\begin{aligned} I &\approx \int dx \exp[[F(x_0) + \tfrac{1}{2}(x - x_0)^2 F''(x_0)]V] \\ &= \left[\frac{2\pi}{V|F''(x_0)|}\right]^{1/2} \exp[F(x_0)V]. \end{aligned} \tag{11.3.10}$$

This is equivalent to

$$-\ln I \approx -F(x_0)V + O(\ln V), \tag{11.3.11}$$

i.e., in this approximation the correction to the leading order term $F(x_0)V$ is of the order $\ln V$ only.

In our case the function $F(x)$ corresponds to the function $f(\alpha) - \beta\alpha$, and eq. (11.3.8) yields

$$\partial f/\partial\alpha = \beta. \tag{11.3.12}$$

Let us compare eq. (11.3.6) with eq. (6.1.9), choosing for $M^\sigma$ the mean value $b$ of the bit-number:

$$\Psi = \beta b - S = \beta\alpha V - S. \tag{11.3.13}$$

We see that $f(\alpha)$ can be regarded as the entropy density $S/V$ of the escort distribution $P$ in the limit $V \to \infty$:

$$\lim_{V\to\infty} (S/V) = f(\alpha). \tag{11.3.14}$$

$\alpha$ may be regarded as the analogue of the mean energy density. From eqs. (11.1.6) and (11.2.1) we obtain

$$\lim_{V\to\infty} (\Psi/V) = \tau(\beta), \tag{11.3.15}$$

where

$$\tau(\beta) = (\beta - 1)D(\beta). \tag{11.3.16}$$

Up to a factor $\beta^{-1}$, the quantity $\tau(\beta)$ is identical with the free energy density. In a numerical experiment it is possible to obtain $\tau(\beta)$ by a box counting algorithm. One simply determines the slope in a $(\ln Z, \ln \varepsilon)$-plot (in higher dimensions other numerical algorithms are more suitable, see, for example, Pawelzik and Schuster (1987)). Then, according to eqs. (11.3.6) and (11.3.15), $f(\alpha)$ can be obtained by a Legendre transformation. Applying the general concept of section 7.1 to $\Psi$ as a function of $\beta$, and $S$ as a function of $b$ we have

$$S(b) = \beta b - \Psi(\beta) \tag{11.3.17}$$

$$d\Psi/d\beta = b, \qquad dS/db = \beta, \tag{11.3.18}$$

which is equivalent to

$$f(\alpha) = \beta\alpha - \tau(\beta) \tag{11.3.19}$$

$$d\tau/d\beta = \alpha, \qquad df/d\alpha = \beta, \tag{11.3.20}$$

provided all occurring functions are differentiable (for the nondifferentiable case, see section 21.1). In this context it is important to notice that the function $\Psi(\beta)$ (or $\tau(\beta)$) is concave for arbitrary probability distributions, as was shown in section 7.4. Hence, the Legendre transformation is well defined. The entropy $S(b)$ (or $f(\alpha)$) is concave as well, for the same reason as in conventional thermodynamics.

Starting from the calculated $\tau(\beta)$ we can use eq. (11.3.20) to obtain $\alpha$ as a function of $\beta$. Then we finally obtain $f(\alpha)$ by use of eq. (11.3.19). Hence, the detailed information about *spectra of scaling exponents* given by $f(\alpha)$ can be obtained from the global scanning quantity $\tau(\beta)$ formed with the escort distribution. It should, however, be clear that both $f(\alpha)$ and $\tau(\beta)$ contain the same amount of information about the system. They are just related to each other by a Legendre transformation.

The advantage of the *spectrum of local dimensions* $f(\alpha)$ is that it has a kind of 'physical' interpretation: loosely speaking, it is the 'fractal dimension' of the subset of points that possesses the local scaling index $\alpha$. This is easily recognized from eq. (11.3.4), which states that the number of boxes with a certain $\alpha$ scales as

$$\gamma(\alpha) \sim \varepsilon^{-f(\alpha)}. \tag{11.3.21}$$

More precisely, $f(\alpha)$ is the fractal dimension of a set of boxes with *averaged* pointwise dimension $\alpha$ (for a critical discussion of this subtlety see Grassberger, Badii and Politi (1988)).

Let us discuss some general features of the $f(\alpha)$ spectrum. Using eq. (11.3.16), we may also write eqs. (11.3.19) and (11.3.20) in the form

$$\alpha(\beta) = D(\beta) + (\beta - 1)D'(\beta) \tag{11.3.22}$$

$$f(\alpha(\beta)) = D(\beta) + \beta(\beta - 1)D'(\beta). \tag{11.3.23}$$

Here $\alpha(\beta)$ denotes the value of $\alpha$ where $\beta\alpha - f(\alpha)$ takes on its minimum. The special cases $\beta = 0$ and $\beta = 1$ yield

$$f(\alpha(0)) = D(0) = \alpha(0) + D'(0), \tag{11.3.24}$$

$$f(\alpha(1)) = D(1) = \alpha(1). \tag{11.3.25}$$

This means that the maximum of $f(\alpha)$, determined by the condition

$$\mathrm{d}f/\mathrm{d}\alpha = 0 = \beta, \tag{11.3.26}$$

is just the capacity $D(0)$. The information dimension $D(1)$ is given by the value of $f(\alpha)$ where $\mathrm{d}f/\mathrm{d}\alpha = 1$.

Fig. 11.1 shows numerical results of both $D(\beta)$ and $f(\alpha)$ for the Feigenbaum attractor. Quite generally, $f(\alpha)$ is a single humped positive function defined on the interval $[\alpha_{\min}, \alpha_{\max}]$.

The minimum scaling index $\alpha_{\min}$, respectively the maximum scaling index $\alpha_{\max}$, is related to the Rényi dimensions $D(\pm\infty)$ by

$$\alpha_{\min} = D(+\infty), \tag{11.3.27}$$

$$\alpha_{\max} = D(-\infty). \tag{11.3.28}$$

This can be seen as follows. Let

$$p_{\max} = \max_i p_i \tag{11.3.29}$$

be the largest one of all the probabilities $p_i$. It corresponds to the smallest scaling index $\alpha_{\min}$ via

$$p_{\max} \sim \exp(-\alpha_{\min} V). \tag{11.3.30}$$

For large $\beta$ the partition function (11.1.7) is dominated by $p_{\max}$. We have

$$\begin{aligned} D(\beta) &= \lim_{V\to\infty} \frac{-1}{(\beta - 1)V} \ln \sum_{i=1}^{r} p_i^{\beta} \\ &\simeq \lim_{V\to\infty} \frac{-1}{(\beta - 1)V} \ln (p_{\max})^{\beta} \\ &= \frac{\beta}{\beta - 1} \alpha_{\min}. \end{aligned} \tag{11.3.31}$$

This, indeed, yields eq. (11.3.27) for $\beta \to \infty$. In an analogous way eq. (11.3.28) can be derived: for $\beta \to -\infty$ the partition function is dominated by

$$p_{\min} \sim \exp(-\alpha_{\max} V). \tag{11.3.32}$$

A great advantage of both the functions $D(\beta)$ and $f(\alpha)$ is that although

Fig. 11.1 (*a*) Rényi dimensions and (*b*) $f(\alpha)$ spectrum of the Feigenbaum attractor.

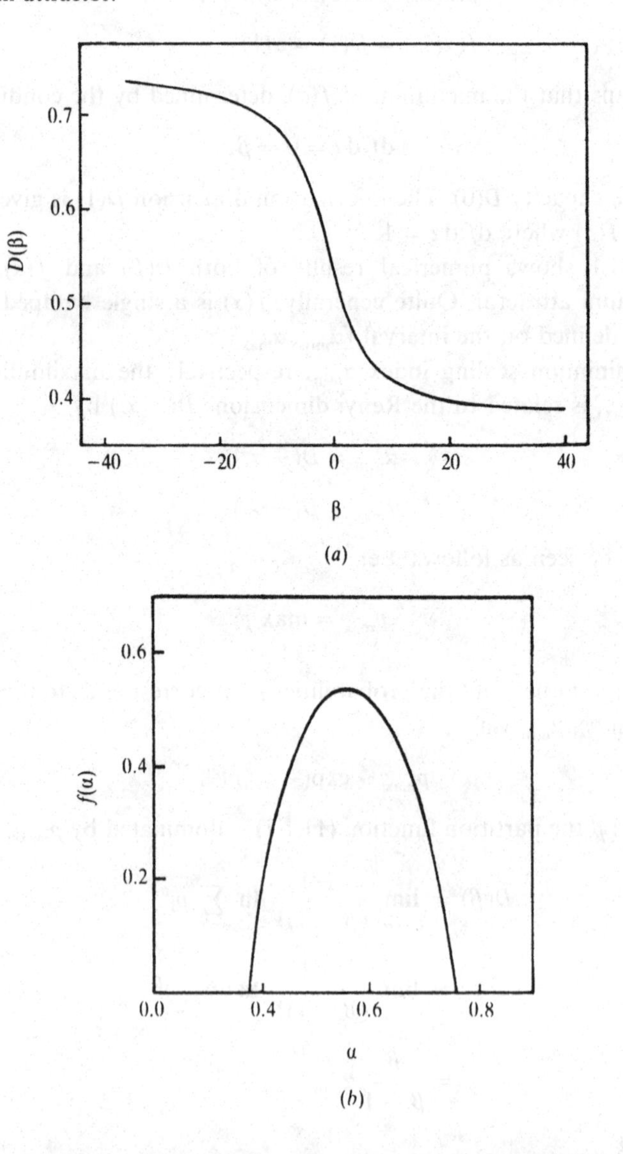

the natural invariant density may have a very complicated singular structure, the functions $D(\beta)$ and $f(\alpha)$ are comparatively smooth. In sections 11.4 and 21.1 we shall consider further examples, where $D(\beta)$ and $f(\alpha)$ can be calculated analytically.

## 11.4 Definition of the Rényi dimensions for cells of variable size

The above definition of the Rényi dimensions is very simple and plain. However, there are some cases (especially for negative $\beta$), where the limit $\varepsilon \to 0$ is ill defined for boxes of equal size. For these cases, a more refined definition of the Rényi dimensions can be given. The relation between this new definition and the old definition is similar to that between Hausdorff dimension and capacity: there is coincidence in most interesting cases.

To give the new definition, let us cover the multifractal by disjoint cells $\sigma_1, \sigma_2, \ldots, \sigma_r$. These covering cells may have different sizes and shapes. Let $p_i$ be the probability attributed to the cell $\sigma_i$. Let each cell $\sigma_i$ be completely covered by a spherical ball of smallest possible radius $l_i$ such that $l_i < l$. A generalized partition function can be defined as follows (Halsey *et al.* 1986)

$$Z(\beta, \tau) = \inf_{\{\sigma\}} \sum_{i=1}^{r} \frac{p_i^\beta}{l_i^\tau} \qquad \text{for } \beta \leqslant 1, \tau \leqslant 0, \tag{11.4.1}$$

respectively

$$Z(\beta, \tau) = \sup_{\{\sigma\}} \sum_{i=1}^{r} \frac{p_i^\beta}{l_i^\tau} \qquad \text{for } \beta > 1, \tau > 0. \tag{11.4.2}$$

The infimum, respectively supremum, is taken over all possible coverings.

In an analogous way as was done for boxes of equal size, we consider a limiting process where the size of all cells goes to zero and, of course, their number $r$ goes to infinity. The Rényi dimension $D(\beta)$ is now defined by the value

$$\tau = (\beta - 1)D(\beta), \tag{11.4.3}$$

where $Z(\beta, \tau)$ neither diverges nor goes to zero in the limit $l \to 0$ (notice the close similarity to the definition of the Hausdorff dimension in section 10.3).

As an example let us consider the two-scale Cantor set with a *two-scale multiplicative measure* (Halsey *et al.* 1986). This is one of the simplest nontrivial examples of a multifractal. At each step of construction of the two-scale Cantor set (see section 10.3) we attribute to the various length scales certain probabilities. These probabilities need not be proportional to the lengths, *a priori* they are independent quantities.

A two-scale Cantor set with multiplicative two-scale measure is defined by the property that we associate with a length scale

$$l_i^{(N)} = a_1^{N-k} a_2^{k} \tag{11.4.4}$$

the probability

$$p_i^{(N)} = w_1^{N-k} w_2^{k} \qquad (w_1 + w_2 = 1), \tag{11.4.5}$$

i.e., the probabilities factorize in the same way as the length scales do. Notice, however, that in general, the shrinking ratios $a_1$, $a_2$ of the length scales are independent of the shrinking ratios $w_1$, $w_2$ of the probabilities. In an analogous way to that in section 10.3 the partition function of eqs. (11.4.1) and (11.4.2) can easily be evaluated. We obtain

$$Z(\beta, \tau) = \left(\frac{w_1^{\beta}}{a_1^{\tau}} + \frac{w_2^{\beta}}{a_2^{\tau}}\right)^N. \tag{11.4.6}$$

This expression stays finite for $N \to \infty$ if

$$\frac{w_1^{\beta}}{a_1^{\tau}} + \frac{w_2^{\beta}}{a_2^{\tau}} = 1. \tag{11.4.7}$$

The last equation implicitly determines the Rényi dimension $D(\beta) = \tau(\beta)/(\beta - 1)$ of the two-scale Cantor set with multiplicative measure. Numerically one may solve this equation for $\tau$. Fig. 11.2 shows $D(\beta)$ for

Fig. 11.2 Rényi dimensions of the two-scale Cantor set with a multiplicative measure ($a_1 = 0.25$, $a_2 = 0.40$, $w_1 = 0.60$, $w_2 = 0.40$).

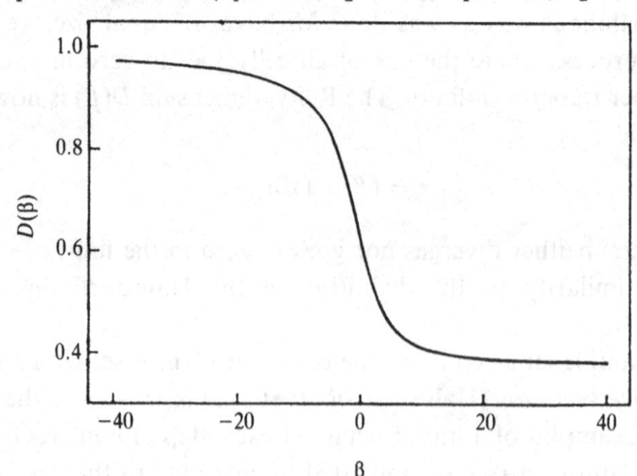

the parameter values $a_1 = 0.25$, $a_2 = 0.4$, $w_1 = 0.6$, $w_2 = 0.4$. Notice that more general fractals can often be asymptotically approximated by two-scale or many-scale Cantor sets.

As another example let us consider the *Feigenbaum attractor*. The Feigenbaum attractor is *not* a two-scale Cantor set, although in a crude approximation it has a certain similarity with such a set. If we consider a trajectory $x_1, x_2, \ldots, x_{2^{N+1}}$ on the attractor, we may cover the attractor by little pieces of length

$$l_i^{(N)} = |x_n - x_{n+2^N}| \qquad n = 1, \ldots, 2^N. \tag{11.4.8}$$

Indeed, this is the best possible covering of the attractor. To each piece we associate the probability

$$p_i^{(N)} = 1/2^N. \tag{11.4.9}$$

Thus the partition function is

$$Z(\beta, \tau) = \sum_i 2^{-N\beta}(l_i^{(N)})^{-\tau}. \tag{11.4.10}$$

The value of $\tau$ where $Z(\beta, \tau)$ neither diverges nor goes to zero yields the dimension spectrum plotted in fig. 11.1(*a*). Indeed, this method yields much better numerical results than a grid of boxes of equal size.

For the Feigenbaum attractor, the largest length scale $l_{\max}{}^{(N)}$ is observed in the vicinity of 0. It scales with $N$ as

$$l_{\max}{}^{(N)} \sim 1/|\alpha|^N, \tag{11.4.11}$$

where $\alpha$ is the Feigenbaum constant (see section 1.3). The smallest length scale $l_{\min}{}^{(N)}$ is observed in the vicinity of 1. It is the image of $l_{\max}{}^{(N)}$. As the map has a maximum of order $z$ at the point $x = 0$, this means

$$l_{\min}{}^{(N)} \sim 1/|\alpha|^{zN} \tag{11.4.12}$$

(see fig. 11.3). For $\beta \to +\infty$ the dominating behaviour is expected to come from the region in the phase space where the measure is most concentrated, i.e., from the smallest length scales. Hence,

$$\begin{aligned} Z(\beta, \tau) &\approx 2^{-N\beta}(l_{\min}{}^{(N)})^{-\tau} \approx 2^{-N\beta}|\alpha|^{zN\tau} \\ &= \exp[(-\beta \ln 2 + z\tau \ln|\alpha|)N] =: 1. \end{aligned} \tag{11.4.13}$$

It follows that

$$-\beta \ln 2 + z(\beta - 1)D(\beta) \ln|\alpha| = 0 \tag{11.4.14}$$

or

$$D(\infty) = \lim_{\beta \to \infty} D(\beta) = \lim_{\beta \to \infty} \frac{\beta}{\beta - 1} \frac{\ln 2}{z \ln|\alpha|} = \frac{\ln 2}{z \ln|\alpha|}. \qquad (11.4.14)$$

Similarly, for $\beta \to -\infty$ we obtain

$$Z(\beta, \tau) \approx 2^{-N\beta} (l^{(N)}{}_{\max})^{-\tau} \qquad (11.4.15)$$

and

$$D(-\infty) = \ln 2/\ln|\alpha|. \qquad (11.4.16)$$

Thus we see that the Feigenbaum constant $\alpha$ is related to special values of the dimension spectrum, namely to the limit dimensions $D(\pm\infty)$.

Fig. 11.3 Scaling of interval lengths in the vicinity of 0 and 1 for a map with a quadratic maximum.

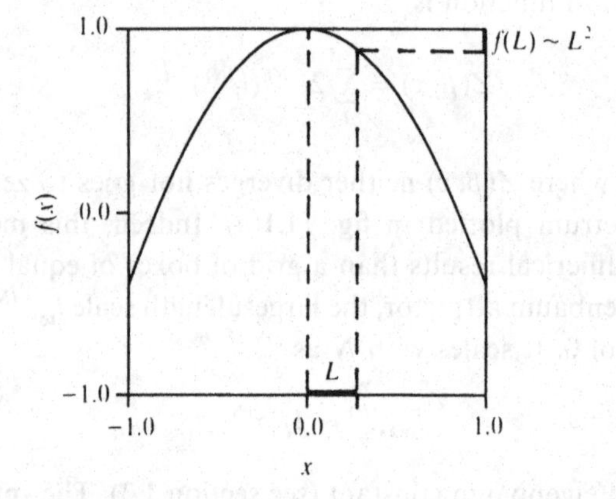

# *12
# Bit-cumulants in multifractal statistics

The fluctuating bit-numbers of multifractals (or of any probability distribution) can also be characterized by the set of all bit-moments or all bit-cumulants. The most interesting cumulant is the second cumulant. It measures the variance of the fluctuating bit-number. In the thermodynamic context this quantity can be regarded as a heat capacity. Quite generally, we shall discuss the role of the bit-variance in thermodynamics and elucidate the sensitivity of this quantity to correlations. We shall then consider the thermodynamic limit of box size going to zero, and present numerical results for the heat capacity of the attractor of the logistic map.

## 12.1 Moments and cumulants of bit-numbers

Consider an arbitrary random quantity $b$ with values $b_i$ in the microstates $i$ of a sample set. The $k$th *moment* of $b$ is defined as the expectation value of the $k$th power of $b$:

$$M_k = \langle b^k \rangle = \sum_i p_i b_i^{\,k}. \tag{12.1.1}$$

A useful tool in probability theory is the so called *generating function* or *generator* of the moments defined by the following expectation value:

$$K(\sigma) = \langle \exp \sigma b \rangle = \sum_{k=0}^{\infty} (\sigma^k/k!) M_k. \tag{12.1.2}$$

Up to the factor $1/k!$, the moments are the expansion coefficients of the generating function. They can easily be obtained from $K(\sigma)$ by differentiation

$$M_k = \left( \frac{\partial^k}{\partial \sigma^k} K(\sigma) \right)_{\sigma=0}. \tag{12.1.3}$$

The set of all moments fully characterizes the random variable $b$ under consideration. In probability theory $K(\mathrm{i}\sigma)$, where i is the imaginary unit, is called the *characteristic function*. The probability distribution of $b$ can be obtained by a Fourier transformation of the characteristic function (see, for example, van Kampen (1981)).

Alternatively, instead of $K(\sigma)$ one may consider another function $G(\sigma)$ defined as

$$G(\sigma) = \ln K(\sigma). \tag{12.1.4}$$

Up to a factor $1/k!$, the *cumulants of order* $k$ are defined to be the coefficients $\Gamma_k$ that occur if $G(\sigma)$ is expanded with respect to the parameter $\sigma$:

$$G(\sigma) = \ln \sum_i p_i \exp(\sigma b_i) = \sum_{k=0}^{\infty} (\sigma^k/k!)\Gamma_k. \tag{12.1.5}$$

The function $G(\sigma)$ is called the *generating function* or the *generator* of the cumulants. Again the cumulants can be obtained from $G(\sigma)$ by differentiation:

$$\Gamma_k = \left( \frac{\partial^k}{\partial \sigma^k} G(\sigma) \right)_{\sigma=0}. \tag{12.1.6}$$

Unlike the moments, the cumulants have a distinctive feature; they become additive for independent systems. That means the following. Let us assume that $(i, j)$ are the microstates of a compound system $\Sigma$, which is the composition of two subsystems $\Sigma^{\mathrm{I}}$ and $\Sigma^{\mathrm{II}}$ with microstates $i$ and $j$, respectively. If the subsystems are statistically independent, the probability of the microstates of the compound system factorizes with respect to the subsystems:

$$p_{ij} = p_i^{\mathrm{I}} p_j^{\mathrm{II}}. \tag{12.1.7}$$

Hence we obtain for the characteristic function $G(\sigma)$ of the entire system

$$\begin{aligned} G(\sigma) &= \ln \sum_{i,j} p_{ij} \exp[\sigma b_{ij}] \\ &= \ln \sum_{i,j} p_i^{\mathrm{I}} p_j^{\mathrm{II}} \exp[\sigma(b_i^{\mathrm{I}} + b_j^{\mathrm{II}})] \\ &= \ln\left(\sum_i p_i^{\mathrm{I}} \exp \sigma b_i^{\mathrm{I}}\right) + \ln\left(\sum_j p_j^{\mathrm{II}} \exp \sigma b_j^{\mathrm{II}}\right) \\ &= G^{\mathrm{I}}(\sigma) + G^{\mathrm{II}}(\sigma). \end{aligned} \tag{12.1.8}$$

Consequently

$$\Gamma_k = \Gamma_k^{\mathrm{I}} + \Gamma_k^{\mathrm{II}}. \tag{12.1.9}$$

So far our consideration has been valid for arbitrary random variables $b$. Let us now choose a special random variable $b$ which is intimately connected to its own probability distribution. For a given probability distribution $p$ we may choose for $b_i$ the bit-number

$$b_i = -\ln p_i. \tag{12.1.10}$$

It is a unique attribute to each event $i$ and its probability $p_i$. This special choice allows us to define the cumulants of the bit-number in just the same way as for a usual random quantity. We call these cumulants the *bit-cumulants.*

According to eq. (12.1.5), the generator $G(\sigma)$ of the bit-cumulants is connected with the Rényi information $I_\beta(p)$ by

$$G(1-\beta) = (\beta - 1)I_\beta(p). \tag{12.1.11}$$

This also confirms the additivity with respect to independent systems. The bit-cumulants $\Gamma_k$ form another interesting set of characteristic quantities for a given probability distribution. Knowing the set of all bit-cumulants is equivalent to knowing the function $G(\sigma)$ or $I_\beta(p)$.

Using eqs. (12.1.5) and (12.1.6), it is easy to evaluate the first few cumulants by differentiation of the generating function. The zeroth cumulant $\Gamma_0$ is zero. The first cumulant $\Gamma_1$ is the mean value $\langle b \rangle$, that is to say the Shannon entropy of the distribution $p$. The second bit-cumulant is the variance of $b$

$$\Gamma_2 = \langle (\Delta b)^2 \rangle = \langle b^2 \rangle - \langle b \rangle^2. \tag{12.1.12}$$

*Remark*: The following fact makes it easier to calculate the cumulants of orders higher than one: The generating function $G_\Delta(\sigma)$ of the fluctuation

$$\Delta b_i = b_i - \langle b \rangle \tag{12.1.13}$$

differs from $G(\sigma)$ by a term linear in $\sigma$:

$$G_\Delta(\sigma) = G(\sigma) - \langle b \rangle \sigma. \tag{12.1.14}$$

Thus the cumulants of higher order are the same for $\Delta b$ and $b$.

## 12.2 The bit-variance in thermodynamics

Before we discuss the bit-variance for chaotic systems, we return to conventional thermodynamics and consider this quantity for the generalized canonical distribution

$$P_i = \exp(\Psi - \beta_\sigma M_i^\sigma). \tag{12.2.1}$$

The corresponding bit-number is

$$B_i = \beta_\sigma M_i^\sigma - \Psi \tag{12.2.2}$$

and we obtain

$$\langle(\Delta B)^2\rangle = \beta_\sigma \beta_\tau \langle \Delta\tilde{M}^\sigma \, \Delta\tilde{M}^\tau \rangle = \beta_\sigma \beta_\tau K^{\sigma\tau}. \tag{12.2.3}$$

Here we have used the correlation matrix of fluctuations $K^{\sigma\tau}$ (see eq. (7.4.9)), which is identical to the susceptibility matrix $Q^{\sigma\tau}$ of eq. (7.4.7). With the help of the entropy $S$ we obtain

$$\langle(\Delta B)^2\rangle = -\beta_\sigma \beta_\tau \frac{\partial M^\tau}{\partial \beta_\sigma} = -\beta_\sigma \frac{\partial S}{\partial M^\tau} \frac{\partial M^\tau}{\partial \beta_\sigma} = -\beta_\sigma \frac{\partial S}{\partial \beta_\sigma}. \tag{12.2.4}$$

where we have used eqs. (7.3.4) and (7.4.2). To follow the conventional notation of thermodynamics, we change from the entropic intensities $\beta_\sigma$ to the energetic intensities

$$y_\sigma = T\beta_\sigma, \tag{12.2.5}$$

where $T$ is the temperature. With

$$\left(\frac{\partial S}{\partial T}\right)_y = \frac{\partial S}{\partial \beta_\sigma}\left(\frac{\partial \beta_\sigma}{\partial T}\right)_y = -\frac{1}{T}\frac{\partial S}{\partial \beta_\sigma}\beta_\sigma, \tag{12.2.6}$$

where we have to sum over all $\sigma$, we find

$$\langle(\Delta B)^2\rangle = T\left(\frac{\partial S}{\partial T}\right)_y = c_y. \tag{12.2.7}$$

The subscript $y$ denotes that all $y_\tau$ are kept fixed when performing the differentiation. The quantity $c_y$ is the heat capacity for the heating process where the $y_\tau$ are kept fixed. For a standard amount of a substance it is called 'specific heat'. For example, for the pressure ensemble, $c_\Pi$ is the specific heat for constant pressure $\Pi$. For the canonical distribution we have $\beta = 1/T$, hence $y = 1$. The condition of keeping $y$ constant is trivially fulfilled. Therefore, in this case $c_y$ is called the specific heat $c_V$ for constant volume $V$, since $V$ is a fixed parameter for the canonical distribution. Applying eq. (12.2.4) to the canonical distribution with energy $E$ we may also write

$$\langle(\Delta B)^2\rangle = -\beta^2(\partial E/\partial \beta)_V = (\partial E/\partial T)_V, \tag{12.2.8}$$

which is another expression for $c_V$. In the case of the grand canonical ensemble, the process of keeping the chemical potential $\mu$ fixed while varying other thermal variables is, in general, connected with a change of the particle number. This, however, implies a change in the amount of substance. Therefore, $c_\mu$ is not a heat capacity for a certain amount of substance at all. It is a new quantity, which nevertheless has a physical meaning and can be expressed with the help of thermal variables. For instance, it appears to be useful for the classification of nonequilibrium phase transitions (Schlögl 1971, 1972, 1983). But we shall not go into details.

## 12.3 The sensitivity to correlations

We have already stressed that if a bit-cumulant $\Gamma_k$ deviates from additivity with respect to some subsystems, this is an indication of correlations between these subsystems. The subsystems may be different spatial parts of a physical system; more generally, they may be different sets of degrees of freedom of the system. The well known dramatic behaviour of the specific heat near a critical phase transition point is caused by the building up of 'critical correlations' in the various spatial regions of a substance.

In broad terms, we can say the following. If in a probability distribution we suppress the correlations between subsystems without changing the probabilities of the single microstates of the subsystems, we lose information, and the entropy is expected to increase. That means the entropy takes on its maximum value for the uncorrelated distribution. If we use an adequate parameter – say $\delta$ – to describe small deviations from the uncorrelated distribution, the entropy $\Gamma_1$ deviates at best in second order of $\delta$ from the uncorrelated value, whereas, in general, all higher bit-cumulants, in particular the bit-variance $\Gamma_2$, will deviate in first order of $\delta$. That means that the bit-variance, like any higher bit-cumulant, is more sensitive to correlations between subsystems than the entropy is.

*Remark*: The reader who is interested in a more detailed proof of this statement is recommended to start with the following scheme. Consider a system that is composed of subsystems $\Sigma_m$. The microstates of each subsystem are labelled by the subscripts

$$i_m = 1, 2, \ldots, R_m. \tag{12.3.1}$$

The probability of the whole system being in a certain microstate is called the compound probability $p(i_1, i_2, \ldots)$. The probability $p_m(i_m)$ of the subsystem $\Sigma_m$ being in a particular state $i_m$ is called a 'marginal' probability and is obtained by performing the sum of all compound probabilities belonging to the same microstate $i_m$. In the special case that the subsystems are statistically independent, the compound probability is

$$p^0(i_1, i_2, \ldots) = p_1(i_1)p_2(i_2) \ldots . \tag{12.3.2}$$

To discuss the onset of correlations small relative variations $\delta(i_1, i_2, \ldots)$ may be introduced by setting

$$p(i_1, i_2, \ldots) = p^0(i_1, i_2, \ldots)[1 + \delta(i_1, i_2, \ldots)]. \tag{12.3.3}$$

This describes the deviation of the correlated distribution from the uncorrelated one. A more detailed analysis (Schlögl 1983) yields in lowest orders of $\delta$

$$\Gamma_1 = \Gamma_1{}^0 - \tfrac{1}{2}\langle\delta^2\rangle^0 + \cdots, \tag{12.3.4}$$

$$\Gamma_2 = \Gamma_2{}^0 + \langle(b^0)^2\delta\rangle^0 + \cdots. \tag{12.3.5}$$

The superscript 0 denotes that the corresponding quantities and mean values $\langle\cdots\rangle^0$ are taken with respect to the uncorrelated distribution $p^0$. Notice that $\Gamma_1 - \Gamma_1{}^0 \sim \delta^2$, whereas $\Gamma_2 - \Gamma_2{}^0 \sim \delta$. This, indeed, implies that $\Gamma_2$ is more sensitive to correlations in the compound distribution than $\Gamma_1$.

## 12.4 Heat capacity of a fractal distribution

Let us now consider an arbitrary, possibly fractal, probability distribution. As before, we may divide the $d$-dimensional phase space into boxes of equal size $\varepsilon$ and attribute to each box the probability

$$p_i = \exp(-b_i). \tag{12.4.1}$$

The corresponding escort distribution is

$$P_i = \exp(\Psi - \beta b_i). \tag{12.4.2}$$

In accordance with eqs. (7.2.7) and (12.2.4), the bit-variance of this canonical distribution is

$$\Gamma_2(P) = -\beta^2 \frac{\partial^2 \Psi}{\partial \beta^2}. \tag{12.4.3}$$

Of special interest is the case $\beta = 1$, because here the escort distribution $P$ approaches the original distribution $p$. The bit-variance of the distribution (12.4.1) is

$$\Gamma_2(p) = (\Gamma_2(P))_{\beta=1} = -(\partial^2\Psi/\partial\beta^2)_{\beta=1}. \tag{12.4.4}$$

Quite generally we have discussed the sensitivity of the bit-variance to correlations between subsystems. But what are the subsystems for a map? For a phase space with dimension $d > 1$ we might think of the different coordinates $x_1, \ldots, x_d$. Another possibility is to consider bit-numbers with respect to the compound probabilities $p(i_1, \ldots, i_N)$ of the symbol sequences $i_1, \ldots, i_N$. We introduced these compound probabilities in section 3.6 and we shall discuss them in more detail in chapter 14. Here let us just deal with the one-point probability $p_i$, which is a marginal distribution of the compound probability. This means it is obtained from $p(i_1, i_2, \ldots)$ by summation over all values of $i_n$, with the exception of the one which occurs as $i$ in the one-point distribution $p_i$. The marginal distribution is not able to describe correlations between the cells. Nevertheless, one can expect that in many cases it is implicitly influenced by these correlations so that a dramatic increase or decrease of correlations will also be reflected in a corresponding behaviour of the natural invariant density. For example, for phase transitions with respect to an external control parameter (see section 21.6), the bit-variance of the marginal distribution is expected to change dramatically.

To discuss the limit of box size $\varepsilon$ going to zero, respectively of $V = -\ln \varepsilon$ going to infinity, we write in leading order

$$b_i = \alpha_i V \tag{12.4.5}$$

$$\Psi = \tau(\beta) V \tag{12.4.6}$$

obtaining

$$\Gamma_2(p) = -(\partial^2 \tau / \partial \beta^2)_{\beta=1} V. \tag{12.4.7}$$

In other words, in the limit $V \to \infty$ the 'bit-variance density', comparable to a specific heat, stays finite:

$$C_2 = \lim_{V \to \infty} \frac{1}{V} \Gamma_2(p) = -(\partial^2 \tau / \partial \beta^2)_{\beta=1}. \tag{12.4.8}$$

According to the definition given by eq. (10.2.13), we may also interpret $C_2$ as a kind of 'dimension' with respect to the bit-variance.

What kind of information is contained in $C_2$? Assuming that the function

$$\tau(\beta) = (\beta - 1) D(\beta) \tag{12.4.9}$$

is analytic, we may expand it in a Taylor series around a fixed value $\beta^*$:

$$\tau(\beta) = \sum_{s=0}^{\infty} \frac{(\beta - \beta^*)^s}{s!} \tau^{(s)}(\beta^*). \tag{12.4.10}$$

Here $\tau^{(s)}(\beta)$ denotes the $s$th derivative of $\tau$. Obviously, knowing the set of all coefficients $\tau^{(s)}(\beta^*)$ is equivalent to knowing the entire function $\tau(\beta)$.

The bit-variance density $C_2$ is the coefficient $\tau^{(2)}(\beta^*)$ for the special choice $\beta^* = 1$. Therefore, $C_2$ contains less information than the entire function $\tau(\beta)$. However, it is of special interest because, in general, the second cumulant is the most interesting cumulant for any fluctuation phenomenon. Moreover, the special value $\beta = \beta^* = 1$ is interesting because it corresponds to the original distribution rather than to an escort distribution with $\beta \neq 1$. For a uniform measure (i.e., all $p_i$ coincide) $\tau(\beta)$ is constant, hence all $\tau^{(s)}(\beta)$ vanish for $s \geqslant 1$. For a nonuniform measure, $C_2$ measures the variance of the fluctuating bit-number.

Let us discuss the relationship between $C_2$ and the $f(\alpha)$ spectrum. Because of

$$\partial\tau/\partial\beta = \alpha, \qquad \partial f/\partial\alpha = \beta, \tag{12.4.11}$$

Fig. 12.1 (*a*) Mean value $C_1$ and (*b*) variance $C_2$ for the attractor of the logistic map as a function of the parameter $r$.

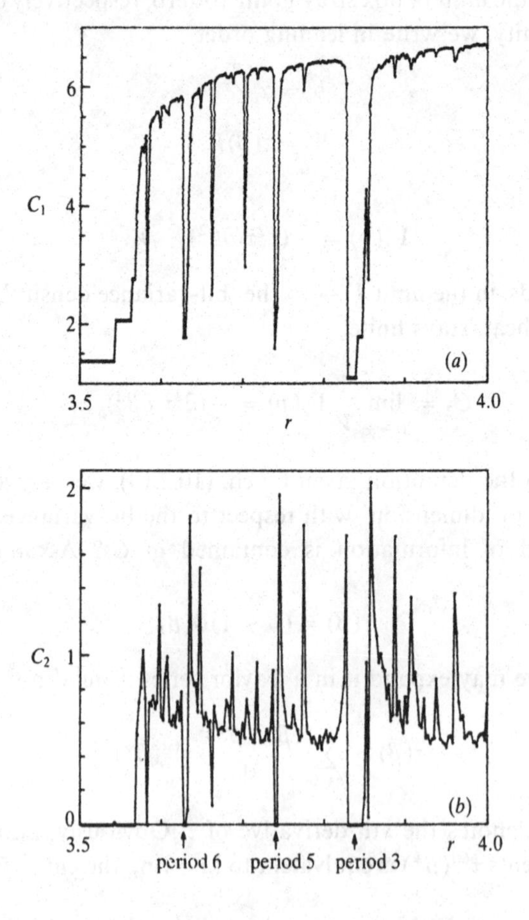

we have

$$\partial^2\tau/\partial\beta^2 = \partial\alpha/\partial\beta, \qquad \partial^2 f/\partial\alpha^2 = \partial\beta/\partial\alpha. \tag{12.4.12}$$

Hence

$$C_2 = -\left(\frac{\partial^2\tau}{\partial\beta^2}\right)_{\beta=1} = -\frac{1}{(\partial^2 f/\partial\alpha^2)_{\alpha=\alpha(1)}}, \tag{12.4.13}$$

i.e., $C_2$ is related to the second derivative of the function $f(\alpha)$ at the fixed point $\alpha(1) = f(\alpha(1)) = D(1)$ (see eq. (11.3.25)).

For an ergodic map, we can express $C_1$ and $C_2$ with the help of the natural invariant density $\rho$ as

$$C_1 = -\langle \ln \rho \rangle \tag{12.4.14}$$

$$C_2 = \langle (\ln \rho)^2 \rangle - \langle \ln \rho \rangle^2. \tag{12.4.15}$$

In fig. 12.1 numerical results for the logistic map

$$x_{n+1} = rx_n(1 - x_n) \tag{12.4.16}$$

are plotted (Schlögl and Schöll 1988).

Besides the first and second cumulant densities one could certainly extend the analysis to all higher cumulant densities

$$C_k = \lim_{V\to\infty} (\Gamma_k/V). \tag{12.4.17}$$

The set of all cumulant densities $C_k$ yields the entire information contained in the function $D(\beta)$.

# *13
# Thermodynamics of finite volume

In this chapter we shall investigate the way a multifractal system approaches the limiting values if the box size decreases. That is to say, we consider small but finite boxes and determine the next order corrections to the Rényi dimensions. Various analogies with traditional thermodynamics will be discussed.

## 13.1 Boxes of finite size

The Rényi dimensions are quantities defined in the limit $\varepsilon \to 0$ or $V = -\ln \varepsilon \to \infty$. In any physical or any computer experiment, however, the measuring device has a finite precision $\varepsilon > 0$ only. Then just the Rényi information per volume $V$

$$\frac{I_\beta}{V} = \frac{1}{(\beta - 1)V} \ln \sum_{i=1}^{r} p_i^\beta \tag{13.1.1}$$

for finite $V$ can be determined instead of the Rényi dimension

$$D(\beta) = -\lim_{V \to \infty} (I_\beta / V). \tag{13.1.2}$$

It is worthwhile studying the dependence of $I_\beta$ on large but finite $V$ because this behaviour is another independent characteristic property of the attractor of a map, or generally of any probability distribution. We may imagine that an experimentalist has measured a probability distribution $p_i$ with an apparatus of restricted precision $\varepsilon$. Then it is indeed interesting for him to know what information he will gain if he doubles the precision by buying a new apparatus.

**The reduced Rényi information** In the limit $V \to \infty$, the Rényi information $I_\beta$ can usually be split into a diverging part plus a constant term. According to eq. (13.1.2), the growth rate of the diverging part is the Rényi dimension. The constant term is called the *reduced Rényi information* (Rényi 1970; Csordás and Szépfalusy 1989; Markosová and Szépfalusy 1991; Kaufmann 1992) and defined as follows:

$$C(\beta) = \lim_{V \to \infty} [-I_\beta - D(\beta)V]. \tag{13.1.3}$$

Provided a series expansion of the Rényi information per volume in powers of $V^{-1}$ exists, the reduced Rényi information corresponds to the leading term in $V^{-1}$.

For nonpathological measures we expect that in the limit $\varepsilon \to 0$ the partition function (11.1.7) scales as

$$Z(\beta) = \sum_{i=1}^{r} p_i^{\beta} = A(\beta)\varepsilon^{(\beta - 1)D(\beta)}, \tag{13.1.4}$$

where $A(\beta)$ is a constant depending on $\beta$ only. In this case the reduced Rényi information is related to $A(\beta)$ by

$$C(\beta) = \frac{1}{1 - \beta} \ln A(\beta), \tag{13.1.5}$$

as can easily be verified by taking the logarithm in eq. (13.1.4). However, there are also exceptional cases where the asymptotic behaviour of $Z(\beta)$ deviates from the simple form of eq. (13.1.4), in particular, if the probabilities $p_i$ do not scale as $\varepsilon^{\alpha_i}$ for $\varepsilon \to 0$, but depend on $\varepsilon$ in a more complicated way. Then eq. (13.1.3) is still formally valid, but $C(\beta)$ may diverge.

The quantity $C(\beta)$ is sometimes used to classify phase transitions of dynamical systems (Csordás and Szépfalusy 1989a), or to characterize the complexity of a measure (Grassberger 1986b). It is quite clear that there may be many different probability distributions $p$ with the same dimension function $D(\beta)$. The function $C(\beta)$ is then an important tool to distinguish between them. On the other hand, it should be clear that the reduced Rényi information $C(\beta)$ does not reach a significance comparable to the Rényi dimension for the characterization of a dynamical system, because $C(\beta)$ is not invariant under smooth coordinate transformations (Kaufmann 1992), whereas $D(\beta)$ is. Nevertheless, $C(\beta)$ is a useful quantity to analyse the properties of a given probability distribution and to estimate finite box size effects.

As an *example* let us consider the velocity distribution function of a damped particle that is kicked by a chaotic force (Beck and Roepstorff 1987a; Beck 1991c). This model has already been described in section 1.4 where we dealt with maps of Kaplan–Yorke type and their physical interpretation. The velocity distribution is just the marginal natural invariant density

$$\rho(y) = \int_X \mathrm{d}x\, \rho(x, y) \tag{13.1.6}$$

of the $y$-variable of the map

$$x_{n+1} = T(x_n), \tag{13.1.7}$$

$$y_{n+1} = \lambda y_n + x_{n+1}. \tag{13.1.8}$$

The map $T$ determines the time evolution of the kicks. In eq. (13.1.6) we integrate over all possible values of $x$, i.e., over the entire phase space $X$ of the map $T$, to obtain the marginal density $\rho(y)$.

As a standard example we may choose for the time evolution of the kicks the map

$$T(x) = 1 - 2x^2. \tag{13.1.9}$$

Fig 13.1 shows histograms of the rescaled velocity $\tau^{1/2} y_n$ for various values of $\lambda$. Increasing $\lambda$ from 0 to 1, we obtain a transition scenario from a complicated singular probability distribution to a smooth Gaussian distribution. The convergence to the Gaussian for $\lambda \to 1$ can be proved rigorously for appropriate classes of maps $T$, see, e.g. Beck (1990b) and references therein. In particular, it can be proved for all maps $T$ conjugated to the Bernoulli shift. To describe the transition scenario in a quantitative way, it is useful to study the dependence of the Rényi dimensions $D(\beta)$ on the parameter $\lambda$, as well as that of the reduced Rényi information $C(\beta)$. Fig. 13.2 shows $D(\beta)$ as a function of $\lambda$ for various values of $\beta$. Fig. 13.3 shows the analogous plot for $C(\beta)$. Notice that for large $\lambda$ the function $D(\beta)$ hardly depends on $\lambda$. But still the 'complexity' of the probability distribution changes quite a lot with $\lambda$. This dependence is well described by the more sensitive function $C(\beta)$.

*Remark*: This example illustrates that it is sometimes useful to study the Rényi dimensions or reduced Rényi informations of a marginal probability distribution (in our case $\rho(y)$) rather than those of the complete attractor (in our case described by the density $\rho(x, y)$). We may call the corresponding quantities 'marginal dimensions' respectively 'marginal reduced informations'. Marginal dimensions and informations are much easier to determine numerically, since the projected phase space has a lower dimension. Moreover, they sometimes have a direct 'physical meaning', as elucidated by the above example of a velocity distribution.

## 13.2 Thermodynamic potentials for finite volume

We may incorporate the concept of the reduced Rényi informations into the thermodynamic formulation. In fact, if one is interested in finite volume effects, we may take into consideration the leading order

Fig. 13.1 Velocity distributions of a kicked damped particle for various values of the parameter $\lambda$: (*a*) $\lambda = 0.1$, (*b*) 0.2, (*c*) 0.3, (*d*) 0.4.

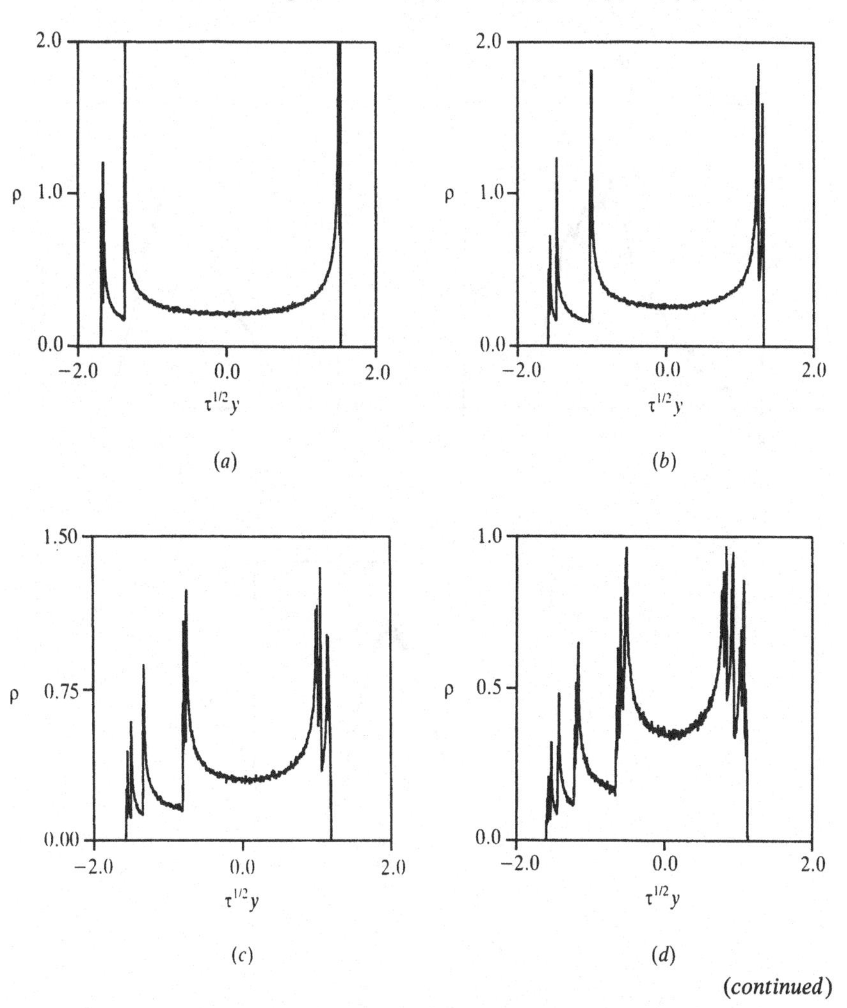


(*continued*)

Fig. 13.1 (*continued*) (*e*) $\lambda = 0.5$, (*f*) 0.6, (*g*) 0.7, (*h*) 0.8, (*i*) 0.9.

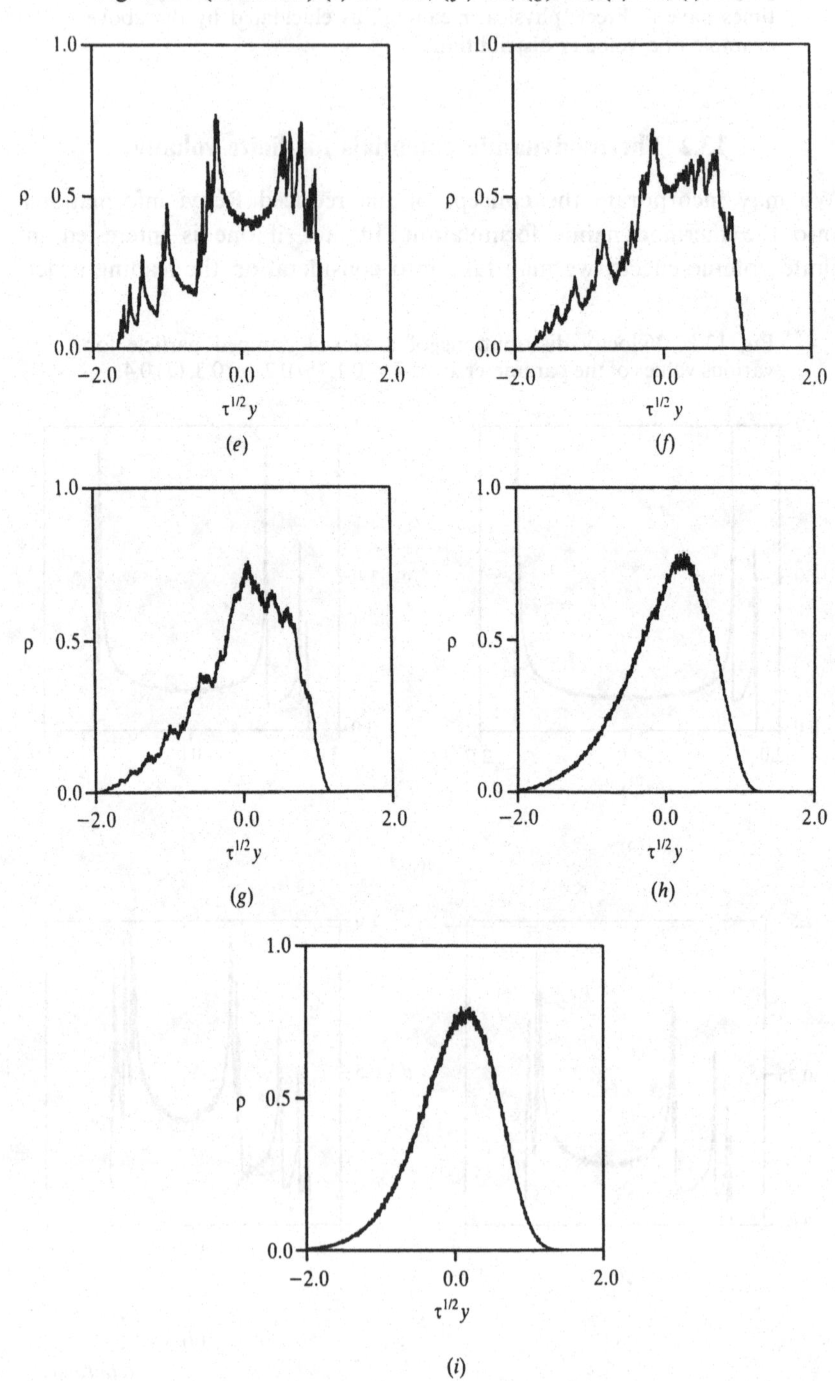

contributions in $V^{-1}$ of the Rényi information per volume rather than just the zeroth order term. To indicate that we take into account the finite volume effects, we mark the corresponding potentials with a tilde, in order to distinguish these potentials from other thermodynamic potentials depending on two parameters that will be introduced in chapter 20.

According to eqs. (9.1.5) and (9.1.6), the Helmholtz free energy is given by

$$\tilde{F}(\beta, V) = \frac{1}{\beta}\Psi(\beta) = -\frac{\beta - 1}{\beta} I_\beta(p). \tag{13.2.1}$$

In traditional thermodynamics, the 'pressure' $\tilde{\Pi}$ is defined as

$$\tilde{\Pi} = -\left(\frac{\partial \tilde{F}}{\partial V}\right)_\beta = \frac{\beta - 1}{\beta}\left(\frac{\partial I_\beta(p)}{\partial V}\right)_\beta. \tag{13.2.2}$$

We shall adopt this definition for a while, although in chapter 20 we shall give a more general definition of pressure. Moreover, the pressure $\tilde{\Pi}$ introduced here should not be confused with the 'topological pressure' of chapter 16. It is clear that we are mainly interested in a region of $V$ where the volume can already be regarded as a quasi-continuous variable

Fig. 13.2 Rényi dimensions of the velocity distributions as a function of $\lambda$. From top to bottom the plot shows $D(0)$, $D(1)$, $D(2)$, . . . , $D(10)$. The dimensions $D(0)$, $D(1)$, $D(2)$ have the constant value 1.

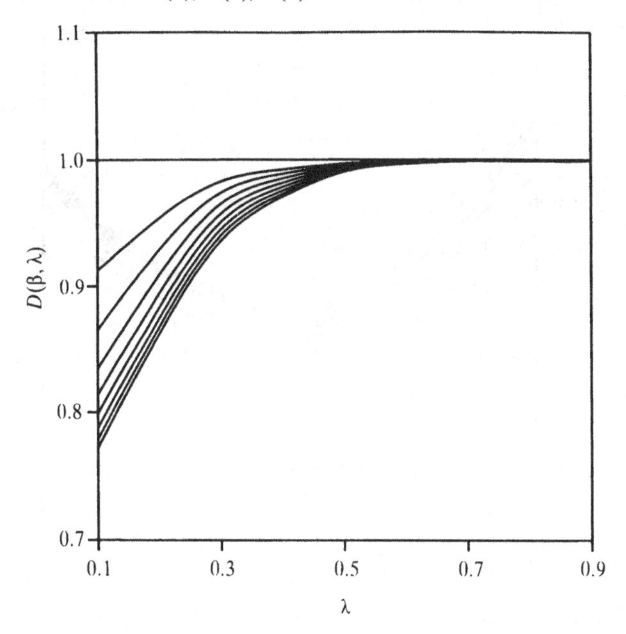

such that the derivative in eq. (13.2.2) makes sense. That is to say, we are interested in values of $V$ such that $R \sim \exp Vd$ is very large.

For nonpathological probability distributions, we expect that the free energy $\tilde{F}$ behaves for large $V$ as

$$\tilde{F}(\beta, V) = a_1(\beta)V + a_0(\beta) + O(V^{-1}), \tag{13.2.3}$$

where $O(V^{-1})$ generally is an abbreviation for 'terms of order $V^{-1}$'. In the following we consider large $V$ only and restrict ourselves to terms of order $V$ and $V^0 = 1$. We may write eq. (13.1.3) in the form

$$-I_\beta(p) = D(\beta)V + C(\beta) + O(V^{-1}). \tag{13.2.4}$$

Comparing eqs. (13.2.3) and (13.2.4), we obtain for large $V$

$$a_1(\beta) = \frac{\beta - 1}{\beta} D(\beta) = -\tilde{\Pi}, \tag{13.2.5}$$

$$a_0(\beta) = \frac{\beta - 1}{\beta} C(\beta). \tag{13.2.6}$$

If we adopt from traditional thermodynamics the definition (6.2.9) of the

Fig. 13.3 Reduced Rényi informations of the velocity distributions as a function of $\lambda$. From top to bottom the plot shows $C(10)$, $C(9), \dots, C(1), C(0)$. As in Fig. 13.2, the curves are spline interpolations of numerical data calculated for $\lambda = 0.1, 0.2, 0.3, \dots, 0.9$.

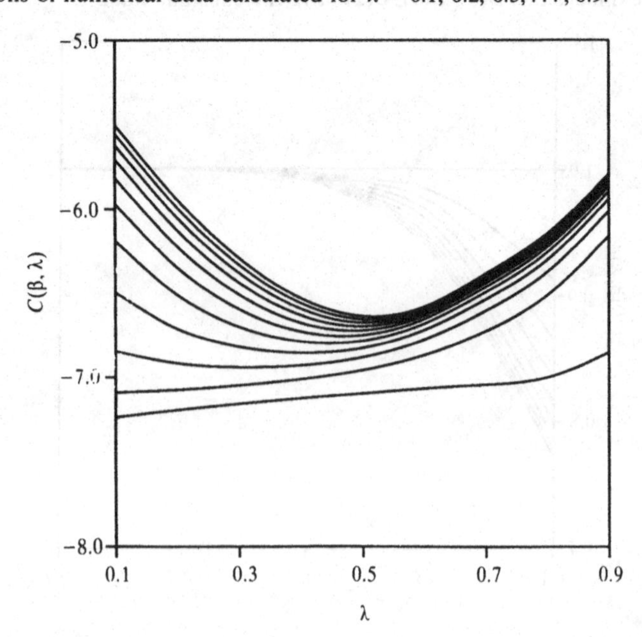

Gibbs free energy, we obtain

$$\tilde{G}(\beta, \Pi) = \tilde{F}(\beta, V) + \tilde{\Pi}V = \frac{\beta - 1}{\beta} C(\beta) + O(V^{-1}). \qquad (13.2.7)$$

As eq. (13.2.7) shows, the Gibbs free energy is finite in the limit of $V \to \infty$.

> *Remark*: The reader should not be confused by the fact that in our formulation the Gibbs free energy $\tilde{G}$ does not grow in proportion to the volume $V$. This has to do with our definition (13.2.2) of the pressure and the special form (13.2.3) of the free energy $\tilde{F}$, which results in the fact that the pressure $\tilde{\Pi}$, according to eq. (13.2.5), does not depend on the volume $V$, but is uniquely determined by the temperature. Hence we consider quite a special situation. In section 20.1 we shall give a more general definition of a pressure.

We remind the reader that in the thermodynamic analogy $T = \beta^{-1}$ is the temperature. The entropy

$$\tilde{S} = -\left(\frac{\partial \tilde{F}}{\partial T}\right)_V = \beta^2 \left(\frac{\partial \tilde{F}}{\partial \beta}\right)_V \qquad (13.2.8)$$

can also be evaluated for large $V$. We obtain

$$\tilde{S}(\beta, V) = f(\alpha)V + \sigma(\beta) + O(V^{-1}). \qquad (13.2.9)$$

The leading part $f(\alpha)$ has already been introduced as the 'spectrum of singularities' in eq. (11.3.14). The term $\sigma(\beta)$ arises from the $\beta$-dependence of $C(\beta)$. We would like to stress once more that the entropy $\tilde{S}$ and the other thermodynamic quantities discussed here are those of the escort distribution, the Rényi information $I_\beta(p)$, however, is that of the original distribution $p$.

Let us summarize. Up to trivial factors, for large $V$ the Gibbs free energy $\tilde{G}$ is the reduced Rényi information $C(\beta)$, the pressure $\tilde{\Pi}$ is the Rényi dimension. The Helmholtz free energy $\tilde{F}$ contains contributions from both. Its volume-independent part is the Gibbs free energy. The pressure $\tilde{\Pi}$ is a measure of how strongly the observed Rényi information (the Helmholtz free energy) depends on the precision of the observation apparatus.

## 13.3 An analytically solvable example

We now consider as a simple example the *Ulam map*, where both $D(\beta)$ and $C(\beta)$ can be calculated in an analytical way. In chapter 2, eq. (2.3.20), we found that the natural invariant density of this map is given by

$$\rho(x) = \frac{1}{\pi(1 - x^2)^{1/2}}. \qquad (13.3.1)$$

Let us partition the phase space, i.e., the interval $X = [-1, 1]$, into a grid of $R$ intervals of equal length $\varepsilon$, the one-dimensional boxes. The boxes are labelled $i = 1, \ldots, R$. The probability of finding an iterate in box $i$ is

$$p_i = \int_{-1+(i-1)\varepsilon}^{-1+i\varepsilon} \mathrm{d}x\, \rho(x). \qquad (13.3.2)$$

For almost all boxes in the interior of the phase space, $\rho(x)$ is smooth and therefore $p_i \sim \varepsilon$. Only for the two boxes at the edges $x = \pm 1$, where $\rho(x)$ diverges, does $p_i$ scale in a different way. There, with $x = \pm(1 - \varepsilon)$ we have $\rho(x) \sim \varepsilon^{-1/2}$ and thus

$$p_1 = p_R \sim \varepsilon^{1/2}. \qquad (13.3.3)$$

The partition function

$$Z(\beta, V) = \sum_{i=1}^{R} p_i^{\beta} \qquad (13.3.4)$$

scales likes

$$Z(\beta, V) \sim (R - 2)\varepsilon^{\beta} + 2\varepsilon^{\beta/2} \sim \varepsilon^{\beta-1} + 2\varepsilon^{\beta/2} \sim \varepsilon^{(\beta-1)D(\beta)}, \qquad (13.3.5)$$

where we have used the fact that $R \sim \varepsilon^{-1}$. Hence we obtain for the Rényi dimensions $D(\beta)$:

$$(\beta - 1)D(\beta) = \min(\beta - 1, \beta/2). \qquad (13.3.6)$$

This means

$$D(\beta) = \begin{cases} 1 & \text{for } \beta \leqslant 2 \\ \dfrac{\beta}{2(\beta - 1)} & \text{for } \beta > 2. \end{cases} \qquad (13.3.7)$$

We notice that at the critical value $\beta_c = 2$ the function $D(\beta)$ is not differentiable. As $\beta^{-1}(\beta - 1)D(\beta)$ is a free energy density, this has similarities with a first order phase transition occurring at the critical temperature $T_c = \beta_c^{-1} = \frac{1}{2}$ in the limit $V \to \infty$ (Ott, Withers and Yorke 1984). We shall discuss phase transitions in the theory of chaotic systems in more detail in chapter 21.

We can also calculate the reduced Rényi informations for the supercritical regime $T > T_c$ if we appeal to a general result of Rényi (1970).

This states that for a given probability density $\rho$ on a phase space $X$, the reduced Rényi information is given by the integral

$$C(\beta) = \frac{1}{1-\beta} \ln \int_X dx \, [\rho(x)]^\beta \tag{13.3.8}$$

for all those values of $\beta$ for which the integral exists. For the Ulam map we obtain

$$C(\beta) = \frac{1}{1-\beta} \ln \left[ \frac{1}{\pi^\beta} \int_{-1}^{+1} dx \, \frac{1}{(1-x^2)^{\beta/2}} \right]. \tag{13.3.9}$$

The integral exists for $\beta < 2$ and yields (Gradshteyn and Ryzhik 1965)

$$C(\beta) = \beta(\beta-1)^{-1} \ln \pi + (1-\beta)^{-1} \ln B(1-\beta/2, 1-\beta/2) + \ln 2. \tag{13.3.10}$$

Here

$$B(x, y) = \frac{\Gamma(x)\Gamma(y)}{\Gamma(x+y)} \tag{13.3.11}$$

denotes the beta function and $\Gamma(x)$ is Euler's gamma function. Hence

$$C(\beta) = \frac{\beta}{\beta-1} \ln \pi + \frac{1}{\beta-1} \left[ 2 \ln \Gamma\left(1-\frac{\beta}{2}\right) - \ln \Gamma(2-\beta) \right] + \ln 2. \tag{13.3.12}$$

From eqs. (13.2.3), (13.2.5), and (13.2.6) it follows that the Helmholtz free energy for large $V$ and $\beta < 2$ is given by

$$\tilde{F}(\beta, V) = \frac{\beta-1}{\beta}(V + \ln 2) + \ln \pi + \frac{1}{\beta} \left[ 2 \ln \Gamma\left(1-\frac{\beta}{2}\right) - \ln \Gamma(2-\beta) \right]. \tag{13.3.13}$$

Part IV · Dynamical analysis of chaotic systems

# 14
# Statistics of dynamical symbol sequences

Whereas in the preceding chapters static properties of dynamical systems were discussed, in this chapter we shall be concerned with dynamical aspects, namely with the analysis of entire sequences of iterates. We shall define appropriate dynamical partition functions and consider their scaling behaviour with respect to an increasing number of iteration steps. To obtain probabilities of enumerable events, we shall use the symbolic dynamics technique of chapter 3. An important set of characteristic quantities of a dynamical system will be introduced: the so called 'Rényi entropies' $K(\beta)$, which can be regarded as the dynamical counterparts of the Rényi dimensions $D(\beta)$. The most important Rényi entropy is the 'Kolmogorov–Sinai entropy' $K(1)$. It describes the average information loss (respectively, increase of information) with respect to the time development of a trajectory.

## 14.1 The Kolmogorov–Sinai entropy

For the dynamical analysis of chaotic systems we need an appropriate partition of the phase space. As already explained in detail in section 3.1, we may choose either a grid of boxes of equal size $\varepsilon$ or a partition into cells $A_i$ of variable size. In either case, we denote the total number of cells of the partition by $R$ and use for the partition the notation

$$\{A\} = \{A_1, A_2, \ldots, A_R\}. \tag{14.1.1}$$

Let us summarize the most important results of chapter 3. Each trajectory $x_0, x_1, x_2, \ldots$ of a map $f$ generates a symbol sequence $i_0, i_1, i_2, \ldots$ determined by the condition

$$x_n \in A_{i_n}, \tag{14.1.2}$$

i.e., the $n$th iterate is found in cell $A_{i_n}$. If we consider a finite symbol sequence $i_0, i_1, \ldots, i_{N-1}$ of $N$ symbols, then there are usually several initial values $x_0$ that generate this sequence. The set $J(i_0, \ldots, i_{N-1})$ of all initial values $x_0$ that generate the sequence $i_0, \ldots, i_{N-1}$ is called an *N-cylinder.*

The probability of a given finite symbol sequence is denoted by $p(i_0, \ldots, i_{N-1})$. It is determined by the probability measure $\sigma$ that is attributed to the $N$-cylinder:

$$p(i_0, \ldots, i_{N-1}) = \int_{J(i_0, \ldots, i_{N-1})} d\sigma(x). \tag{14.1.3}$$

The initial distribution $\sigma$ can, in principle, be chosen arbitrarily. In an experiment, $\sigma$ determines the ensemble of symbol sequences with which we are concerned. However, as already mentioned in section 2.2, for an ergodic map the iterates of a generic trajectory are distributed according to the natural invariant density $\rho(x)$. Hence it is reasonable to choose for the initial distribution the natural invariant measure $\mu$. This means

$$p(i_0, \ldots, i_{N-1}) = \int_{J(i_0, \ldots, i_{N-1})} dx\, \rho(x) = \int_{J(i_0, \ldots, i_{N-1})} d\mu(x). \tag{14.1.4}$$

The advantage of this choice is that the corresponding stochastic process generated by the map is 'stationary', i.e., the probabilities $p(i_0, \ldots, i_{N-1})$ do not change with time.

Let us now measure the information that is contained in the probabilities $p(i_0, \ldots, i_{N-1})$. According to chapter 4, an appropriate information measure is the Shannon information

$$I(p) = \sum_{i_0, \ldots, i_{N-1}} p(i_0, \ldots, i_{N-1}) \ln p(i_0, \ldots, i_{N-1}). \tag{14.1.5}$$

Here the sum runs over all allowed symbol sequences $i_0, \ldots, i_{N-1}$, i.e., over events with nonzero probability. According to the interpretation given in section 4.2, the negative dynamical Shannon information (the dynamical Shannon entropy)

$$H = -I(p) \tag{14.1.6}$$

measures the lack of knowledge about the question of which symbol sequence is to be expected. Notice that for a given map $f$ this entropy depends on the following quantities:

(1) the probability distribution $\sigma$ of the initial values;
(2) the partition $\{A\}$ of the phase space;
(3) the length $N$ of the symbol sequence.

We indicate this by writing

$$H = H(\sigma, \{A\}, N). \tag{14.1.7}$$

We are now interested in the asymptotic behaviour of the dynamical Shannon entropy if we let the length $N$ of the symbol sequence go to infinity. In this case $H$ will usually diverge. What, however, stays finite is the dynamical Shannon entropy per time

$$\begin{aligned} h(\sigma, \{A\}) &= \lim_{N\to\infty} (H/N) \\ &= -\lim_{N\to\infty} (1/N) \sum_{i_0,\ldots,i_{N-1}} p(i_0, \ldots, i_{N-1}) \ln p(i_0, \ldots, i_{N-1}). \end{aligned} \tag{14.1.8}$$

This quantity still depends on the measure $\sigma$ and the partition $\{A\}$.

In order to construct a fundamental quantity that can uniquely be attributed to a map $f$ and that is independent of the arbitrarily chosen partition $\{A\}$, we take a supremum with respect to all possible partitions of the phase space and define

$$h(\sigma) = \sup_{\{A\}} h(\sigma, \{A\}). \tag{14.1.9}$$

This quantity is called the *Kolmogorov–Sinai entropy* (KS entropy, metric entropy) of a map with respect to the measure $\sigma$ (Kolmogorov 1958; Sinai 1959). Usually, one takes for $\sigma$ the natural invariant measure $\mu$. Then $h = h(\mu)$ is just a number uniquely attributed to the map. If we do not mention the measure explicitly, we always mean the KS entropy with respect to the natural invariant measure.

The supremum over all partitions can be rather cumbersome for practical applications: how can we find the partition that yields the maximum value? There is, however, an important simplification if a *generating partition* $\{A\}$ of the phase space is known (see section 3.3). For such a partition it can be proved that $h(\sigma, \{A\})$ takes on its maximum value, i.e., we can omit the supremum over partitions (Cornfeld *et al.* 1982).

The KS entropy has turned out to be an extremely useful quantity in nonlinear dynamics. It is invariant under changes of coordinates (see, for example, section 2.3). Quite generally, it is often used to *define chaos*: a dynamical system is said to be chaotic if it possesses a positive KS entropy. Further, we can measure the 'degree of chaoticity' in a quantitative way: the larger the KS entropy, the 'stronger' are the chaotic properties of the system under consideration. This has to do with the fact that the KS entropy measures the average production (respectively, the average loss) of information per iteration step. To see this, remember that, according to the interpretation given in chapter 4, the Shannon entropy $H(\sigma, \{A\}, N)$

is the missing information needed in order to locate the system on a certain event (in this case, on a certain symbolic trajectory $i_0{}^*, i_1{}^*, \ldots, i_{N-1}{}^*$ of length $N$). Thus

$$\Delta H_N = H(\sigma, \{A\}, N + 1) - H(\sigma, \{A\}, N) \qquad (14.1.10)$$

is the additional missing information if we want to predict the next cell $i_N{}^*$ of the symbolic trajectory provided we know that so far we have had the sequence $i_0{}^*, \ldots, i_{N-1}{}^*$. Thus the average loss of information per iteration step is

$$\begin{aligned}\overline{\Delta H} = \lim_{M\to\infty} \frac{1}{M} \sum_{N=1}^{M} \Delta H_N &= \lim_{M\to\infty} \frac{1}{M} \sum_{N=1}^{M} [H(\sigma, \{A\}, N + 1) - H(\sigma, \{A\}, N)] \\ &= \lim_{M\to\infty} \frac{1}{M} [H(\sigma, \{A\}, M + 1) - H(\sigma, \{A\}, 1)] \\ &= h(\sigma, \{A\}) \qquad (14.1.11)\end{aligned}$$

provided $H(\sigma, \{A\}, 1)$ is finite.

It is quite clear that among all possible partitions $\{A\}$ we should choose the one that yields the most effective description. This means we have to take a supremum over all $\{A\}$. Moreover, a stationary state is properly described by the natural invariant measure $\mu$ rather than any measure $\sigma$. Hence

$$h = \sup_{\{A\}} h(\mu, \{A\}) \qquad (14.1.12)$$

is an appropriate measure of the *information loss* per time. Of course, we can also look at that problem the other way round: the more symbols we observe, the more information we get about the precise position of the initial value $x_0$. In this sense a chaotic system *produces* information, and $h$ measures the average *increase of information* rather than the average loss. We shall come back to this question in section 15.2.

## 14.2 The Rényi entropies

As we have described in chapter 5, besides the Shannon information there are further important information measures, the Rényi informations. Of course, we can also measure the information loss associated with symbol sequences by these Rényi informations of order $\beta$. The corresponding generalizations of the Kolmogorov–Sinai entropy are called 'Rényi entropies'. (To distinguish these special quantities attributed to symbol sequences from the negative Rényi informations as general information measures in statistics, a better name might be 'dynamical Rényi entropies', but it has become common to skip the attribute 'dynamical'.)

As already mentioned in chapter 3, for a given dynamical system there are allowed and forbidden symbol sequences. Let us restrict ourselves to allowed sequences, i.e., sequences with nonzero probabilities $p(i_0, \ldots, i_{N-1})$. For a fixed $N$, let us denote by $\omega(N)$ the number of allowed sequences. It is useful to label each of these allowed sequences by a single index $j$ that runs from 1 to $\omega(N)$. The corresponding probability of the $j$th symbol sequence is denoted by

$$p_j^{(N)} = p(i_0, \ldots, i_{N-1}) \qquad j = 1, 2, \ldots, \omega(N). \tag{14.2.1}$$

With these probabilities we can now do thermostatistics in just the same way as we did for the multifractal distributions in chapters 9 and 11. First of all, we may define dynamical escort distributions by

$$P_j^{(N)} = \frac{(p_j^{(N)})^\beta}{\sum_{j'} (p_{j'}^{(N)})^\beta}. \tag{14.2.2}$$

Next, we introduce the dynamical partition function as

$$Z_N^{\mathrm{dyn}}(\beta) = \sum_{j=1}^{\omega} (p_j^{(N)})^\beta. \tag{14.2.3}$$

The Helmholtz free energy, up to a trivial factor, is identical with the dynamical Rényi information $I_\beta$ of the original distribution $p$:

$$I_\beta(p) = \frac{1}{\beta - 1} \ln Z_N^{\mathrm{dyn}}(\beta) = \frac{1}{1 - \beta} \Psi(\beta). \tag{14.2.4}$$

In a similar way to that in the previous section, we may proceed to the negative dynamical Rényi information

$$H_\beta = H_\beta(\sigma, \{A\}, N) = -I_\beta \tag{14.2.5}$$

and consider the limit $N \to \infty$

$$h_\beta(\sigma, \{A\}) = \lim_{N \to \infty} \frac{H_\beta}{N} = \lim_{N \to \infty} \frac{1}{N} \frac{1}{1 - \beta} \ln Z_N^{\mathrm{dyn}}(\beta). \tag{14.2.6}$$

Notice the fundamental difference compared to the introduction of the Rényi dimensions in section 11.2: There we considered a grid of boxes of equal size $\varepsilon$ and studied the scaling behaviour of the static partition function in the limit $\varepsilon \to 0$. Here, however, we consider a finite partition $\{A\}$ of the phase space. The cells may have variable finite size. We keep the partition constant and study the scaling behaviour of the dynamical partition function *with respect to the time* $N$, not with respect to the box size $\varepsilon$.

To get rid of the partition, we may proceed (at least for $\beta \geqslant 0$) in an analogous way to that used for the KS entropy: we take a supremum over all partitions. Moreover, one usually restricts oneself to the natural invariant measure $\sigma = \mu$. The *Rényi entropies of order* $\beta$ are then defined as

$$K(\beta) = \sup_{\{A\}} h_\beta(\mu, \{A\}). \tag{14.2.7}$$

Of course, for practical applications the supremum over all partitions is quite cumbersome and therefore often omitted. This is not a problem if the partition of the phase space is a generating one. Then the supremum over partitions is already reached. In this case we can write

$$K(\beta) = \lim_{N \to \infty} \frac{1}{1-\beta} \frac{1}{N} \ln \sum_{j=1}^{\omega} (p_j^{(N)})^\beta, \tag{14.2.8}$$

where the $p_j^{(N)}$ is related to the $N$-cylinder $J_j^{(N)}$ by

$$p_j^{(N)} = \mu(J_j^{(N)}) = \int_{J_j^{(N)}} \mathrm{d}\mu(x) \tag{14.2.9}$$

(see section 3.6).

If no generating partition is known, one usually uses a grid of $R$ boxes of equal size $\varepsilon$ and performs the limit $\varepsilon \to 0$ *after* the limit $N \to \infty$ has been performed. That is, for each $R$ we consider symbol sequence probabilities $p(i_0, \ldots, i_{N-1}) = p_j^{(N)}$ for the sequence $j$ and define

$$K(\beta) = \lim_{\varepsilon \to 0} \lim_{N \to \infty} \frac{1}{1-\beta} \frac{1}{N} \ln \sum_{j=1}^{\omega} (p_j^{(N)})^\beta. \tag{14.2.10}$$

Notice that the limit $\varepsilon \to 0$ is just taken to get rid of possible finite size effects of the partition. The important limit is the limit $N \to \infty$.

The difference between eq. (14.2.7) and eq. (14.2.10) is similar to the difference between the capacity and the Hausdorff dimension. From a rigorous mathematical point of view, definition (14.2.7) is preferable. For a computer experiment, however, the equal size definition (14.2.10) might be more suitable. In most cases of physical relevance, we expect the two definitions to coincide.

Two important special cases of the Rényi entropies are the following:

**The topological entropy $K(0)$** For $\beta = 0$ the partition function $Z_N(\beta)$ is equal to the number $\omega(N)$ of allowed symbol sequences of length $N$. For large $N$,

$$\omega(N) \sim \exp[NK(0)]. \tag{14.2.11}$$

Thus $K(0)$ describes the growth rate of the number of allowed symbol sequences with $N$.

**The Kolmogorov–Sinai entropy $K(1)$** A Taylor expansion of $K(\beta)$ around $\beta = 1$ (analogous to eqs. (5.1.3), (5.1.4)) shows that under the assumption of smoothness with respect to $\beta$ the Rényi entropy $K(1)$ is identical with the KS entropy $h$.

Let us now determine the Rényi entropies for a very simple example, the *binary shift map.* This map has already been considered in section 3.2. The natural invariant density is $\rho(x) = 1$, which can be deduced by the same arguments that we have used for the tent map in section 2.2 in the context of eq. (2.2.16). For the binary shift map a generating partition exists. It consists of two cells, namely the intervals 'left' $[0, \frac{1}{2})$, and 'right' $[\frac{1}{2}, 1)$, as described in section 3.2. Hence, for each cell

$$p_i = \tfrac{1}{2}, \qquad i = 0, 1 \tag{14.2.12}$$

and for the symbol sequence of length $N$

$$p(i_0, \ldots, i_{N-1}) = p_j^{(N)} = \int_{J_j^{(N)}} dx = (\tfrac{1}{2})^N, \tag{14.2.13}$$

(see section 3.5). Notice that $f(x)$ has a constant slope 2, which means that each interval expands with the constant factor 2 per iteration step. In the forward direction of time, the dynamical bit-number $b_j^{(N)}$ of $p_j^{(N)}$ increases by one bit. As $b_j^{(N)}$ is a bit-number the observer is missing, this increase means an information loss of one bit per step. In the backward direction, this corresponds to an increase of information of one bit per step, namely ln 2. The number $\omega(N)$ of possible symbol sequences $i_0, \ldots, i_{N-1}$ is $2^N$. Hence we obtain

$$K(\beta) = \lim_{N\to\infty} \frac{1}{1-\beta} \frac{1}{N} \ln\left[2^N \cdot 2^{-\beta N}\right] = \ln 2. \tag{14.2.14}$$

Notice that we do not need to perform the supremum over partitions, as the partition is a generating one. This example is somewhat trivial, as all Rényi entropies have a constant $\beta$-independent value.

For more complicated dynamics, $K(\beta)$ will depend on $\beta$. From the general arguments presented in section 5.3 it follows that for arbitrary dynamics $K(\beta)$ is a monotonically decreasing function of $\beta$ for all real $\beta$, whereas $(\beta - 1)\beta^{-1}K(\beta)$ is monotonically increasing for $\beta \in (-\infty, 0)$ and $\beta \in (0, \infty)$ (Beck 1990a). All the general inequalities presented at the end of section 11.2 for the function $D(\beta)$ are valid for $K(\beta)$ as well.

As an example of a map with a nontrivial spectrum of Rényi entropies let us now consider the *asymmetric triangular map*. It is a generalization of the symmetric tent map and defined on the interval $X = [0, 1]$ as follows:

$$f(x) = \begin{cases} \dfrac{x}{w} & \text{for } 0 \leqslant x \leqslant w \\[2ex] \dfrac{1-x}{1-w} & \text{for } w \leqslant x \leqslant 1. \end{cases} \tag{14.2.15}$$

Here $w \in (0, 1)$ is an arbitrary parameter. Fig. 14.1 shows the graph of this function.

For an arbitrary interval of length $l_2$ and its preimages of length $l_0$ and $l_1$ we have

$$\frac{l_2}{l_0} = \frac{1}{w}, \qquad \frac{l_2}{l_1} = \frac{1}{1-w}. \tag{14.2.16}$$

Hence

$$l_0 + l_1 = wl_2 + (1-w)l_2 = l_2, \tag{14.2.17}$$

i.e., the map conserves the total length of arbitrary intervals. In other words, the Lebesgue measure is again the natural invariant measure for this map. It is obvious that a generating partition for this map is a partition consisting of the two intervals

$$A_0 = [0, w), \qquad A_1 = [w, 1). \tag{14.2.18}$$

Fig. 14.1 Conservation of interval lengths by the asymmetric triangular map.

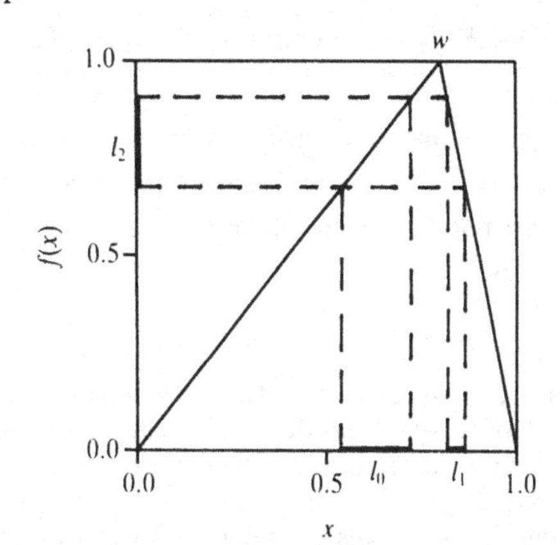

Again we may attribute to $A_0$ the symbol $L$ (left) and to $A_1$ the symbol $R$ (right). All symbol sequences are allowed. An $N$-cylinder corresponding to a certain symbol sequence of length $N$ has the length

$$l_j^{(N)} = w^K(1-w)^{N-K}, \tag{14.2.19}$$

where $K$ denotes the number of symbols $L$ in the sequence. As we attribute to each cylinder the Lebesgue measure, we have for the probability $p_j^{(N)}$ to observe the symbol sequence $j$

$$p_j^{(N)} = l_j^{(N)}. \tag{14.2.20}$$

Hence

$$\begin{aligned} Z_N{}^{\text{dyn}}(\beta) &= \sum_j (p_j{}^{(N)})^\beta \\ &= \sum_K \binom{N}{K} w^{\beta K}(1-w)^{\beta(N-K)} \\ &= [w^\beta + (1-w)^\beta]^N \end{aligned} \tag{14.2.21}$$

and

$$\begin{aligned} K(\beta) &= \lim_{N\to\infty} \frac{1}{N}\frac{1}{1-\beta} \ln Z_N(\beta) \\ &= \frac{1}{1-\beta}\ln[w^\beta + (1-w)^\beta]. \end{aligned} \tag{14.2.22}$$

In particular, we obtain for the topological entropy

$$K(0) = \ln 2 \tag{14.2.23}$$

and for the KS entropy

$$K(1) = \lim_{\beta\to 1} K(\beta) = -w\ln w - (1-w)\ln(1-w) \tag{14.2.24}$$

as can be easily seen from a Taylor expansion around $\beta = 1$ in eq. (14.2.22). Notice that for this simple example the KS entropy is equal to the usual Shannon entropy of a two-event probability distribution with $p_1 = w$ and $p_2 = 1 - w$.

*Remark*: In Beck and Graudenz (1992) it has been shown that successive measurements on a quantum mechanical system with a two-dimensional state space produce information in just the same way as the asymmetric triangular map does. That is, the Rényi entropies associated with the measurement process coincide with those of the chaotic map. This consideration can be generalized to more general cases and yields an interesting connection between the quantum mechanics of measurements and chaos theory.

## 14.3 Equivalence with spin systems

For the probabilities of sequences $i_0, \ldots, i_{N-1}$ of $N$ symbols

$$p(i_0, \ldots, i_{N-1}) = \exp[-H(i_0, \ldots, i_{N-1})] \tag{14.3.1}$$

we may write the escort distributions as

$$P(i_0, \ldots, i_{N-1}) = \exp[\Psi - \beta H(i_0, \ldots, i_{N-1})]. \tag{14.3.2}$$

We have written $H$ for the 'dynamical' bit-number, instead of $b$. Each of the indices $i_n$ can take on the possible values

$$i = 1, 2, \ldots, R. \tag{14.3.3}$$

Formally we observe that our system is equivalent to a spin model of statistical mechanics, namely a Potts model of $R$th order (see chapter 8). The dynamical bit-number $H$ is the analogue of the energy of the spin system. In general, it depends on *all* spins $i_0, \ldots, i_{N-1}$. For the binary shift map or the triangular map we have $i_n = 0$ or 1, and the spin system is an Ising model. All spin configurations are allowed in this case. However, for $w \neq \frac{1}{2}$ they have different probabilities. In general, the probabilities $P(i_0, \ldots, i_{N-1})$ do not factorize and there is a complicated interaction between the spins. We shall come back to this problem later.

As in statistical mechanics, in the theory of dynamical systems one is especially interested in the limit $N \to \infty$, that is to say, in the long-time behaviour, corresponding to the thermodynamic limit for the spin system. Thus we consider a one-dimensional lattice with infinitely many spins. In this limit the Helmholtz free energy per spin is

$$\varphi(\beta) = \lim_{N\to\infty} \frac{-1}{N\beta} \ln \sum_{i_0,\ldots,i_{N-1}} \exp[-\beta H(i_0, \ldots, i_{N-1})]. \tag{14.3.4}$$

The possibility of forbidden sequences of the dynamical system means that there are forbidden configurations of the spin systems. $\varphi(\beta)$ is the dynamical free energy per iteration step. It is related to the Rényi entropies $K(\beta)$ by

$$\varphi(\beta) = \frac{\beta - 1}{\beta} K(\beta). \tag{14.3.5}$$

In particular, we notice that according to the general considerations of section 5.3 $\varphi(\beta)$ is a monotonically increasing function of $\beta$ for arbitrary dynamics (except at the singular point $\beta = 0$).

## 14.4 Spectra of dynamical scaling indices

In analogy to the consideration in section 11.3 we may pass from the sum

$$\Psi(\beta) = -\ln \sum_{i_0,\ldots,i_{N-1}} p(i_0, \ldots, i_{N-1})^\beta \tag{14.4.1}$$

to an integral

$$\Psi(\beta) \sim -\ln \int d\gamma\, W(\gamma) \exp(-\beta N\gamma). \tag{14.4.2}$$

Here $\gamma = H/N$ is a dynamical crowding index, which is expected to stay finite in the limit $N \to \infty$, and $W(\gamma)\, d\gamma$ is the number of symbol sequences $i_0, \ldots, i_{N-1}$ with values of $\gamma$ in the range between $\gamma$ and $\gamma + d\gamma$. Now the time $N$ is the analogue of $V = -\ln \varepsilon$ in eq. (11.3.3). Thus we write for large $N$

$$W(\gamma) \sim \exp[Ng(\gamma)], \tag{14.4.3}$$

and obtain

$$\Psi \sim -\ln \int d\gamma \exp\{[g(\gamma) - \beta\gamma]N\}. \tag{14.4.4}$$

Again assuming that the maximum of the integrand is unique and sharply peaked, for large $N$ we may replace the integral by the integrand at its maximum value, obtaining

$$\lim_{N\to\infty} (\Psi/N) = \beta\varphi = \beta\gamma - g(\gamma), \tag{14.4.5}$$

where $\gamma = \gamma(\beta)$ denotes the value of $\gamma$ where the integrand in eq. (14.4.4) takes the maximum. The spectrum $g(\gamma)$ of the dynamical crowding indices $\gamma$ is the dynamical counterpart of the spectrum $f(\alpha)$ of static crowding indices $\alpha$ (Eckmann and Procaccia 1986). It is obtained via

$$\frac{\partial}{\partial\beta}(\beta\varphi) = \gamma \tag{14.4.6}$$

from the dynamical free energy by a Legendre transformation.

Also all other considerations described in chapters 12 and 13 can easily be generalized to the dynamical case: we may study the set of all cumulant densities of the dynamical bit-numbers as an alternative to the function $K(\beta)$. Moreover, we may consider leading order corrections to the Rényi entropies for finite $N$, described by the reduced dynamical Rényi information.

# 15 Expansion rate and information loss

In this chapter we shall deal with the time evolution of little phase space elements. Typically, such an element expands in certain directions and shrinks in other directions. That is to say, it changes its shape when a map is iterated. This is also true for Hamiltonian systems, although there we have the condition of constraint that the total volume of the phase space element is conserved (but not its shape). Average expansion and contraction rates are quantitatively described by the Liapunov exponents, which we shall define in this chapter. We shall discuss the relation between Liapunov exponents and the loss of information. Moreover, we shall consider fluctuations of expansion rates. These are described by the so called 'spectra of local Liapunov exponents'. For simplicity we shall first restrict ourselves to one-dimensional maps, and afterwards generalize to the $d$-dimensional case.

## 15.1 Expansion of one-dimensional systems

Let us consider a one-dimensional map $f$ on a one-dimensional phase space $X = [a, b]$

$$x_{n+1} = f(x_n). \tag{15.1.1}$$

In the following we restrict ourselves to differentiable or at least piecewise differentiable functions $f(x)$. Consider a small interval $I_0$ of length $\Delta_0$. After one iteration it is transformed into an interval $I_1$ of length

$$\Delta_1 = |f'(x_0)|\Delta_0, \tag{15.1.2}$$

where $x_0$ is an appropriate value in $I_0$, and $f'$ denotes the derivative of $f$. For $|f'(x_0)| > 1$, the transformation is a genuine spatial expansion, for $|f'(x_0)| < 1$ it is a contraction, a negative expansion.

We can extend this consideration to the expansion of $I_0$ during $N$ iteration steps in the following way. We write

$$x_N = f(x_{N-1}) = f^2(x_{N-2}) = \cdots = f^N(x_0), \tag{15.1.3}$$

i.e., we consider the $N$-times iterated map $f^N(x)$. After $N$ iterations, the small interval $I_0$ is transformed into an interval of length

$$\Delta_N = |f^{N\prime}(x_0)|\Delta_0, \tag{15.1.4}$$

where $f^{N\prime}$ is the derivative of $f^N$, and $x_0$ is again an appropriate value in $I_0$. As a consequence of the chain rule we have

$$f^{N\prime}(x_0) = \prod_{n=0}^{N-1} f'(x_n). \tag{15.1.5}$$

A *local expansion rate* $E_N(x_0)$ is defined by

$$\Delta_N = \exp[NE_N(x_0)]\Delta_0. \tag{15.1.6}$$

Obviously

$$E_N(x_0) = (1/N)\ln|f^{N\prime}(x_0)| = (1/N)\sum_{n=0}^{N-1} \ln|f'(x_n)|. \tag{15.1.7}$$

For finite $N$, the quantity $E_N$ is dependent on the initial value $x_0$. However, assuming that the map is ergodic, the time average

$$\lambda = \lim_{N\to\infty} E_N(x_0) \tag{15.1.8}$$

exists and is independent of $x_0$ (up to a set of measure zero). Remember that for ergodic maps arbitrary time averages $\bar{Q}$ of observables $Q(x)$ can be expressed as ensemble averages $\langle Q\rangle$ (see section 2.2):

$$\bar{Q} = \lim_{N\to\infty} \frac{1}{N}\sum_{n=0}^{N-1} Q(x_n) = \int \mathrm{d}x\, \rho(x)Q(x) = \langle Q\rangle. \tag{15.1.9}$$

Here $\rho(x)$ denotes the natural invariant density of $f$. In particular, we can choose $Q(x) = \ln|f'(x)|$ obtaining

$$\lambda = \int \mathrm{d}x\, \rho(x)\ln|f'(x)|. \tag{15.1.10}$$

$\lambda$ is called the *Liapunov exponent* of the one-dimensional map $f$. It describes the average expansion rate. A positive Liapunov exponent means that small intervals (small uncertainties of the initial condition) expand on average, a negative Liapunov exponent means that intervals shrink on average. In other words, a positive Liapunov exponent implies sensitive dependence on initial conditions.

**Example** The Ulam map

$$x_{n+1} = 1 - 2x_n^{\,2} \tag{15.1.11}$$

is known to be ergodic (Ulam and von Neumann 1947). Hence we can use eq. (15.1.10) for the calculation of $\lambda$. The derivative of $f$ is $f'(x) = -4x$, the natural invariant density was derived in section 2.3 to be

$$\rho(x) = \frac{1}{\pi(1 - x^2)^{1/2}}. \tag{15.1.12}$$

Hence we obtain

$$\lambda = \int_{-1}^{+1} \mathrm{d}x \frac{\ln(4|x|)}{\pi(1 - x^2)^{1/2}} = \ln 2. \tag{15.1.13}$$

The Liapunov exponent is positive, i.e., we have sensitive dependence on initial conditions.

## 15.2 The information loss

We may give an interpretation to the expansion rate in terms of the information concept. An iteration step of the one-dimensional map causes the expansion of a small interval $I_n$ containing $x_n$ into $I_{n+1}$ containing $x_{n+1}$. This is connected with a change of knowledge about the question of where the iterate is after the step if we only know that it was in $I_n$ before the step, but do not know the precise position of $x_n$. It is a prognostic question, and the change is a loss of knowledge if the expansion is positive. It is, however, a gain of knowledge with respect to the retrospective question, namely of where the iterate was before the step, if we only know that after the step it is in $I_{n+1}$. In the literature these two different aspects give rise to different names for the change of knowledge, namely either a 'loss' or an 'increase' of information.

Let us assume that we do not know anything about the system except that at time $n$ the iterate is in $I_n$ and at time $n + 1$ it is in $I_{n+1}$. The smaller the interval length, the more information we have on the position of the iterate. As a measure of our knowledge on the position we may regard the bit-number $b_n$ associated with the relative size of the interval $I_n$:

$$\Delta_n/L = \exp(-b_n). \tag{15.2.1}$$

Here $L$ is the total size of the phase space. Expressed in bit-units ln 2, the information loss is the difference of the bit-numbers before and after the iteration step:

$$b_n - b_{n+1} = \ln \Delta_{n+1} - \ln \Delta_n = \ln|f'(x_n)|. \tag{15.2.2}$$

As already said, this difference can be interpreted either as a loss or as an increase of information. With respect to the prognostic question, we shall call it the *information loss*. According to eq. (15.1.8), the Liapunov exponent $\lambda$ is its time average.

In the previous chapter we derived that quite generally, for arbitrary phase space dimensions, the average information loss is measured by the KS entropy $K(1)$. Thus we see that obviously for one-dimensional maps the Liapunov exponent and the KS entropy coincide provided the above heuristic argumentation is correct. In fact, in section 19.3 we shall derive $\lambda = K(1)$ for appropriate classes of one-dimensional maps, using thermodynamic tools. In general, however, only the inequality

$$K(1) \leqslant \lambda \tag{15.2.3}$$

can be proved rigorously. We would like to emphasize that the generalization of this relationship between information loss and expansion rate to higher dimensions is nontrivial and will be discussed in section 15.4. In general, in higher dimensions the KS entropy rather than the Liapunov exponent measures the loss of information.

## *15.3 The variance of the loss

In the previous section we made it plausible that

$$J(x_n) = \ln|f'(x_n)| \tag{15.3.1}$$

is the change of information caused by the $n$th iteration step $n \to n+1$ of a one-dimensional map, the 'information loss' about the question where the trajectory will go to. $J$ may also be regarded as a fluctuating bit-number. We can define cumulants $\Gamma_k(J)$ of $J$ by

$$\ln\langle\exp(\sigma J)\rangle = \sum_{k=0}^{\infty} \frac{\sigma^k}{k!} \Gamma_k(J). \tag{15.3.2}$$

Here $\langle\cdots\rangle$ denotes the ensemble average with respect to the natural invariant density $\rho(x)$. It is equal to the time average over a generic trajectory provided the map is ergodic:

$$\langle\exp[\sigma J(x)]\rangle = \lim_{N\to\infty} \frac{1}{N} \sum_{n=0}^{N-1} |f'(x_n)|^\sigma = \int dx\, \rho(x)|f'(x)|^\sigma \tag{15.3.3}$$

The first cumulant is the average of $J$

$$\Gamma_1(J) = \langle J \rangle. \tag{15.3.4}$$

It is identical with the Liapunov exponent. Explicitly written

$$\langle J \rangle = \lim_{N \to \infty} N^{-1} \sum_{n=0}^{N-1} \ln|f'(x_n)| = \int \mathrm{d}x\, \rho(x) \ln|f'(x)|. \tag{15.3.5}$$

The second cumulant is the variance of information loss for a single iteration step, sometimes also called 'nonuniformity factor':

$$\Gamma_2(J) = \langle J^2 \rangle - \langle J \rangle^2. \tag{15.3.6}$$

The importance of this quantity was first emphasized in Nicolis, Mayer-Kress, and Haubs (1983) and Fujisaka (1983).

In the following we shall discuss simple standard examples of one-dimensional maps and determine the corresponding first two cumulants.

**The triangular map** For the map

$$f(x) = \begin{cases} \dfrac{x}{w} & \text{for } x \in [0, w] \\ \dfrac{1-x}{1-w} & \text{for } x \in [w, 1] \end{cases} \tag{15.3.7}$$

already introduced in section 14.2, the quantity $J = \ln|f'(x)|$ is independent of $x$ for each of the two intervals $[0, w]$ and $[w, 1]$. The natural invariant density is $\rho(x) = 1$. Hence the probability that an iterate is in the first interval $[0, w]$ is $w$. From this we obtain

$$\Gamma_1(J) = -w \ln w - (1-w)\ln(1-w) \tag{15.3.8}$$

and

$$\Gamma_2(J) = w(1-w)\{\ln[(1-w)/w]\}^2. \tag{15.3.9}$$

**The logistic map** In fig. 15.1 numerical results for the map

$$f(x) = rx(1-x) \tag{15.3.10}$$

are plotted (Schlögl 1987). Fig. 15.1($a$) shows the first cumulant $\Gamma_1(J)$ and fig. 15.1($b$) the second cumulant $\Gamma_2(J)$ depending on the parameter $r$. To compare the two values at singular peaks, in fig. 15.1($c$) the ratio

$$Q = |\Gamma_1(J)|/[\Gamma_2(J)]^{1/2} \tag{15.3.11}$$

is plotted as well. The first four cusps of $Q$ pertaining to superstable periods are marked as dots. The corresponding parameter values $r_k$ of $r$ converge to the accumulation point of period doubling for $k \to \infty$ (see section 1.3). The fact that $Q$ remains finite for the values of $r$ where $\Gamma_1(J)$ and $\Gamma_2(J)$ diverge shows that $\Gamma_2(J)$ is more sensitive to changes of $r$ than $\Gamma_1(J)$.

Fig. 15.1 (*a*) Liapunov exponent $\Gamma_1(J)$, (*b*) variance $\Gamma_2(J)$, and (*c*) the ratio $Q = |\Gamma_1|/(\Gamma_2)^{1/2}$ for the logistic map as a function of the parameter $r$. The first four cusps of $Q$ belonging to superstable periods of lengths 2, 4, 8, 16 are marked by dots.

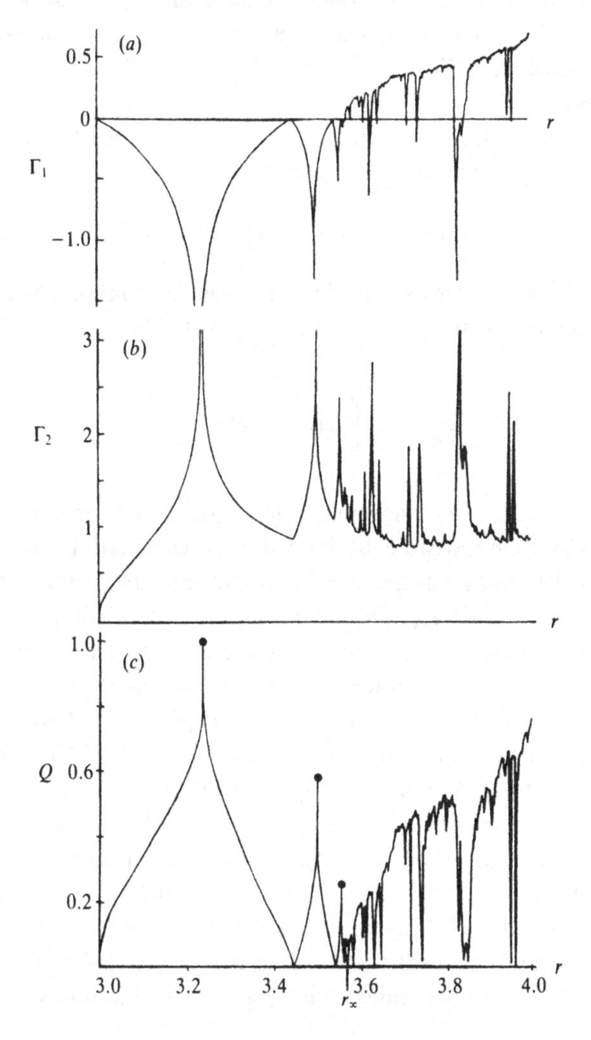

## 15.4 Spectra of local Liapunov exponents

According to eq. (15.1.7), the Liapunov exponent $\lambda$ can be regarded as the time average of the local observable $\ln|f'(x_n)|$. For finite $N$ this average still depends on the initial value $x_0$. If we do not perform the limit $N \to \infty$ in eq. (15.1.8), we may write

$$\lambda(x_0, N) = E_N(x_0) = (1/N) \sum_{n=0}^{N-1} \ln|f'(x_n)|. \tag{15.4.1}$$

This quantity is usually called the *local Liapunov exponent* of the point $x_0$. It describes a local expansion or contraction rate. Local Liapunov exponents are fluctuating quantities, and therefore it is possible to apply thermostatistics to them in a way similar to that used for static and dynamical crowding indices in chapters 11 and 14. This will be worked out in the following.

According to eq. (15.1.4), a small interval $I_0$ containing the initial value $x_0$ expands after $N$ iterations by the factor

$$L_N(x_0) = |f^{N\prime}(x_0)| = \exp[NE_N(x_0)]. \tag{15.4.2}$$

For a fixed finite $N$, let us consider an *ensemble of initial values* $x_0$ and determine the moments of $L_N$ with respect to the natural invariant measure $\mu$:

$$\langle L_N{}^\beta \rangle = \int_X \mathrm{d}\mu(x_0)|f^{N\prime}(x_0)|^\beta. \tag{15.4.3}$$

Here $\beta$ is an arbitrary real parameter chosen such that the integral exists. In eq. (15.4.3) each possible initial value is weighted by the natural invariant density, and we determine the ensemble average of the $\beta$th power of the expansion factor $L_N$. Of course, as we choose for $\mu$ the natural invariant measure, for ergodic maps $f$ the ensemble average is equal to the time average of $L_N{}^\beta$. Notice that for the special choice $N = 1$, the quantity $\langle L_N{}^\beta \rangle$ is identical to the quantity $\langle \exp(\sigma J) \rangle$ that we studied in the previous section. This can be seen from eq. (15.3.3). Now, however, we consider arbitrary $N$ and thus a generalization of the above concept. We are no longer concerned with the expansion behaviour of a single iteration step, but with that of a large number of iterations.

We may write eq. (15.4.3) in a form that reminds us of the partition function $Z_N(\beta)$ of a thermodynamic system. To distinguish this partition function from the dynamical partition function of eq. (14.2.3), we denote it by $Z_N{}^{\mathrm{Lia}}(\beta)$, where the index 'Lia' stands for 'Liapunov exponent'.

Using eq. (15.1.7) we obtain

$$Z_N^{\mathrm{Lia}}(\beta) \equiv \langle L_N{}^{\beta} \rangle = \int \mathrm{d}\mu(x_0) \exp[\beta N E_N(x_0)]$$

$$= \int \mathrm{d}\mu(x_0) \exp\left[ \beta \sum_{n=0}^{N-1} \ln|f'(x_n)| \right]. \qquad (15.4.4)$$

The only difference compared to previously introduced partition functions is that now we have an integral over all possible states $x_0$, rather than a sum. Moreover, each state $x_0$ is not only weighted by the 'Boltzmann factor' $\exp[\beta N E_N(x_0)]$, but also by the natural invariant measure $\mathrm{d}\mu(x_0) = \rho(x_0)\,\mathrm{d}x_0$. In the following we shall call $N$ a particle number (just a formal analogy). The free energy per particle (up to a trivial minus sign) is given by

$$\Lambda(\beta, N) = (1/\beta N) \ln Z_N^{\mathrm{Lia}}(\beta). \qquad (15.4.5)$$

In particular, we are interested in the free energy per particle in the limit $N \to \infty$

$$\Lambda(\beta) = \lim_{N \to \infty} \Lambda(\beta, N). \qquad (15.4.6)$$

This quantity $\Lambda(\beta)$ is called *generalized Liapunov exponent of order* $\beta$ (Fujisaka 1983). It is a generalization of the ordinary Liapunov exponent in just the same sense as the Rényi dimensions $D(\beta)$ are generalizations of the Hausdorff dimension $D(0)$. Each possible local expansion rate is 'scanned' by the 'inverse temperature' $\beta$ and, in addition, weighted by the invariant measure. A nontrivial dependence of $\Lambda(\beta)$ on the scanning parameter $\beta$ indicates that there is a nontrivial spectrum $\phi(E)$ of local expansion rates $E$ – in just the same way as a nontrivial dependence of $D(\beta)$ on $\beta$ indicates a nontrivial spectrum $f(\alpha)$ of crowding indices $\alpha$. In just the same way as explained in section 11.3 or 14.4, we may proceed from the order-$\beta$ Liapunov exponents $\Lambda(\beta)$ to the spectrum $\phi(E)$ by a Legendre transformation.

When a generating partition exists, the cylinder lengths $l_j^{(N)}$ are inversely proportional to the expansion factors $L_N(x_0^{(j)})$, where $x_0$ is an appropriate initial value inside the $j$th $N$-cylinder. Each cylinder is weighted with the probability $p_j^{(N)}$ corresponding to the natural invariant measure. Hence in this case we can write

$$\Lambda(\beta) = \lim_{N \to \infty} \frac{1}{\beta N} \ln \sum_j p_j^{(N)} (l_j^{(N)})^{-\beta}. \qquad (15.4.7)$$

For ergodic maps, for arbitrary $N$ the limit $\beta \to 0$ of $\Lambda(\beta, N)$ is identical with the ordinary Liapunov exponent $\lambda$. This can be seen as follows. For small $\beta$ and finite $N$

$$\begin{aligned}\Lambda(\beta, N) &= (\beta N)^{-1} \ln \int d\mu(x_0) \exp(\beta \ln|f^{N\prime}(x_0)|) \\ &= (\beta N)^{-1} \ln \int d\mu(x_0)(1 + \beta \ln|f^{N\prime}(x_0)| + O(\beta^2)) \\ &= (\beta N)^{-1} \ln\left(1 + \beta \int d\mu(x_0) \ln|f^{N\prime}(x_0)| + O(\beta^2)\right) \\ &\to N^{-1} \int d\mu(x_0) \ln|f^{N\prime}(x_0)| \qquad (\beta \to 0). \end{aligned} \tag{15.4.8}$$

We obtain $1/N$ times the Liapunov exponent of the $N$-times iterated map $f^N$. This, of course, is equal to the Liapunov exponent of $f$ for arbitrary $N$. Thus indeed

$$\lim_{\beta \to 0} \Lambda(\beta, N) = \int d\mu(x_0) \ln|f'(x_0)| = \lambda. \tag{15.4.9}$$

Once more let us mention that all ensemble averages such as $\langle L_N{}^\beta \rangle$ can be expressed as time averages for ergodic maps. Thus the generalized Liapunov exponent contains information about *temporal* fluctuations of expansion rates.

**Example** As a simple example, let us determine the quantity $\Lambda(\beta, N)$ for the Ulam map

$$x_{n+1} = 1 - 2x_n{}^2 \qquad x \in [-1, 1]. \tag{15.4.10}$$

Setting

$$x_0 = \cos \pi u \qquad u \in [0, 1] \tag{15.4.11}$$

we can write

$$x_1 = 1 - 2\cos^2 \pi u = -\cos \pi 2u. \tag{15.4.12}$$

Applying this formula $N$ times we obtain

$$x_N = -\cos \pi 2^N u. \tag{15.4.13}$$

Thus

$$x_N = f^N(x_0) = -\cos(2^N \arccos x_0), \tag{15.4.14}$$

$$f^{N\prime}(x_0) = -\frac{2^N}{(1 - x_0{}^2)^{1/2}} \sin(2^N \arccos x_0). \tag{15.4.15}$$

The natural invariant density has already been derived in section 2.2 to be

$$\rho(x) = \frac{1}{\pi(1 - x^2)^{1/2}}. \tag{15.4.16}$$

Hence

$$\begin{aligned} Z_N^{\text{Lia}}(\beta) &= \int_{-1}^{+1} dx\, \rho(x) |f^{N\prime}(x)|^\beta \\ &= \frac{2^{N\beta}}{\pi} \int_{-1}^{+1} dx \frac{|\sin(2^N \arccos x)|^\beta}{(1 - x^2)^{(\beta+1)/2}}. \end{aligned} \tag{15.4.17}$$

Substituting $x = \cos \pi u$ we obtain

$$Z_N^{\text{Lia}}(\beta) = 2^{N\beta} \int_0^1 du \frac{|\sin 2^N \pi u|^\beta}{|\sin \pi u|^\beta} \tag{15.4.18}$$

and

$$\Lambda(\beta, N) = \ln 2 + (N\beta)^{-1} \ln \int_0^1 du \left| \frac{\sin 2^N \pi u}{\sin \pi u} \right|^\beta. \tag{15.4.19}$$

Hence $\Lambda(\beta, N)$ is equal to the Liapunov exponent $\lambda = \ln 2$ plus a correction term depending on $\beta$ and $N$. For the special values $\beta = 0$, $\beta = 1$, and $\beta = 2$ we obtain (Gradshteyn and Ryzhik 1965)

$$\Lambda(0, N) = \ln 2, \tag{15.4.20}$$

$$\begin{aligned} \Lambda(1, N) = \ln 2 + \frac{1}{N} \ln \Bigg\{ \frac{4}{\pi} \sum_{k=0}^{2^{N-1}-1} \frac{1}{2k+1} \sum_{j=0}^{2^{N-1}-1} \\ \times \left[ \sin(2k+1)(2j+1) \frac{\pi}{2^N} - \sin(2k+1)2j \frac{\pi}{2^N} \right] \Bigg\}, \end{aligned} \tag{15.4.21}$$

$$\Lambda(2, N) = \tfrac{3}{2} \ln 2. \tag{15.4.22}$$

## 15.5 Liapunov exponents for higher-dimensional systems

Let us now discuss the expansion and contraction behaviour in higher dimensions. Suppose we have a $d$-dimensional map $f$. The first part of the considerations in section 15.1 can easily be generalized if we replace the derivative $f'(x)$ by the Jacobi determinant

$$U(x) = \det Df(x) = \det(\partial f^\alpha / \partial x^\beta). \tag{15.5.1}$$

After one iteration, a small $d$-dimensional phase space element $I_0$ of volume $\Delta_0$ expands to the size

$$\Delta_1 = |U(x_0)|\Delta_0, \tag{15.5.2}$$

where $x_0$ is appropriately chosen in $I_0$. After $N$ iterations it expands to a volume of size

$$\Delta_N = |U_N(x_0)|\Delta_0, \tag{15.5.3}$$

where

$$U_N(x_0) = \det Df^N(x_0) \tag{15.5.4}$$

is the Jacobi determinant of the $N$-times iterated map $f^N$. As

$$Df^N(x_0) = \prod_{n=0}^{N-1} Df(x_n), \tag{15.5.5}$$

we can write

$$U_N(x_0) = \prod_{n=0}^{N-1} U(x_n). \tag{15.5.6}$$

A local volume expansion rate $E_N(x_0)$ can be defined by

$$\Delta_N = \Delta_0 \exp[NE_N(x_0)] \tag{15.5.7}$$

or

$$E_N(x_0) = (1/N) \sum_{n=0}^{N-1} \ln|U(x_n)|. \tag{15.5.8}$$

This expansion rate, however, in most cases is not the quantity in which we are interested. For example, if we have a Hamiltonian system (see section 1.6), the Jacobi determinant $U(x)$ has the constant value 1. But in spite of that, Hamiltonian maps can exhibit extremely complicated chaotic behaviour, as the cells may expand in some directions and shrink in others, leading in total to a conservation of the volume. For dissipative systems, the local volume expansion rate is more interesting, but here also one is mainly interested in a more detailed description, namely in expansion rates with respect to the various phase space directions.

To obtain this more detailed information, we study the time development of $d$-dimensional vectors $y$ in the so called *tangent space*, as will be explained in the following. Consider two vectors $x_0$ and $\tilde{x}_0$ in the phase space $X$ that differ only by an infinitesimally small amount. After one iteration $x_0$ is mapped onto

$$x_1 = f(x_0) \tag{15.5.9}$$

and $\tilde{x}_0$ onto

$$\tilde{x}_1 = f(\tilde{x}_0) = f(x_0) + Df(x_0)(\tilde{x}_0 - x_0). \tag{15.5.10}$$

Hence the difference $y_1 = \tilde{x}_1 - x_1$ obeys the linearized evolution equation

$$y_1 = \tilde{x}_1 - x_1 = Df(x_0)(\tilde{x}_0 - x_0) = Df(x_0)y_0. \tag{15.5.11}$$

$y_1$ and $y_0$ are vectors in the tangent space. Notice that the dynamics in this tangent space is governed by a *linear* equation. Applying eq. (15.5.11) $N$ times we obtain

$$y_N = \left(\prod_{n=0}^{N-1} Df(x_n)\right)y_0 = Df^N(x_0)y_0. \tag{15.5.12}$$

As for any linear problem, it is reasonable to look for eigenvectors and eigenvalues of the matrix $Df^N(x_0)$. We denote the eigenvalues by $\sigma_\alpha(N, x_0)$. If the initial vector $y_0$ points into the direction of the $\alpha$th eigenvector, we have

$$y_N = \sigma_\alpha y_0. \tag{15.5.13}$$

Hence the length $|y_0|$ expands by the factor $|\sigma_\alpha|$ during $N$ iteration steps. The Liapunov exponents are defined as

$$\lambda_\alpha(x_0) = \lim_{N\to\infty} \frac{1}{N} \ln|\sigma_\alpha(N, x_0)| \qquad \alpha = 1, \ldots, d. \tag{15.5.14}$$

They describe the average expansion rate in the direction of the $\alpha$th eigenvector of the matrix

$$Df^N(x_0) = Df(x_{N-1})\cdots Df(x_1)Df(x_0). \tag{15.5.15}$$

For ergodic maps $f$, the Liapunov exponents are independent of the initial value $x_0$ (for almost all $x_0$, i.e., up to a set of measure zero) and are independent of the norm chosen, i.e., how we measure the length of the vectors. This is a consequence of the celebrated 'multiplicative ergodic theorem' proved by Oseledec (1968). This theorem is an extension of the Birkhoff ergodic theorem mentioned in section 2.2. A generalization of the Birkhoff ergodic theorem is necessary due to the fact that in eq. (15.5.12) the quantities $Df(x_n)$ are noncommuting matrices rather than scalars.

Let us order the Liapunov exponents $\lambda_1, \lambda_2, \ldots, \lambda_d$ according to their size. The most important Liapunov exponent is certainly the largest Liapunov exponent $\lambda_1$. A generic initial vector $y_0$ in the tangent space will possess a component pointing into the direction of the eigenvector belonging to $\lambda_1$. This component will expand with the largest rate $\lambda_1$ and thus yield the dominating behaviour for large $N$. The second largest Liapunov exponent $\lambda_2$ describes the separation rate of much more special initial conditions $y_0$, namely those that do not possess a component in the direction of the eigenvector corresponding to $\lambda_1$. Similarly, the exponent $\lambda_j$ is of relevance if one considers initial vectors $y_0$ that do not possess components pointing into the directions of the eigenvectors of $\lambda_1, \ldots, \lambda_{j-1}$. This, of course, is quite difficult to realize in a numerical experiment.

Liapunov exponents have turned out to be extremely useful for the quantitative analysis of nonlinear dynamics. They are invariant under a smooth change of coordinates. Positive Liapunov exponents correspond to expanding directions in the phase space, negative Liapunov exponents to contracting directions. The existence of just one positive Liapunov exponent already implies the exponential increase of small uncertainties of the initial data. Indeed, often a map is said to be chaotic if at least one Liapunov exponent is positive. Notice that in spite of this the total volume of arbitrary phase space elements may shrink everywhere. The essential point for chaotic behaviour is the existence of at least *one direction* in which phase space elements expand on average. With respect to the total volume this expansion may be compensated by the contracting directions.

**Example** Let us consider as a simple example the map

$$f(x, y) = (ax - \lfloor ax \rfloor, by + x), \qquad (0 < b < 1). \tag{15.5.16}$$

Here $\lfloor \cdots \rfloor$ denotes the integer part, and $a$ and $b$ are parameters. We have

$$Df = \begin{pmatrix} a & 0 \\ 1 & b \end{pmatrix}, \tag{15.5.17}$$

which is independent of $x_0$ in this special case. Moreover,

$$(Df)^N = \begin{pmatrix} a^N & 0 \\ \ldots & b^N \end{pmatrix}. \tag{15.5.18}$$

Hence,

$$\sigma_1(N) = a^N, \qquad \sigma_2(N) = b^N. \tag{15.5.19}$$

This yields the Liapunov exponents

$$\lambda_1 = \ln|a|, \qquad \lambda_2 = \ln|b|. \tag{15.5.20}$$

For $|a| > 1$ the map exhibits chaotic behaviour. A small uncertainty of the initial value $x_0$ will grow fast, indeed exponentially fast. Nevertheless, the volume expansion coefficient is

$$E = \lim_{N \to \infty} \frac{1}{N} \sum_{n=0}^{N-1} \ln|\det Df| = \lambda_1 + \lambda_2, \tag{15.5.21}$$

which is negative for $|a| < 1/b$.

As another example let us consider *Arnold's cat map* defined by

$$f: \begin{cases} x_{n+1} = (x_n + y_n) \bmod 1 \\ y_{n+1} = (x_n + 2y_n) \bmod 1. \end{cases} \tag{15.5.22}$$

The symbol 'mod 1' means that we take the fractional part, i.e.,

$$x \bmod 1 = x - \lfloor x \rfloor. \tag{15.5.23}$$

The phase space is the unit square $X = [0, 1]^2$. For this map we have

$$Df = \begin{pmatrix} 1 & 1 \\ 1 & 2 \end{pmatrix} \tag{15.5.24}$$

and

$$\det Df = 1. \tag{15.5.25}$$

Hence it is a Hamiltonian system. The eigenvalues of $Df$ are

$$\sigma_{1/2} = \tfrac{1}{2}(3 \pm 5^{1/2}). \tag{15.5.26}$$

Because $Df$ is a constant matrix, we have for the eigenvalues $\sigma_{1/2}(N)$ of $Df^N$

$$\sigma_{1/2}(N) = \sigma_{1/2}{}^N. \tag{15.5.27}$$

Hence the Liapunov exponents are

$$\lambda_1 = \ln\left(\frac{3 + 5^{1/2}}{2}\right) > 0 \tag{15.5.28}$$

and

$$\lambda_2 = \ln\left(\frac{3 - 5^{1/2}}{2}\right) < 0. \tag{15.5.29}$$

There is one expanding and one contracting direction. Because the map is a Hamiltonian dynamical system, the total volume expansion rate is zero. Indeed we observe

$$\lambda_1 + \lambda_2 = 0. \tag{15.5.30}$$

**Information loss in higher dimensions** The average information loss of a $d$-dimensional map is measured by the KS entropy $K(1)$ introduced in section 14.1. In section 15.2 we made it plausible that typically $K(1) = \lambda$ for one-dimensional systems. One might conjecture that for higher-dimensional systems also there is a relation between the KS entropy and the Liapunov exponents. If we intuitively try to generalize the arguments of section 15.2 to the $d$-dimensional case, we could argue as follows. The information loss is produced by the expanding directions. It cannot be restored by the contracting directions. Hence, the total information loss is given by the sum of *all positive* Liapunov exponents:

$$K(1) = \sum_{\lambda_\alpha > 0} \lambda_\alpha. \tag{15.5.31}$$

This equation can be proved for certain classes of dynamical systems and is called *Pesin's identity* (Pesin 1977; Ruelle 1978b; Eckmann and Ruelle 1985; Paladin and Vulpiani 1987). In particular, it is valid for the so called *hyperbolic* dynamical systems. This notion will be explained in the next section. In general, however, only the inequality

$$K(1) \leqslant \sum_{\lambda_\alpha > 0} \lambda_\alpha \tag{15.5.32}$$

is valid.

**Generalized Liapunov exponents in higher dimensions** Not too much is known of generalized Liapunov exponents for higher-dimensional maps. It is certainly reasonable to consider fluctuations of local expansion rates in higher dimensions as well. To define a set of generalized Liapunov exponents for $d$-dimensional maps we may project onto the various eigenvectors of the matrix $Df^N(x_0)$. This can be achieved by replacing the expansion factor $L_N$ in eqs. (15.4.2) and (15.4.3) by the eigenvalue $\sigma_\alpha(N, x_0)$. For two-dimensional maps with one expanding and one contracting direction in the phase space we may simply consider moments of expansion rates of little pieces of the unstable manifold (the notion of stable and unstable manifold will be explained in the next section). We shall not go into details, but refer the reader to some of the standard literature on the subject (Grassberger 1986a; Paladin and Vulpiani 1987; Tél 1990).

## 15.6 Stable and unstable manifolds

Suppose the $d$-dimensional map $f$ has a fixed point $x^*$, i.e.,

$$f(x^*) = x^*. \tag{15.6.1}$$

In section 1.2 we characterized the stability properties of fixed points by the eigenvalues $\eta_\alpha(x^*)$ of the Jacobi matrix $Df(x^*)$. Equivalently, we may characterize the fixed point by the set of its Liapunov exponents $\lambda_\alpha(x^*)$, $\alpha = 1, \ldots, d$. This makes no difference, since

$$Df^N(x^*) = [Df(x^*)]^N \tag{15.6.2}$$

and thus, according to eq. (15.5.14),

$$\lambda_\alpha(x^*) = \ln|\eta_\alpha(x^*)|. \tag{15.6.3}$$

The fixed point is called *hyperbolic* if all its Liapunov exponents are nonzero. It is called *attracting* if all $\lambda_\alpha < 0$, and *repelling* if all $\lambda_\alpha > 0$. If some $\lambda_\alpha$ are positive and others are negative, the fixed point is said to be of *saddle type*. For example, the fixed point $x^* = (0, 0)$ of the cat map is a hyperbolic fixed point of saddle type, because $\lambda_1 > 0$ and $\lambda_2 < 0$. As $Df^N(x^*)$ is a matrix acting in the tangent space, it describes the behaviour of vectors $x$ that differ from $x^*$ by an infinitesimal amount only, i.e., we consider a small vicinity of $x^*$. In the vicinity of $x^*$, the subspace spanned by the eigenvectors corresponding to negative $\lambda_\alpha$ is called the *local stable manifold of* $x^*$, and that corresponding to positive $\lambda_\alpha$ the *local unstable manifold of* $x^*$. We may also define a *global stable manifold* $W^s(x^*)$ of $x^*$ as follows:

$$W^s(x^*) = \left\{ y\colon \lim_{n\to\infty} f^n(y) = x^* \right\}. \tag{15.6.4}$$

It is the set of all points $y$ in the phase space that are attracted by the fixed point $x^*$. In contrast to the local manifold, the global manifold also contains points $y$ that are 'far away' from $x^*$.

The definition of the *global unstable manifold* $W^u(x^*)$ of $x^*$ is easy if the map is a diffeomorphism, i.e., if a unique differentiable $f^{-1}$ exists (see section 1.5). Then we simply define $W^u(x^*)$ to be the global stable manifold of the inverse map $f^{-1}$:

$$W^u(x^*) = \left\{ y\colon \lim_{n\to\infty} f^{-n}(y) = x^* \right\}. \tag{15.6.5}$$

That is to say, the global unstable manifold contains all points that are attracted by $x^*$ under time reversal.

The definition of $W^u(x^*)$ can be easily generalized to noninvertible maps, but we shall not go into these technical details. From now on we shall skip the attribute 'global', and simply talk of stable and unstable manifolds, always meaning the global manifolds.

In practice, one very often considers two-dimensional maps. In this case the stable and unstable manifold of a fixed point of saddle type are simply 'lines' that are folding and accumulating in a very complicated way. Locally, these lines are smooth. An easy method to calculate them (in the case of a diffeomorphism) is the following one. We cover the fixed point by a small ball containing a large number of initial values. Then we iterate this ensemble of initial values. Under successive iterations the ball is deformed into a long line, which is just the unstable manifold $W^u(x^*)$. If we iterate $f^{-1}$ instead of $f$, we obtain the stable manifold $W^s(x^*)$ by the same method.

Stable and unstable manifolds can also be defined for periodic orbits rather than fixed points. If the periodic orbit has the period length $L$, one simply replaces the map $f$ by the $L$-times iterated map $f^L$. More generally, we may even introduce stable and unstable manifolds for an arbitrary point $x$ in the phase space. In this case one defines

$$W^s(x) = \left\{ y\colon \lim_{n\to\infty} \frac{1}{n} \ln|f^n(x) - f^n(y)| < 0 \right\}. \tag{15.6.6}$$

That is to say, the stable manifold of an arbitrary point $x$ consists of all those points $y$ for which the distance between the orbits of $x$ and $y$ decays as

$$|f^n(x) - f^n(y)| \sim \exp(\lambda n), \qquad \lambda < 0, \tag{15.6.7}$$

i.e., small distances contract on the stable manifold. Similarly, we can define (for a diffeomorphism) the unstable manifold of $x$ by

$$W^u(x) = \left\{ y\colon \lim_{n\to\infty} \frac{1}{n} \ln|f^{-n}(x) - f^{-n}(y)| < 0 \right\}, \tag{15.6.8}$$

i.e., small distances contract in reverse time, which means that they expand in the original time direction.

In chapter 1 we mentioned that typically a strange attractor of a two-dimensional chaotic map has the structure of a 'line' that is curled up and folded in a very complicated way (see figs. 1.7 and 1.8). It turns out that this is just a visualization of the *unstable* manifold of an arbitrary point $x$ on the attractor. Quite generally one can prove that the unstable manifold of $x$ is contained in the attractor (see, for example, Eckmann and Ruelle (1985)). Moreover, the shape of the unstable manifold of an arbitrary unstable fixed point or periodic orbit of the map is usually an indistinguishably close approximation of the strange attractor of the map (Guckenheimer and Holmes 1983); it gives us information on the geometrical structure of the attractor.

If we take a little piece of the unstable manifold, the images of this little piece will forever stay on the unstable manifold under successive iterations of the map $f$. The same is true for the stable manifold. This makes these manifolds such helpful tools for the analysis of higher-dimensional dynamical systems. The expansion and contraction behaviour on these manifolds is under full control.

**Homoclinic points** In typical cases the stable and unstable manifolds have a tendency to fold and accumulate in a very complicated way. In particular, the manifolds $W^s(x^*)$ and $W^u(x^*)$ can intersect. Such an intersection point is called a 'homoclinic point'. As images of little pieces of $W^s(x^*)$ and $W^u(x^*)$ will again be pieces of $W^s(x^*)$ and $W^u(x^*)$, the existence of one homoclinic point, in fact, implies the existence of infinitely many. We notice that the manifolds must be interwoven in a very complex way.

Homoclinic points are of utmost interest in chaos theory, because their existence can be used as a criterion for the existence of strongly chaotic motion (see, for example, Guckenheimer and Holmes (1983) for more details). In this sense unstable fixed points $x^*$ of the map yield information on strange attractors.

**Hyperbolic dynamical systems** Of great interest for the rigorous mathematical formulation of the thermodynamic formalism (Ruelle 1978a) are the so called 'hyperbolic dynamical systems'. A *map* is called 'hyperbolic' if the stable and unstable manifolds of arbitrary unstable fixed points (including all periodic orbits) always cross each other transversally. They should never touch each other in a tangential way. Both situations are sketched in fig. 15.2. Most rigorous results of the thermodynamic formalism can only be proved for hyperbolic systems. On the other hand, generic nonlinear maps (such as the logistic map, maps of Kaplan–Yorke type, and the Hénon map) appear to be nonhyperbolic. The situation here is no different from that in other fields of physics: most systems of real interest do not fit the conditions necessary for a rigorous treatment.

**Expanding dynamical systems** It may happen that there is no contracting direction in the phase space at all. A $d$-dimensional map $f$ on the phase space $X$ is called 'expanding' if there are constants $\varepsilon > 0$ and $\alpha > 1$ such that

$$|f(x) - f(y)| > \alpha|x - y| \tag{15.6.9}$$

for all $x \in X$ with $|x - y| < \varepsilon$. In other words, at each point of the phase space small phase elements expand into all directions. As this condition must be satisfied for all $x \in X$, a somewhat more precise name for this property is *everywhere expanding*. However, the 'everywhere' is usually skipped.

In the case of a one-dimensional differentiable map the condition (15.6.9) reduces to

$$|f'(x)| > 1 \tag{15.6.10}$$

for *all* $x \in X$. If $f$ is just piecewise differentiable, eq. (15.6.10) has to be valid for all $x$ in the intervals of differentiability. In the one-dimensional case, the expanding condition is almost identical with the hyperbolicity condition: a one-dimensional map is called *hyperbolic* if there is an $N \geqslant 1$ such that

$$|f^{N\prime}(x)| > 1 \tag{15.6.11}$$

for *all* $x \in X$, i.e., the $N$-times iterated map is expanding.

To elucidate these definitions, let us consider a few examples. The tent map is both expanding and hyperbolic, since $f(x)$ is piecewise differentiable and $|f'(x)| > 1$. The logistic map is neither expanding nor hyperbolic, since $f'(x) = 0$ for $x = 0$. The cat map is hyperbolic but not expanding: the map has a constant Jacobi matrix $Df$ with one positive eigenvalue $\lambda_1$ and one negative $\lambda_2$. In this simple case the unstable and the stable manifolds correspond to the constant directions parallel to the corresponding eigenvectors. They cross each other in a transversal way, i.e., the map is hyperbolic, but since $\lambda_2 < 0$, it is not expanding. The Hénon map is neither hyperbolic nor expanding. This can be guessed since this

Fig. 15.2 Stable and unstable manifold of (*a*) a hyperbolic system and (*b*) a nonhyperbolic system.

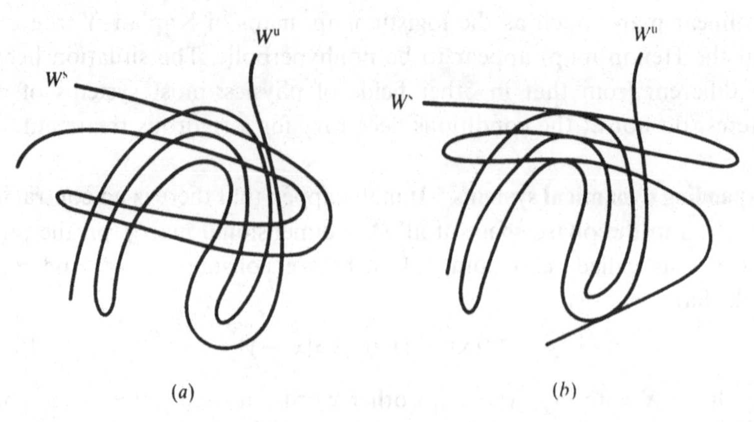

map is a kind of two-dimensional extension of the logistic map, and the logistic map is neither hyperbolic nor expanding. In fact, for the Hénon map homoclinic tangencies are observed. The reader may, for example, numerically determine the stable and unstable manifolds of the fixed point $x^* = (\bar{x}, b\bar{x})$, where $\bar{x} = \{b - 1 + [(b-1)^2 + 4a]^{1/2}\}/2a$. The manifolds touch each other in a tangential way, similar to fig. 15.2(*b*).

# 16
# The topological pressure

So far we have introduced three important different types of 'free energies': the Rényi dimensions, describing scaling properties of the natural invariant measure with respect to space, the dynamical Rényi entropies, describing scaling properties of symbol sequence probabilities with respect to time, and the generalized Liapunov exponents of order $\beta$ containing information about expansion rates and their temporal fluctuations. All the above free energies had in common that the natural invariant measure $\mu$ entered into the partition function, in the sense that the cells, respectively the initial values, were weighted by $\mu$ in an appropriate way. In many cases, however, the natural invariant measure is not known analytically, but one still wants to analyse the dynamical system by thermodynamical methods. Then a very useful type of partition function can be defined that contains information about expansion rates and their fluctuations, but *without* weighting them with respect to the natural invariant measure. Rather, one takes into account geometrical properties of the map such as cylinder sizes or unstable periodic orbits, weighting them with the same *a priori* probability. The corresponding free energy is called the 'topological pressure' for historical reasons. As in the previous chapter we shall first restrict the discussion to one-dimensional maps and afterwards generalize to higher-dimensional cases.

## 16.1 Definition of the topological pressure

As we have discussed in chapter 15, the local expansion rate of a one-dimensional map $f$ is defined as

$$E_N(x_0) = (1/N) \sum_{n=0}^{N-1} \ln|f'(x_n)|. \quad (16.1.1)$$

It depends on the initial value $x_0$. Only in the limit $N \to \infty$ and for ergodic maps does it lose this dependence. An important question concerning a chaotic map $f$ is the following: which local expansion rates occur and how do they fluctuate for various $x_0$? If we weight each initial value with the natural invariant density, this question leads us to the spectra of local Liapunov exponents introduced in section 15.4. However, these spectra are influenced by two different effects:

(1) the natural invariant density;
(2) the local expansion rates.

In many cases it is advantageous to investigate pure expansion effects. That means, we are interested in appropriate ensembles of initial values $x_0$ where each $x_0$ has the same *a priori* probability. For practical applications, it has turned out to be very useful to consider an appropriately chosen discrete set $H_N$ of initial values $x_0^{(k)}$, $k = 1, 2, \ldots, K(N)$ that can be easily constructed from topological properties of the map and that leads to a fast convergence of the free energy density in the limit $N \to \infty$.

Which set $H_N$ of initial values should be chosen to provide a 'good' description of the system? The answer is not unique and, in fact, there are different choices of $H_N$ used in the literature. The most common choices are as follows.

(1) *The set of N-cylinders.* Suppose we have a generating partition of the phase space (see section 3.3). After $N$ backward iterations of the map there may be $K(N)$ cylinders. The sizes of the cylinders shrink to zero in the limit $N \to \infty$. We then choose one initial value $x_0^{(k)}$, $k = 1, 2, \ldots, K(N)$ in each cylinder. This defines our ensemble $H_N$. Of course, for finite $N$ the cylinders have finite sizes, and the position of $x_0^{(k)}$ inside the $k$th cylinder is not uniquely determined. However, in the limit $N \to \infty$ the concrete choice of $x_0^{(k)}$ inside the cylinder is expected to be irrelevant for the free energy density (this can be rigorously proved for hyperbolic maps).

(2) *The set of unstable fixed points of $f^N$.* Usually a chaotic map $f$ has a large number of unstable periodic orbits, i.e., there are initial values $x_0$ such that

$$f^N(x_0) = x_0. \tag{16.1.2}$$

Here we again denote by $f^N$ the $N$-times iterated function $f$. Consider an arbitrary number $N$. Then we seek for the solutions $x_0$ that satisfy eq. (16.1.2). Choose for $H_N$ the set of these $x_0$. Of course, as we assume that $f$ is chaotic, these periodic orbits must be unstable, i.e., they do not attract a large number of trajectories.

Nevertheless, the set $H_N$ of unstable fixed points of $f^N$ provides us with an appropriate ensemble of initial values to characterize local expansion rates. The idea behind this is as follows. Although the periodic orbits are unstable, a typical chaotic orbit will approximately follow an unstable periodic orbit if $N$ is large enough. Hence, for large $N$, the ensemble of unstable periodic orbits somehow represents properties of typical chaotic orbits.

(3) *The set of preimages of an appropriate point $y^* \in X$ under $f^N$. A priori* we have the freedom to consider other representative points rather than the fixed points for our ensemble $H_N$. In fact, given an appropriate point $y^*$ in the phase space, we may simply choose for $H_N$ the set of points that are mapped onto $y^*$ after $N$ iterations. If $f$ is a single humped one-dimensional map, a common choice for $y^*$ is the $y$ value where $f(y)$ takes its extremum.

Whatever ensemble $H_N$ of initial values is chosen, we define a partition function $Z_N{}^{\text{top}}(\beta)$ by

$$\begin{aligned} Z_N{}^{\text{top}}(\beta) &= \sum_{k=1}^{K} \exp[-\beta N E_N(x_0{}^{(k)})] \\ &= \sum_{k=1}^{K} \exp\left[-\beta \sum_{n=0}^{N-1} \ln|f'(x_n{}^{(k)})|\right] \\ &= \sum_{k=1}^{K} |f^{N\prime}(x_0{}^{(k)})|^{-\beta}. \end{aligned} \tag{16.1.3}$$

We have written $Z_N{}^{\text{top}}$ in order to distinguish this partition function from the dynamical partition function $Z_N{}^{\text{dyn}}$ of section 14.2 and the partition function $Z_N{}^{\text{Lia}}$ for the generalized Liapunov exponent of section 15.4. $Z_N{}^{\text{top}}$ has the form of a canonical partition function where the energy is replaced by $N$-times the local expansion rate $E_N$. Again we identify the parameter $\beta$ with an inverse temperature. $x_n{}^{(k)}$ denotes the $n$th iterate of the initial value $x_0{}^{(k)}$. The quantities

$$P(x_0{}^{(k)}) = \frac{1}{Z_N{}^{\text{top}}(\beta)} \exp[-\beta N E_N(x_0{}^{(k)})] \tag{16.1.4}$$

have the form of an escort distribution of the distribution

$$p(x_0{}^{(k)}) = \frac{1}{Z_N{}^{\text{top}}(1)} \exp[-N E_N(x_0{}^{(k)})]. \tag{16.1.5}$$

Now, however, it is an escort distribution with respect to the ensemble

$H_N$. A probabilistic interpretation of eqs. (16.1.4) and (16.1.5) in terms of the so called Gibbs and SRB measures will be given in section 19.2.

Let us now consider the limit $N \to \infty$. Up to a trivial factor, the *free energy per particle* associated with the partition function (16.1.3) is

$$\mathscr{P}(\beta) = \lim_{N \to \infty} \frac{1}{N} \ln Z_N^{\text{top}}(\beta). \tag{16.1.6}$$

For historical reasons, this free energy is called the *topological pressure.*

*Remark*: Something should be said about the name 'topological pressure'. We notice that

$$F_N(\beta) = -\beta^{-1} \ln Z_N^{\text{top}}(\beta) \tag{16.1.7}$$

is a Helmholtz free energy. The most natural analogy to thermodynamics is to regard the positive integer $N$ as the analogue of the particle number in statistical mechanics and to call

$$\mu = \partial F_N / \partial N \tag{16.1.8}$$

a 'chemical potential'. Of course, we could also again regard $N$ as the analogue of a volume, notwithstanding that this has already been done with $-\ln \varepsilon$. This would mean that $-\mu$ is not called a chemical potential but a 'pressure'. Thus – ignoring the trivial factor $\beta$ – the following quantity would be a 'pressure' for infinite volume

$$\hat{\mathscr{P}}(\beta) = \lim_{N \to \infty} \frac{\partial}{\partial N} \ln Z_N^{\text{top}}(\beta). \tag{16.1.9}$$

It is expected to coincide with

$$\mathscr{P}(\beta) = \lim_{N \to \infty} \frac{1}{N} \ln Z_N^{\text{top}}(\beta). \tag{16.1.10}$$

This may serve as an explanation for the (somewhat misleading) name 'topological pressure'. In any case, it is more natural to regard $\mathscr{P}(\beta)$ as a free energy density.

The topological pressure can be interpreted as a free energy density associated with an ensemble of uniformly distributed trajectories in a suitably defined 'trajectory space'. *A priori*, this free energy density can depend on the ensemble $H_N$ that is chosen to define it. Moreover, for the choice (1) of $N$-cylinders it can also depend on the precise position of $x_0^{(k)}$ inside the $N$-cylinder $J_k^{(N)}$. However, these difficulties do not occur if the map $f$ is hyperbolic. As already mentioned in section 15.6, a one-dimensional map $f$ is called hyperbolic if at arbitrary positions in the phase space $X$ little intervals expand under the action of $f^N$ ($N$ large enough). Provided $f^N$ is piecewise differentiable, this property is equivalent to the condition

$$|f^{N\prime}(x)| > 1 \tag{16.1.11}$$

for all $x \in X$ where the derivative exists. For hyperbolic maps it can be shown that the topological pressure is independent of the choice (1)–(3) of $H_N$ (Bowen 1975; Ruelle 1978a; Bohr and Rand 1987; Bohr and Tél 1988). We shall call $-\beta^{-1}\mathscr{P}(\beta)$ a *topological expansion free energy*, as the topological pressure contains information about local expansion rates and their fluctuations with respect to the topological ensemble $H_N$. Moreover, we may call $\beta^{-1}(\beta-1)D(\beta)$ a 'static free energy', $\beta^{-1}(\beta-1)K(\beta)$ a 'dynamic free energy', and $-\Lambda(\beta)$ a 'metric expansion free energy'. Static, dynamic, and expansion free energies contain the most interesting information about a chaotic dynamical system. In general, they are independent quantities. Only for special classes of maps is one type of free energy sufficient to derive the others. This will be discussed in more detail in sections 19.3–19.5.

## 16.2 Length scale interpretation

We now present a nice geometrical rather than dynamical interpretation of the topological pressure, which, in particular, is suited to the analysis of fractal sets (Bohr and Tél 1988; Tél 1988). The application to fractals will be worked out in more detail in chapter 18.

We assume that there is a generating partition of the phase space that consists of a finite number of cells $i$ of size $L_i$. We remind the reader that an $N$-cylinder is the set $J[i_0, i_1, \ldots, i_{N-1}]$ of all initial points that generate the symbol sequence $i_0, i_1, \ldots, i_{N-1}$ (see section 3.5). After $N-1$ iteration steps the trajectory arrives in cell $i_{N-1}$. Let us use the notation already introduced in section 14.2: For a given $N$, all $N$-cylinders are labelled by the index $j$ and denoted by $J_j^{(N)}$. The 1-cylinders $J_j^{(1)}$ are just the cells of the generating partition. In the one-dimensional case they are intervals of length $L_j$, whereas the $N$-cylinders are intervals with length $l_j^{(N)}$. After $N$ iterations, each cylinder expands to a certain cell $i_N$ of the partition. In the case of a generating partition, for sufficiently large $N$ the $N$-cylinders become infinitesimally small, and their expansion can be properly described with the help of the local expansion rate. The length of the cell of the generating partition onto which the cylinder $J_j^{(N)}$ is mapped may be called $L_{i(j)}$. The index $i$ is a function of the cylinder index $j$. We obtain

$$l_j^{(N+1)}/L_{i(j)} = \exp[-NE_N(x_0^{(j)})]. \tag{16.2.1}$$

Here $x_0^{(j)}$ is some initial value in the cylinder $J_j^{(N+1)}$.

We may now write the partition function of eq. (16.1.3) in the form

$$Z_N^{\text{top}}(\beta) = \sum_j (l_j^{(N+1)}/L_{i(j)})^\beta. \tag{16.2.2}$$

Notice that the lengths $L_{i(j)}$ are bounded between fixed constants $c_1$ and $c_2$, namely the smallest and the largest cell size of the generating partition:

$$c_1 \leqslant L_{i(j)} \leqslant c_2. \tag{16.2.3}$$

Hence for $\beta > 0$

$$c_2^{-\beta} \sum_j (l_j^{(N+1)})^\beta \leqslant Z_N^{\text{top}}(\beta) \leqslant c_1^{-\beta} \sum_j (l_j^{(N+1)})^\beta. \tag{16.2.4}$$

This means

$$\begin{aligned} \frac{1}{N+1} \ln c_2^{-\beta} + \frac{1}{N+1} \ln \sum_j (l_j^{(N+1)})^\beta & \\ \leqslant \frac{1}{N+1} \ln Z_N^{\text{top}}(\beta) & \\ \leqslant \frac{1}{N+1} \ln c_1^{-\beta} + \frac{1}{N+1} \ln \sum_j (l_j^{(N+1)})^\beta. & \end{aligned} \tag{16.2.5}$$

In the limit $N \to \infty$ the terms involving the constants $c_1, c_2$ converge to zero, and we obtain

$$\lim_{N \to \infty} \frac{1}{N} \ln \sum_j (l_j^{(N)})^\beta = \mathscr{P}(\beta). \tag{16.2.6}$$

The analogous consideration applies for $\beta < 0$. We obtain the result that instead of the partition function (16.2.2) we may consider the partition function

$$\tilde{Z}_N^{\text{top}}(\beta) = \sum_j (l_j^{(N)})^\beta, \tag{16.2.7}$$

which is obtained by just raising all cylinder lengths to the power $\beta$. The function $\tilde{Z}_N^{\text{top}}(\beta)$ yields the same free energy density $\mathscr{P}(\beta)$ in the limit $N \to \infty$ as $Z_N^{\text{top}}(\beta)$ does. Therefore, in future we shall not distinguish between both partition functions as long as we are concerned with the thermodynamic limit $N \to \infty$.

We may interpret eq. (16.2.7) in a geometrical way. The contribution of the $N$-cylinders to $\tilde{Z}_N^{\text{top}}(\beta)$ consists of the sum of length scales of the cylinders raised to the power $\beta$. The topological pressure contains information about 'geometrical' properties of the cylinders in the thermodynamic limit $N \to \infty$.

To see the connection with statistical mechanics, we remember that the index $j$ stands for an entire sequence of $N$ symbols $i_0, i_1, \ldots, i_{N-1}$. Hence we may regard

$$H_j = -\ln l_j^{(N)} \tag{16.2.8}$$

as the energy of the spin configuration $i_0, i_1, \ldots, i_{N-1}$. The energy in general depends on *all* spins. This is a similar interpretation to that in section 14.3. Now, however, the energy is determined by the lengths of the cylinders rather than the symbol sequence probabilities $p(i_0, i_1, \ldots, i_{N-1})$. Both approaches are possible. They coincide if the invariant measure is the Lebesgue measure.

As explained in section 8.3, the interaction energy is determined by conditional probabilities. Hence, to extract the underlying interaction of our spin system, we have to analyse the ratios

$$q(i_N \mid i_0, \ldots, i_{N-1}) = \frac{l(i_0, \ldots, i_{N-1}, i_N)}{l(i_0, \ldots, i_{N-1})}. \tag{16.2.9}$$

In nonlinear dynamics these ratios are often called *scaling functions* (Feigenbaum 1980; Jensen, Kadanoff, and Procaccia 1987). In general, the scaling function $q(i_N \mid i_0, \ldots, i_{N-1})$ depends on all spins $i_0, \ldots, i_N$, not only on nearest neighbours. Hence the underlying interaction is a complicated $N$-spin interaction. In general, it cannot be reduced to a pair interaction. In fact, the interactions occurring for nonlinear dynamical systems (especially for nonhyperbolic systems) do *not* correspond to standard examples of interactions studied in conventional statistical mechanics.

**Spectra of local cylinder expansion rates** Analogous to the spectrum $f(\alpha)$ of local dimensions defined in section 11.3, or the spectrum $g(\gamma)$ of dynamical crowding indices defined in section 14.4, a spectrum $S(E)$ of local cylinder expansion rates $E_j = H_j/N$ can be introduced. Again it is obtained simply by a Legendre transformation of the topological pressure

$$\mathscr{P}(\beta) = S - \beta E, \qquad \beta = \mathrm{d}S/\mathrm{d}E. \tag{16.2.10}$$

## 16.3 Examples

We shall now determine the topological pressure for some very simple examples, where it is possible to evaluate $\mathscr{P}(\beta)$ directly from the definition, without any further tools. These are the binary shift map and the Ulam map. A more complicated example is the Julia set of the logistic map. Further examples will be treated in section 17.5.

**The binary shift map** Let us again consider the map

$$f(x) = 2x - \lfloor 2x \rfloor \qquad x \in [0, 1] \tag{16.3.1}$$

that we have already discussed in section 3.2, where also the symbol $\lfloor \cdots \rfloor$ was explained. As $|f'(x)| = 2$, the map is hyperbolic. The generating partition consists of the two intervals $[0, \frac{1}{2})$ and $[\frac{1}{2}, 1)$. The lengths $L_i$ of these cells are equal: $L_0 = L_1 = \frac{1}{2}$. To a given $N$, the number of the $N$-cylinders is $2^N$. The length of each $N$-cylinder is

$$l_j^{(N)} = 2^{-N}. \tag{16.3.2}$$

Hence

$$Z_N^{\text{top}}(\beta) = \sum_j [l_j^{(N)}]^\beta = 2^N 2^{-\beta N} = \exp[(1-\beta)N \ln 2] \tag{16.3.3}$$

and

$$\mathscr{P}(\beta) = (1 - \beta) \ln 2. \tag{16.3.4}$$

**The Ulam map** A less trivial example is provided by the map

$$f(x) = 1 - 2x^2, \qquad x \in [-1, 1]. \tag{16.3.5}$$

This map is nonhyperbolic, as $f'(0) = 0$. Nevertheless, it is analytically treatable, since the map is topologically conjugated to the tent map. The corresponding coordinate transformation is

$$x = -\cos \pi y, \tag{16.3.6}$$

as explained in section 2.3. The generating partition consists of the two intervals $[-1, 0)$, $[0, 1)$ with lengths $L_i = 1$. The $N$-cylinders of the tent map are the $2^N$ intervals bounded by the points

$$y^{(j)} = 2^{-N} j, \qquad j = 0, 1, \ldots, 2^N. \tag{16.3.7}$$

The corresponding values of the variable $x$ are

$$x^{(j)} = -\cos \pi 2^{-N} j. \tag{16.3.8}$$

(Notice that $x_0^{(j)}$ in eq. (16.2.1) was an appropriate point inside the $N$-cylinder, whereas $x^{(j)}$ in eq. (16.3.8) denotes an edge point of an

$N$-cylinder.) The length of the $N$-cylinder $J_j^{(N)}$ is

$$l_j^{(N)} = |x^{(j+1)} - x^{(j)}| \tag{16.3.9}$$

(see fig. 3.2($c$) for a plot).

For all values $j$ other than 0 and $2^N - 1$, the length $l_j^{(N)}$ is proportional to $2^{-N}$ for $N \to \infty$. This means the lengths $l_j^{(N)}$ of $2^N - 2$ intervals scale with $N$ as $2^{-N}$. The only exceptions are the intervals with $j = 0$ and $j = 2^N - 1$, which are sited at the edges of the phase space $X = [-1, 1]$. As $\cos \varepsilon \approx 1 - \varepsilon^2/2$ for small $\varepsilon$, these two interval lengths scale as $2^{-2N}$. Thus the partition function scales as follows

$$Z_N^{\text{top}}(\beta) = \sum_j [l_j^{(N)}]^\beta \sim 2^N \times 2^{-N\beta} + 2 \times 2^{-2N\beta} \sim \exp[N\mathscr{P}(\beta)]. \tag{16.3.10}$$

Hence

$$\mathscr{P}(\beta) = \max\{(1 - \beta) \ln 2, \ -2\beta \ln 2\} \tag{16.3.11}$$

or

$$\mathscr{P}(\beta) = \begin{cases} -2\beta \ln 2 & \text{for } \beta \leqslant -1 \\ (1 - \beta) \ln 2 & \text{for } \beta \geqslant -1. \end{cases} \tag{16.3.12}$$

It is interesting to notice that $\mathscr{P}(\beta)$ is not differentiable at $\beta = -1$ (Bohr and Rand 1987). This is a 'phase transition' phenomenon. It will be analysed in more detail in section 21.3.

*Remark*: Strictly speaking, not only the intervals corresponding to $j = 0$ and $j = 2^N - 1$ scale as $4^{-N}$ in the limit $N \to \infty$, but all intervals with index $j$ in the range $[0, K]$ and $[2^N - K, 2^N - 1]$, where $K$ is finite. This, however, does not change the result (16.3.12), since $K$ is a finite number.

**Julia set of the logistic map** This example is somewhat more advanced, since we are dealing with a mapping in the complex plane. In section 10.4 we mentioned that the Julia set of the complex logistic map

$$x_{n+1} = f(x_n) = x_n^2 + C \tag{16.3.13}$$

is intimately related to the branching process generated by the inverse map

$$y_{n+1} = \pm(y_n - C)^{1/2}. \tag{16.3.14}$$

In fact, the Julia set is obtained if we determine all complex preimages of an appropriate point $x^*$ under the map $f^N$, letting $N \to \infty$. Usually one chooses for $x^*$ the unstable fixed point

$$x^* = \tfrac{1}{2} + (\tfrac{1}{4} - C)^{1/2} \tag{16.3.15}$$

of $f$, but any other generic initial value can be taken as well. The sequence of successive symbols $\pm$ in eq. (16.3.14) allows us to code the various preimages.

After $N$ iterations of eq. (16.3.14) we have $2^N$ complex preimages $y$, which we code according to the recurrence relation

$$y(i_1, \ldots, i_N) = \pm[y(i_2, \ldots, i_N) - C]^{1/2}. \tag{16.3.16}$$

The symbol $i_1$ takes on the value $+1$ if the positive branch is chosen, and $-1$ if the negative branch is chosen. It is convenient to define a variable $t$ taking values in $[0, 1]$ as follows

$$t = \sum_{k=1}^{N} \left(\frac{1 + i_k}{2}\right) 2^{-k}. \tag{16.3.17}$$

The dual representation of $t$ provides us with a certain symbol sequence. Thus we may regard the various preimages $y$ as a function of the variable $t$.

Let us now introduce cylinder lengths as follows (Jensen, Kadanoff, and Procaccia 1987):

$$l^{(N)}(i_1, \ldots, i_N) = |y(t + 2^{-N}) - y(t)|. \tag{16.3.18}$$

Notice that here we determine (in the complex plane) the distance between those preimages $y$ whose symbol sequences just differ by the last symbol $i_N$, whereas all previous symbols $i_1, \ldots, i_{N-1}$ coincide. The distances $l^{(N)}(i_1, \ldots, i_N)$ are, in fact, the nearest-neighbour distances of the points forming the $N$th order approximation of the Julia set. Having defined the length scales $l^{(N)}(i_1, \ldots, i_N)$, we can now proceed to the topological pressure via the length scale definition

$$\mathscr{P}(\beta) = \lim_{N \to \infty} \frac{1}{N} \ln \sum_{i_1, \ldots, i_N} [l^{(N)}(i_1, \ldots, i_N)]^{\beta}. \tag{16.3.19}$$

For general parameter values $C$, this function has to be evaluated numerically, although a perturbation treatment of the Julia set for $|C| \to 0$ and $|C| \to \infty$ is possible (Ruelle 1982; Widom *et al.* 1983; Michalski 1990). We shall further analyse the special case $C = \frac{1}{4}$ in section 21.3.

## *16.4 Extension to higher dimensions

For $d$-dimensional dynamical systems there are expanding and contracting directions in the phase space. To define a topological pressure for these systems, one usually projects onto the expanding direction. Typically, for $d = 2$ there is just one expanding and one contracting direction in the phase space. For the ensemble $H_N$ of initial values we can choose the set of unstable fixed points of $f^N$. For each $x_0^{(k)}$ we then determine the largest eigenvalue $\sigma_1(x_0^{(k)}, N)$ of the matrix

$$Df^N(x_0^{(k)}) = Df(x_{N-1}^{(k)}) \cdots Df(x_1^{(k)}) Df(x_0^{(k)}) \tag{16.4.1}$$

(more precisely, the eigenvalue with the maximum absolute value). The partition function is then introduced as

$$Z_N^{\text{top}}(\beta) = \sum_k |\sigma_1(x_0^{(k)}, N)|^{-\beta}, \tag{16.4.2}$$

and the topological pressure is defined as

$$\mathscr{P}(\beta) = \lim_{N \to \infty} \frac{1}{N} \ln Z_N^{\text{top}}(\beta). \tag{16.4.3}$$

Notice that compared to the one-dimensional case we have now replaced the derivative $f^{N\prime}(x_0^{(k)})$ by $\sigma_1(x_0^{(k)}, N)$, i.e., by the eigenvalue corresponding to the expanding direction.

Also in the higher-dimensional case a length scale interpretation of the topological pressure can be given, namely in terms of length scales of little preimage intervals of the unstable manifold (see Tél (1990) for details). Moreover, the topological pressure is again well defined, i.e., independent of the choice of the ensemble $H_N$, for *hyperbolic* maps (see section 15.6 for the general definition of hyperbolicity). Alternatively, a rigorous mathematical treatment is also possible for *expanding* maps in arbitrary dimensions (Ruelle 1989).

## *16.5 Topological pressure for arbitrary test functions

We may write for the topological pressure of a one-dimensional map

$$\mathscr{P}([\phi]) = \lim_{N \to \infty} \frac{1}{N} \sum_j \exp[S_N \phi(x_0^{(j)})], \tag{16.5.1}$$

where again $x_0^{(j)}$ denotes some point of the ensemble $H_N$, and we have used the abbreviations

$$\phi(x) = -\beta \ln|f'(x)| \tag{16.5.2}$$

$$S_N \phi(x) = \phi(x) + \phi(f(x)) + \cdots + \phi(f^{N-1}(x)). \tag{16.5.3}$$

$\phi$ may be regarded as a test function of the iterates. It is straightforward to generalize the concept of a topological pressure and to define a topological pressure via eq. (16.5.1) for other test functions $\phi(x)$. This concept turns out to be fruitful for the rigorous mathematical foundations of the thermodynamic formalism. Several theorems can be proved for entire classes of test functions $\phi$. The special choice of eq. (16.5.2) is not necessary. A general test function can also be introduced for higher-dimensional maps. We shall return to the generalized topological pressure in chapter 19.

*Remark*: Without proof we mention the following. A new physical application of the concept of a topological pressure for arbitrary test functions has been reported in Beck (1991a). In this paper the thermodynamic formalism is applied to maps $f$ of the form

$$f: \begin{cases} x_{n+1} = T(x_n) \\ y_{n+1} = y_n + \sigma^{-1}(\hbar\tau/m)^{1/2}x_n. \end{cases} \tag{16.5.4}$$

$T$ is a map with certain strong mixing properties. These are, for example, satisfied for the Ulam map

$$T(x) = 1 - 2x^2. \tag{16.5.5}$$

$\sigma$ is a constant depending on $T$ (for example, $\sigma^2 = \frac{1}{2}$ for the Ulam map). The parameter $\tau$ is a small time constant, $m$ is a mass parameter and $\hbar$ is Planck's constant. It turns out that if a partition function is defined for this map $f$ with the test function

$$\phi(x, y) = -(\tau/\hbar)V(y), \tag{16.5.6}$$

then in an appropriate scaling limit the topological pressure converges to the ground state energy of a quantum mechanical system with potential $V$. Moreover, the partition function converges to the Green's function.

# 17
# Transfer operator methods

The time evolution of densities in the phase space can be described by an operator, the so called Perron–Frobenius operator. The fixed points of this operator are invariant densities. They will be evaluated for some examples. A generalization of the Perron–Frobenius operator is the transfer operator for a given inverse temperature $\beta$. Its largest eigenvalue is related to the topological pressure. This provides us with a powerful method for calculating the topological pressure.

## 17.1 The Perron–Frobenius operator

Usually, if we iterate a map $f$ on a computer, we start with a single initial value $x_0$ and observe the time evolution of the trajectory $x_0, x_1, x_2, \ldots$. We may, however, also consider the evolution of a large ensemble of initial values $x_0$ described by a probability distribution $\rho_0(x)$, and ask for its evolution in time. That means, we ask how the ensemble changes into $\rho_n(x)$ after $n$ iteration steps. Of course, in a computer experiment we cannot observe the function $\rho_n(x)$ precisely but have to take a large number $S$ of initial values $x_0$, observing the number $S_A$ of iterates $x_n$ in a certain part $A$ of the phase space $X$. The relative frequency of the iterates $x_n$ in the subset $A$ is

$$S_A/S = \int_A \mathrm{d}x \, \rho_n(x), \qquad (17.1.1)$$

where the number $S$ is assumed to be sufficiently large. This is just the intuitive meaning of the densities $\rho_n$ (see section 2.2). Let us now define by

$$\rho_{n+1} = L\rho_n \qquad (17.1.2)$$

the 'Perron–Frobenius operator' $L$. This determines the time evolution of the density $\rho_n$. The operator $L$ is in a certain sense analogous to the Liouville operator $\hat{L}$ of classical Hamiltonian dynamics (the index $n$ corresponds to the time and the operator $L$ corresponds to the operator $\exp(-\mathrm{i}\hat{L}\tau)$, where i is the imaginary unit and $\tau$ the time interval between $n$ and $n+1$). The members of the ensemble develop in time independently from one another. Therefore, the operator $L$ is linear in the space of all densities $\rho$, even in a more extended space of functions $\phi$ on the phase space. It depends on the question under consideration for which class of functions $\phi$ the operator $L$ is defined.

The linearity of $L$ in the function space should not be confused with the nonlinearity of the map $f(x)$ in the phase space $X$. We notice that a nonlinear evolution law in the phase space can always be formulated as a linear problem in a function space. However, this linearity is dearly purchased by the infinite dimension of the function space.

Let us seek the explicit form of the operator $L$ for a given map $f$. For an arbitrary subset $A$ of the phase space, the densities $\rho_n$ and $\rho_{n+1}$ must satisfy

$$\int_A \mathrm{d}x\, \rho_{n+1}(x) = \int_{A'} \mathrm{d}x\, \rho_n(x) \tag{17.1.3}$$

(see also eq. (2.2.4)). The integral on the left hand side is taken over the set $A$, that on the right hand side over the set $A' = f^{-1}(A)$ of all points that are mapped onto $A$ after one iteration step. In other words, $A'$ is the *preimage* of $A$. Eq. (17.1.3) simply expresses the conservation of probability. If we know the probability that at time $n$ the trajectory is found in the region $A'$, then we know that with the same probability it will be found in the region $A$ at time $n+1$.

Now all we have to do is write the condition (17.1.3) in a more convenient way. For the moment let us restrict ourselves to one-dimensional maps. We shall extend the results to higher dimensions afterwards. Assume that $A$ is an interval of the $x$-axis of the form $A = [u, w]$. Then we may write eq. (17.1.3) as

$$\int_u^w \mathrm{d}x\, \rho_{n+1}(x) = \sum_\sigma \left| \int_{\chi_\sigma(u)}^{\chi_\sigma(w)} \mathrm{d}x\, \rho_n(x) \right|. \tag{17.1.4}$$

Here

$$\chi_\sigma(y) \equiv f_\sigma^{-1}(y) \tag{17.1.5}$$

are the preimages of $y$ under $f$. Notice that, in general, these preimages are not unique. For example, as already explained in the context of eq. (2.2.16), for the tent map we always have two possible preimages

$\chi_1(y) = y/2$ and $\chi_2(y) = 1 - y/2$ for a given $y$. Or for the logistic map

$$f(x) = 1 - \mu x^2 \tag{17.1.6}$$

we have the two preimages

$$\chi_{1,2}(y) = \pm[(1 - y)/\mu]^{1/2}. \tag{17.1.7}$$

We distinguish the various preimages of $y$ by the subscript $\sigma$. Now, in the $\sigma$th integral on the right hand side of eq. (17.1.4) let us substitute the variable $x$ by $\chi_\sigma(y)$. The derivative of $\chi_\sigma(y)$ is given by

$$\chi_\sigma'(y) = (f_\sigma^{-1})'(y) = \frac{1}{f'(f_\sigma^{-1}(y))} = \frac{1}{f'(\chi_\sigma(y))}. \tag{17.1.8}$$

Hence

$$\int_u^w \mathrm{d}x\, \rho_{n+1}(x) = \sum_\sigma \int_u^w \mathrm{d}y\, \rho_n(\chi_\sigma(y))|f'(\chi_\sigma(y))|^{-1}. \tag{17.1.9}$$

This equation has to be valid for arbitrary values $u$, $w$. Thus the densities $\rho_{n+1}$ and $\rho_n$ must satisfy

$$\rho_{n+1}(y) = \sum_\sigma \frac{\rho_n(\chi_\sigma(y))}{|f'(\chi_\sigma(y))|}. \tag{17.1.10}$$

We may write this in the form of eq. (17.1.2), where the operator $L$ is given by

$$L\rho(y) = \sum_\sigma \frac{\rho(\chi_\sigma(y))}{|f'(\chi_\sigma(y))|}. \tag{17.1.11}$$

The summation runs over all preimages $\chi_\sigma(y)$ of $y$. A somewhat more compact and commonly used notation for eq. (17.1.11) is

$$L\rho(y) = \sum_{x \in f^{-1}(y)} \frac{\rho(x)}{|f'(x)|}. \tag{17.1.12}$$

Eq. (17.1.12) yields the explicit expression for the Perron–Frobenius operator of a one-dimensional map $f$. The generalization to a $d$-dimensional map is straightforward. In this case $L$ is given by

$$L\rho(y) = \sum_{x \in f^{-1}(y)} \frac{\rho(x)}{|U(x)|}, \tag{17.1.13}$$

where $U(x)$ is the Jacobi determinant of $f(x)$.

## 17.2 Invariant densities as fixed points of the Perron–Frobenius operator

The linear operator $L$ can be applied not only to normalized probability densities but to quite a general class of functions $\phi(x)$. Any function $\phi$ that is invariant under the action of the operator is called a *fixed point* of $L$. It should be clear that these are fixed points in a function space, i.e., entire fixed point functions. They should not be confused with the fixed points of the map $f$ in the phase space. Of utmost interest in nonlinear dynamics are positive and normalizable fixed points $\rho$. They can be interpreted as probability densities and fulfil

$$L\rho(x) = \rho(x). \tag{17.2.1}$$

As already mentioned in chapter 2, $\rho(x)$ is called an *invariant density*. It is invariant under the dynamics and describes a stationary state of the system. Instead of invariant densities $\rho$ one can also deal with *invariant measures* defined by

$$\mu(A) = \int_A \mathrm{d}x\, \rho(x). \tag{17.2.2}$$

A measure is a function of a set (in this case $A$) rather than a function of a variable $x$ such as the density $\rho$.

In section 2.2 we stressed that, in general, there may exist several invariant densities for an ergodic map. But only one invariant density is really important, the natural invariant density. In the following, we shall determine the Perron–Frobenius operator and the corresponding natural invariant density for some examples.

**The tent map** This is given by

$$f(x) = \begin{cases} 2x & \text{for } x \leqslant \frac{1}{2} \\ 2(1-x) & \text{for } x > \frac{1}{2} \end{cases} \tag{17.2.3}$$

on the phase space $X = [0, 1]$. We have already determined the invariant density by a geometrical consideration in section 2.2. Let us now do it with the Perron–Frobenius approach. We have $|f'(x)| = 2$. The preimages $x_\sigma$ of a given $y$ are $y/2$ and $1 - y/2$. Thus eq. (17.1.11) takes on the form

$$L\rho(y) = \tfrac{1}{2}\rho\left(\frac{y}{2}\right) + \tfrac{1}{2}\rho\left(1 - \frac{y}{2}\right). \tag{17.2.4}$$

This yields for the invariant density the functional equation

$$\rho(y) = \tfrac{1}{2}\rho\left(\frac{y}{2}\right) + \tfrac{1}{2}\rho\left(1 - \frac{y}{2}\right). \tag{17.2.5}$$

This is solved by $\rho(y) = 1$. This is the invariant density we found in section 2.2. But now we have found it in a more systematic, general way.

**The Ulam map** This is the mapping

$$f(x) = 1 - 2x^2 \tag{17.2.6}$$

on the phase space $X = [-1, 1]$. We have $|f'(x)| = 4|x|$. The preimages of a given $y$ are $\chi_\pm(y) = \pm[(1-y)/2]^{1/2}$. Hence

$$L\rho(y) = \frac{1}{4}\left(\frac{2}{1-y}\right)^{1/2}\left\{\rho\left[\left(\frac{1-y}{2}\right)^{1/2}\right] + \rho\left[-\left(\frac{1-y}{2}\right)^{1/2}\right]\right\}. \tag{17.2.7}$$

For the invariant density the left hand side has to be equal to $\rho(y)$. It is easily verified that the normalized density

$$\rho(y) = \frac{1}{\pi(1-y^2)^{1/2}} \tag{17.2.8}$$

fulfils this condition. It is indeed the density we obtained in eq. (2.3.20).

**The continued fraction map** If we use the symbol $\lfloor 1/x \rfloor$ for the integer part of $1/x$, the mapping is given by

$$f(x) = \frac{1}{x} - \left\lfloor \frac{1}{x} \right\rfloor. \tag{17.2.9}$$

The phase space is the unit interval [0, 1]. We have

$$|f'(x)| = x^{-2}. \tag{17.2.10}$$

In the expression

$$y = \frac{1}{x} - \left\lfloor \frac{1}{x} \right\rfloor = \frac{1}{x} - n \tag{17.2.11}$$

the number $n$ is a positive integer. Hence the preimages of $y \in [0, 1]$ are

$$\chi_n(y) = \frac{1}{y+n} \qquad (n = 1, 2, \ldots). \tag{17.2.12}$$

Notice that now there are infinitely many preimages. We obtain

$$L\rho(y) = \sum_{n=1}^{\infty} \frac{1}{(y+n)^2} \rho\left(\frac{1}{y+n}\right). \tag{17.2.13}$$

The fixed point condition $L\rho = \rho$ is satisfied by

$$\rho(y) = \frac{1}{(1+y)\ln 2}. \tag{17.2.14}$$

**The cusp map** By this we mean the mapping

$$f(x) = 1 - 2|x|^{1/2} \tag{17.2.15}$$

on the phase space $X = [-1, 1]$. We have

$$|f'(x)| = \frac{1}{|x|^{1/2}} \tag{17.2.16}$$

and

$$\chi_{\pm}(y) = \pm\left(\frac{1-y}{2}\right)^2. \tag{17.2.17}$$

The fixed point condition

$$\rho(y) = \frac{1-y}{2}\left\{\rho\left[\frac{(1-y)^2}{4}\right] + \rho\left[-\frac{(1-y)^2}{4}\right]\right\} \tag{17.2.18}$$

is satisfied by the normalized density

$$\rho(y) = \frac{1-y}{2}. \tag{17.2.19}$$

In general, there is no simple recipe for finding the fixed point of the Perron–Frobenius operator. In fact, only very few invariant densities of chaotic maps are known analytically.

The Perron–Frobenius operator is a very useful tool for one-dimensional maps. It provides us with a systematic method for determining the natural invariant density. In the higher-dimensional case, however, the usefulness of this operator is somewhat restricted. As an example let us consider the Hénon map. This map possesses only one preimage and the constant Jacobi determinant $-b$. The operator $L$ given by eq. (17.1.13) does not help us to explain the complicated structure of the attractor and to find the natural invariant measure on the attractor.

## 17.3 Spectrum of the Perron–Frobenius operator

We may regard the invariant density as an eigenfunction of the Perron–Frobenius operator $L$ for the eigenvalue 1. In general, it is interesting to study the entire set of eigenfunctions $\phi_\alpha$ of the linear operator $L$. The eigenfunctions satisfy the equation

$$L\phi_\alpha(x) = \eta_\alpha \phi_\alpha(x) \tag{17.3.1}$$

for $\phi_\alpha$ chosen in an appropriate function space. The eigenvalues are denoted by $\eta_\alpha$. The eigenfunctions are not necessarily densities because they may take on negative values as well. Moreover, they may not be normalizable. But under certain assumptions a density can be expressed as a linear combination of the eigenfunctions.

The natural invariant density is the eigenfunction $\rho = \phi_0$ corresponding to the largest eigenvalue $\eta_0 = 1$. Also the second largest eigenvalue $\eta_1$ (more precisely, the eigenvalue with the second largest absolute value) has a distinguished 'physical' meaning. It is responsible for the asymptotic approach to the fixed point $\phi_0$. To see this, we assume that there is a complete set of eigenfunctions $\phi_\alpha$ of the operator $L$ and a gap in the spectrum between $\eta_0 = 1$ and $\eta_1$. This statement can be proved for special classes of maps and appropriately chosen function spaces. An arbitrary initial distribution $\rho_0(x)$ can then be expressed as a linear combination of the eigenfunctions

$$\rho_0(x) = \sum_\alpha c_\alpha \phi_\alpha(x), \tag{17.3.2}$$

where the $c_\alpha$ are appropriate coefficients. Applying $L$ $N$-times we obtain

$$\rho_N(x) = L^N \rho_0(x) = \sum_\alpha \eta_\alpha{}^N c_\alpha \phi_\alpha(x) = c_0\phi_0(x) + R_N, \tag{17.3.3}$$

where $R_N \to 0$ for $N \to \infty$ due to the fact that $|\eta_1| < \eta_0 = 1$. Because of normalization we must have $c_0 = 1$. The rest term $R_N$ decays as

$$R_N \sim \eta_1{}^N \tag{17.3.4}$$

for $N \to \infty$. Hence $\eta_1$ determines the approach to the equilibrium state $\phi_0$. In general, depending on the function space, the operator $L$ may also have a continuum in its spectrum. However, as long as there is a gap between $\eta_1$ and $\eta_0$, the same consideration applies.

In section 2.2 we stated that a map is called *mixing* if an arbitrary smooth initial density converges to the natural invariant density $\phi_0$ for $N \to \infty$. Mixing is a stronger property than ergodicity, and like ergodicity it is very difficult to prove in general. A map is called *exponentially mixing* if the asymptotic approach to the invariant density $\phi_0$ is exponentially fast:

$$|\rho_N - \phi_0| \sim \exp(-\gamma N). \tag{17.3.5}$$

Then $\gamma^{-1} > 0$ is called '*relaxation time*'. We notice that $\gamma$ is related to the second largest eigenvalue $\eta_1$ by $\exp(-\gamma) = |\eta_1|$. Usually the spectrum of eigenvalues of $L$ is dependent on the class of functions for which the operator $L$ is defined. So far, the complete spectrum of $L$ has been analytically determined for only a very few examples of maps. But at least the first few eigenvalues are sometimes known (Dörfle 1985; Mayer and Roepstorff 1987, 1988). For a more detailed introduction to the subject, see Lasota and Mackey (1985).

## 17.4 Generalized Perron–Frobenius operator and topological pressure

In the thermodynamical formalism of dynamical systems, it is useful to generalize the concept of the Perron–Frobenius operator in a way that is similar to the introduction of escort distributions for a given probability distribution. We use a parameter $\beta$, which again may be regarded as an inverse temperature, and define the generalized Perron–Frobenius operator $L_\beta$, also called the *transfer operator*, by

$$L_\beta \rho(y) = \sum_{x \in f^{-1}(y)} \rho(x)|f'(x)|^{-\beta} \tag{17.4.1}$$

(Ruelle 1968; Szépfalusy and Tél 1986; Bohr and Tél 1988; Bedford, Keane and Series 1991; Feigenbaum 1988). Here we have restricted ourselves to one-dimensional maps (possible extensions will be discussed later). Let us compare this with the introduction of escort distributions of a probability distribution $p$ in chapter 9. There the variation of $\beta$ corresponds to a variation of the weights of the different microstates $i$. Hence, the variation of temperature allows us to scan the influence of the different microstates $i$ of the distribution $p$. Something similar is true with respect to the operator $L_\beta$ for different values of $\beta$. Here the variation of the temperature allows us to scan the influence of the different preimages on the operator $L$; in other words, it allows us to scan the different channels which lead from the density $\rho_n$ to $\rho_{n+1}$.

Let us apply $L_\beta$ $N$-times to some smooth initial function $\rho_0(x)$:

$$\rho_N(y) := L_\beta{}^N \rho_0(y) = \sum_k \rho_0(x^{(k)})|f^{N\prime}(x^{(k)})|^{-\beta}. \tag{17.4.2}$$

The sum now runs over all $x^{(k)}$ that are preimages of $y$ after $N$ iterations, i.e., over all $x^{(k)} = f^{-N}(y)$. In particular, if we apply $L_\beta$ to the initial function $\rho_0(x) = 1$, we obtain

$$\rho_N(y) = \sum_k |f^{N\prime}(x_0^{(k)})|^{-\beta} = \sum_k \exp\left[ -\beta \sum_{n=0}^{N-1} \ln|f'(x_n^{(k)})| \right]. \quad (17.4.3)$$

Whereas the first sum runs over all initial values $x_0^{(k)}$ that are preimages of $y$ after $N$ iterations, the second sum runs over the iterates $x_n^{(k)}$ of $x_0^{(k)}$. The expression (17.4.3) is just the partition function $Z_N^{\text{top}}(\beta)$ of eq. (16.1.3) that we have introduced to define the topological pressure $\mathscr{P}(\beta)$. We have chosen possibility (3) for the set $H_N$, and $y^*$ corresponds to $y$ (see section 16.1). In fact, we expect that the growth rate of $\rho_N(y)$ does not depend on $y$ for $N \to \infty$, and is given by $\mathscr{P}(\beta)$ in the form

$$\rho_N \sim \rho_0 \exp[N\mathscr{P}(\beta)] \qquad \text{for } N \to \infty. \quad (17.4.4)$$

All that is left to be done is to prove the mathematical details such as the independence of the precise form of the ensemble $H_N$ for $N \to \infty$.

Indeed, it has been rigorously proved that for one-dimensional hyperbolic systems we do have this relation between the topological pressure and the generalized Perron–Frobenius operator (Ruelle 1978a; Walters 1981). Let us denote the eigenvalues of $L_\beta$ by $\eta_\alpha^{(\beta)}$. Then the largest eigenvalue $\eta_0^{(\beta)}$ is related to the topological pressure by

$$\eta_0^{(\beta)} = \exp[\mathscr{P}(\beta)]. \quad (17.4.5)$$

For hyperbolic systems, from the topological pressure one may pass on to the Rényi dimensions, Rényi entropies, and generalized Liapunov exponents (see sections 19.3–19.5).

Eqs. (17.4.3) and (17.4.5) provide us with a powerful numerical tool for calculating the topological pressure. For a given value of $\beta$, we may simply evaluate the right hand side of eq. (17.4.3) for subsequent values of $N$. We then obtain the topological pressure as

$$\mathscr{P}(\beta) = \ln \rho_N(y) - \ln \rho_{N-1}(y) \qquad N \to \infty \quad (17.4.6)$$

where the $y$-dependence disappears for large $N$. Typically, it is sufficient to proceed to $N \approx 10$ to obtain an accurate result (Tél 1990).

Quite generally, the rigorous proof for the relationship between the topological pressure and the largest eigenvalue of the transfer operator can be extended to $d$-dimensional expanding maps and general test functions $\phi$ (Ruelle 1989). It should, however, be clear that this extension holds for expanding or hyperbolic maps only and not for generic cases such as the Hénon map. For general test functions $\phi$ the transfer operator is defined as

$$L_\phi \rho(y) = \sum_{x \in f^{-1}(y)} \rho(x) \exp \phi(x) \qquad (17.4.7)$$

(just replace $-\beta \ln|f'(x)|$ by $\phi(x)$ in eq. (17.4.1)). In fact, for expanding maps several rigorous statements on the spectrum of transfer operators can be proved. In rare cases it is possible to determine explicitly the entire spectrum of $L_\beta$ by analytical means, or to obtain at least some estimates of it (Bohr and Tél 1988; Feigenbaum, Procaccia and Tél 1989; Ruelle 1989; Mayer 1990).

## *17.5 Connection with the transfer matrix method of classical statistical mechanics

The concept of the generalized Perron–Frobenius operator is the analogue of the transfer matrix method of classical statistical mechanics. There the free energy of a spin system with finite-range interaction can be obtained from the largest eigenvalue of a matrix, the so called 'transfer matrix' (see section 8.4). In nonlinear dynamics, the situation is slightly more complicated because here we may have infinite-range interactions between the 'spins' (remember that in eq. (16.2.9) the conditional probability depends on all symbols $i_0, \ldots, i_N$, not only on nearest neighbours). This results in the fact that, in general, one has to determine eigenvalues of operators in an infinite-dimensional space, indeed, in a function space, rather than eigenvalues of a low-dimensional matrix. For hyperbolic maps $f$, the interaction decreases exponentially, and many rigorous results from classical statistical mechanics can be used in this case (Mayer 1980).

In certain simple cases the generalized Perron–Frobenius operator actually reduces to a finite-dimensional matrix. As these cases illustrate very directly the connection with classical statistical mechanics, we shall treat them in some detail in this section.

First let us just consider the special value $\beta = 1$, i.e., the usual Perron–Frobenius operator $L$ given by eq. (17.1.12). We would like to find a matrix approximation of this operator for a given map $f$ on the interval $X = [a, b]$. For this purpose we choose an arbitrary partition of the interval $X$ into cells $A_i$, $i = 1, \ldots, R$ of finite size (see section 3.1). We then define an $R \times R$ matrix $L$ as follows

$$L_{ij} = \frac{\nu(A_j \cap f^{-1}(A_i))}{\nu(A_j)} \qquad i, j = 1, \ldots, R. \qquad (17.5.1)$$

Here $\nu$ denotes the Lebesgue measure. $A_j \cap f^{-1}(A_i)$ means 'intersection of the set $A_j$ and the set $f^{-1}(A_i)$'. It is simply the set of all points $x \in X$ which satisfy both, namely lying in the interval $A_j$ *and* lying in the preimage of the interval $A_i$. We may give a probabilistic interpretation to the matrix $L$ as follows. We notice that $A_j \cap f^{-1}(A_i)$ is just the set of all initial values that are in cell $A_j$ at time 0 *and* in cell $A_i$ at time 1 (after one iteration step). Hence the equation

$$\nu(A_j)L_{ij} = \nu(A_j \cap f^{-1}(A_i)) \tag{17.5.2}$$

can also be written as

$$q(j)q(i|j) = q(j, i) \tag{17.5.3}$$

where $q(j)$ is the probability of the system to be in cell $A_j$ at time 0, $q(j, i)$ is the probability to be first in cell $A_j$ and then in cell $A_i$, and $q(i|j)$ is the conditional probability to be in cell $A_i$ after one step, provided the system was in cell $A_j$ before.

Notice that we have weighted the events with respect to the Lebesgue measure $\nu$. Thus we see that the matrix $L$ can be interpreted in terms of a conditional probability, where the *a priori* probability distribution of the events is the uniform distribution.

*Remark*: Notice that

$$\sum_i q(i|j) = 1, \tag{17.5.4}$$

i.e., the sum of all column entries of $L$ is equal to 1. The transposed matrix $L^T$ is a so called 'stochastic matrix'. The theory of these matrices is well developed and several useful general results are known (see, for example, Gantmacher (1959) and Fritz, Huppert, and Willems (1979)).

Let us now apply the matrix $L$ to $R$-dimensional vectors $u$. It is convenient to write the components of these vectors as

$$u_i = \rho_i \nu(A_i) \qquad i = 1, \ldots, R. \tag{17.5.5}$$

This notation is useful since it allows us to obtain statements on probability *densities* $\rho$ rather than probabilities.

Applying the matrix $L$ to the vector $u$ is a discrete version of applying the Perron–Frobenius operator to some density function $\rho$. Namely, after one iteration step the probability of being in cell $A_i$ is equal to the sum of probabilities of being in the cells $A_j$, weighted with the conditional probability $L_{ij}$ of proceeding from cell $A_j$ to cell $A_i$. We may look for fixed point vectors $u^*$, i.e., vectors satisfying

$$Lu^* = u^*. \tag{17.5.6}$$

The entries $\rho_i$ of the fixed point vector $u^*$ will be denoted by $\rho_i^*$. To obtain better and better approximations, one can perform a limit $R \to \infty$, i.e., a partition into infinitely many cells such that the size of each cell goes to zero. A famous theorem of Li (1976) states that in this limit (for expanding maps) the step function $\hat{\rho}^*(x)$, defined as

$$\hat{\rho}^*(x) = \rho_i^* \qquad x \in A_i, \tag{17.5.7}$$

converges to the invariant density $\rho^*(x)$ of the map $f$:

$$\lim_{R \to \infty} \hat{\rho}^*(x) = \rho^*(x). \tag{17.5.8}$$

In certain cases the matrix $L$ provides us with the *exact* invariant density even for finite $R$. This is the case for piecewise linear expanding maps with a Markov partition. These examples are analytically treatable, and directly allow us to interpret $L$ as the transfer matrix of a Potts model of order $R$ of classical statistical mechanics. This will be worked out in the following. A piecewise linear expanding map is defined by the property

$$|f'(x)| = c_i > 1 \qquad \text{for } x \in A_i. \tag{17.5.9}$$

Here the $c_i$ are appropriate constants. An example is shown in fig. 17.1. Let us denote the edges of the intervals $A_i$ by $a_i$:

$$A_i = [a_{i-1}, a_i], \qquad i = 1, \ldots, R, \qquad a_0 = a, \qquad a_R = b. \tag{17.5.10}$$

Fig. 17.1 Example of a piecewise linear map with a Markov partition.

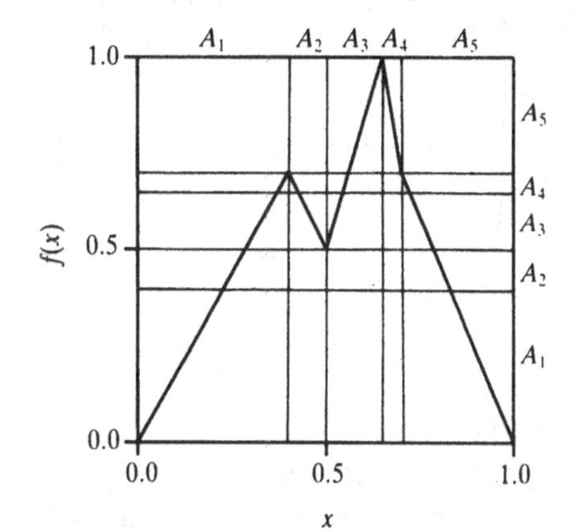

We assume that the partition is a Markov partition, defined by the property

$$f(a_i) \in \{a_0, a_1, \dots, a_R\} \qquad \text{for all } i, \tag{17.5.11}$$

i.e., edges of the partition are mapped onto edges again (see section 3.6).

To determine the matrix $L$, we have to determine the conditional probabilities $q(i\,|\,j) = L_{ij}$ of being in cell $i$ after one iteration step, provided the system was in cell $j$ before, where the relevant measure is the uniform distribution in each cell. Looking at fig. 17.1 we notice that typically some cells $A_i$ are reachable from cell $A_j$, others are not. If cell $A_i$ is not reachable from cell $A_j$, we have

$$q(i\,|\,j) = 0. \tag{17.5.12}$$

Otherwise, we have

$$q(j, i) = \nu(A_i)\frac{1}{c_j} \tag{17.5.13}$$

$$q(j) = \nu(A_j) \tag{17.5.14}$$

and therefore

$$q(i\,|\,j) = \frac{q(j, i)}{q(j)} = \frac{1}{c_j}\frac{\nu(A_i)}{\nu(A_j)}. \tag{17.5.15}$$

Using for the interval length the shorthand notation $l_i = \nu(A_i)$ we obtain for the example of fig. 17.1.

$$L = \begin{pmatrix} \frac{1}{c_1} & 0 & 0 & 0 & \frac{l_1}{c_5 l_5} \\ \frac{l_2}{c_1 l_1} & 0 & 0 & 0 & \frac{l_2}{c_5 l_5} \\ \frac{l_3}{c_1 l_1} & \frac{l_3}{c_2 l_2} & \frac{1}{c_3} & 0 & \frac{l_3}{c_5 l_5} \\ \frac{l_4}{c_1 l_1} & \frac{l_4}{c_2 l_2} & \frac{l_4}{c_3 l_3} & 0 & \frac{l_4}{c_5 l_5} \\ 0 & 0 & \frac{l_5}{c_3 l_3} & \frac{l_5}{c_4 l_4} & 0 \end{pmatrix}. \tag{17.5.16}$$

To determine the invariant density of this map, we have to solve $Lu^* = u^*$ for $u^*$, and thus the problem is reduced to just a set of $R$ linear equations. The invariant density is a step function on the intervals $A_i$.

Multiplying the conditional probabilities of successive events, we obtain probabilities $q(i_0, i_1, \ldots, i_{N-1})$ for strings of symbols $i_0, i_1, \ldots, i_{N-1}$:

$$q(i_0, i_1, \ldots, i_{N-1}) = q(i_0)q(i_1 \mid i_0)q(i_2 \mid i_1)\cdots q(i_{N-1} \mid i_{N-2}). \quad (17.5.17)$$

This simple factorization in terms of conditional probabilities $q(i_k \mid i_{k-1})$ only appears due to the fact that we have a *Markov partition*. In general, the conditional probabilities would depend on the entire history of symbols. Remember that *a priori* we weight the events with respect to the Lebesgue measure. Therefore, using the notation of chapter 16, we can identify $q(i_0, \ldots, i_{N-1})$ with the length $l_i^{(N)}$ of the $N$-cylinder $J_j^{(N)}$ (the index $j$ represents the entire string of symbols).

We are now in a position to see directly the connection with the transfer matrix method of classical statistical mechanics (see section 8.4). Introducing the partition function

$$Z_N^{\text{top}}(\beta) = \sum_j (l_j^{(N)})^\beta = \sum_{i_0,\ldots,i_{N-1}} q^\beta(i_0, \ldots, i_{N-1})$$

$$= \sum_{i_0,\ldots,i_{N-1}} q^\beta(i_0)q^\beta(i_1 \mid i_0)\cdots q^\beta(i_{N-2} \mid i_{N-1}) \quad (17.5.18)$$

we observe that our partition function factorizes in just the same way as the partition function of a classical spin system with nearest-neighbour interaction, as given by eq. (8.4.3). The quantities $v(s, s')$ in eq. (8.4.3) can be identified with the $\beta$th power of the conditional probability,

$$q^\beta(s' \mid s) = v(s, s'). \quad (17.5.19)$$

The Markov property of the partition corresponds to a nearest-neighbour interaction of the spin system. To obtain the free energy, respectively the topological pressure of our system, we just have to determine the largest eigenvalue $\eta_0$ of the transfer matrix

$$L^{(\beta)} = \{L_{ij}{}^\beta\} = \begin{cases} 0 & \text{if } i \text{ is not reachable from } j, \\ \dfrac{l_i^\beta}{c_j^\beta l_j^\beta} & \text{otherwise,} \end{cases} \quad (17.5.20)$$

obtaining

$$\mathscr{P}(\beta) = \ln \eta_0. \quad (17.5.21)$$

Thus, for the class of piecewise linear maps considered here, the topological pressure can be very easily calculated by determining the largest eigenvalue of a finite-dimensional matrix.

# 18
# Repellers and escape

In this chapter we shall deal with so called 'repellers'. Repellers are fractal objects that survive under the iteration of a map $f$ that maps a large number of points outside the phase space $X$. We shall incorporate this concept into the thermodynamic formalism. Trajectories coming close to repellers are responsible for the phenomenon of *transient chaos*, i.e., for chaotic behaviour that takes place for a long but finite time scale only.

## 18.1 Repellers

So far we have dealt with the situation that the map $f$ maps a finite phase space $X$ onto itself. But we may also consider a more general class of mappings. Suppose $f$ maps part of the phase space to infinity. An example is shown in fig. 18.1. It is the map

$$f(x) = \begin{cases} \mu - 1 - \mu x^2 & \text{for } x \in A \text{ or } x \in C \\ \infty & \text{for } x \in B. \end{cases} \tag{18.1.1}$$

We consider it for $\mu > 2$. The phase space is the interval $X = [-1, 1]$. It consists of three regions $A, B, C$. Points in $B$ are mapped to infinity, they escape, whereas points in $A$ and $C$ are mapped onto $X$. The edges of the interval $B$ are determined by the condition $f(x) = 1$, which leads to $A = [-1, -\delta]$, $B = [-\delta, +\delta]$, $C = [\delta, 1]$, $\delta = (1 - 2/\mu)^{1/2}$. It is clear that if such a map is iterated several times, a complicated dynamical problem arises. In particular, the following question is of interest: What does the set of points, that have not escaped after $N$ iterations, look like? Obviously this set is just given by the preimages of the regions $A$, $C$ under the map $f^N$. Slightly generalizing our previous definition, these preimages (or, in general, the preimages of those parts of the phase space that are not mapped to infinity) are again called $N$-cylinders. We may introduce a topological partition function and a topological pressure for these $N$-cylinders in just the same way as was done in chapter 16.

Actually, we need not consider the somewhat artificially defined function $f$ of eq. (18.1.1). Instead of $f$ we can simply iterate the map

$$\tilde{f}(x) = \mu - 1 - \mu x^2 \qquad \mu > 2 \tag{18.1.2}$$

to obtain a system with escape. As soon as an orbit of $\tilde{f}$ reaches region $B$, it will be mapped to infinity under successive future iterations of $\tilde{f}$. In fact, the form of the map in region $B$ is irrelevant: the only important thing is that points entering region $B$ will never return, they are absorbed.

If the number $N$ of iterations goes to infinity, there will be a limit set of points that never escape, i.e., that will always stay in the phase space. This set is called the *repeller*. Typically it has a fractal structure. Let us consider as a very simple example the *classical Cantor set* introduced in section 10.1. We may regard it as the repeller of the following map:

$$f(x) = \begin{cases} 3x - \lfloor 3x \rfloor & \text{for } x \in [0, \frac{1}{3}] \text{ or } x \in [\frac{2}{3}, 1] \\ \infty & \text{else.} \end{cases} \tag{18.1.3}$$

It is plotted in fig. 18.2. Fig. 10.2 shows the corresponding $N$-cylinders. We recognize the construction rule of the classical Cantor set. The topological partition function is defined as

$$Z_N^{\text{top}}(\beta) = \sum_j (l_j^{(N)})^\beta, \tag{18.1.4}$$

where $l_j^{(N)} = (\frac{1}{3})^N$ are the cylinder lengths. We obtain

$$Z_N^{\text{top}}(\beta) = 2^N(\tfrac{1}{3})^{N\beta} = \exp[(\ln 2 - \beta \ln 3)N] \tag{18.1.5}$$

Fig. 18.1 The logistic map with escape.

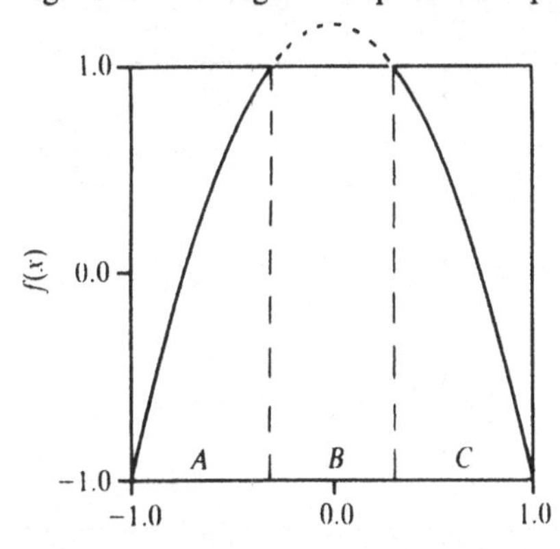

and for the topological pressure

$$\mathscr{P}(\beta) = \lim_{N\to\infty} \frac{1}{N} \ln Z_N^{\text{top}}(\beta) = \ln 2 - \beta \ln 3. \tag{18.1.6}$$

We notice that the fractal dimension (Hausdorff dimension)

$$D(0) = \ln 2/\ln 3 \tag{18.1.7}$$

of the classical Cantor set is given by the value of $\beta$ for which the topological pressure vanishes. This, in fact, is a more general result that we shall derive for entire classes of maps in section 19.5.

A slightly more advanced, but still analytically solvable example is that of the asymmetric triangular map with escape. The graph of this map is plotted in fig. 18.3. In the limit $N \to \infty$, the $N$-cylinders of this map form a two-scale Cantor set with shrinking ratios $a_1, a_2$ (see section 10.3). According to eq. (10.3.8), the topological partition function is given by

$$Z_N^{\text{top}}(\beta) = \sum_i (l_i^{(N)})^\beta = (a_1^\beta + a_2^\beta)^N. \tag{18.1.8}$$

This yields for the topological pressure

$$\mathscr{P}(\beta) = \ln(a_1^\beta + a_2^\beta). \tag{18.1.9}$$

Generally, we notice that the length scale interpretation of the topological pressure allows us to apply this concept quite generally to fractals whose symbolic dynamics is understood and that are thus organized by a certain 'code' (a string of symbols). The code allows us to select arbitrary

Fig. 18.2 The map (18.1.3). Its repeller is the classical Cantor set.

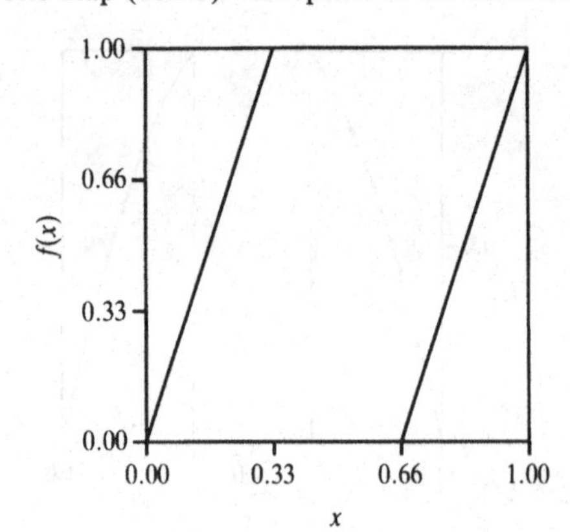

pieces of the fractal object. As soon as we have a code, we can define a partition function $Z_N^{\mathrm{top}}(\beta)$ by taking the sum over the diameters of all coded pieces raised to the power $\beta$.

**The natural invariant measure on a repeller** Trajectories starting on the repeller will never leave it. It is intuitively clear that this defines a probability distribution: the natural invariant probability distribution on a repeller tells us how often the various parts of the repeller are visited by trajectories that never escape.

As soon as we have a measure, we can perform a multifractal analysis of the repeller, in just the same way as we did for attractors in section 11.4.

**Repellers in higher dimensions** Repellers certainly exist for higher-dimensional maps as well. A repeller in higher dimensions typically possesses a fractal structure along both the stable *and* the unstable directions of the system. This clearly distinguishes it from an attractor, which is 'smooth' along the unstable direction. The reader interested in more details may consult Tél (1990) for an introduction to the subject.

## 18.2 Escape rate

For many maps, and also in many experimental situations, one often observes trajectories that apparently exhibit chaotic behaviour on a rather long time scale, but then suddenly the trajectory switches to a different type of behaviour. For example, it may finally approach a stable periodic

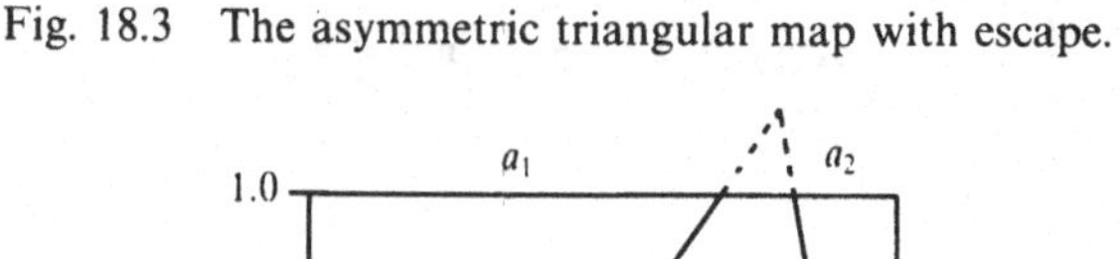

Fig. 18.3 The asymmetric triangular map with escape.

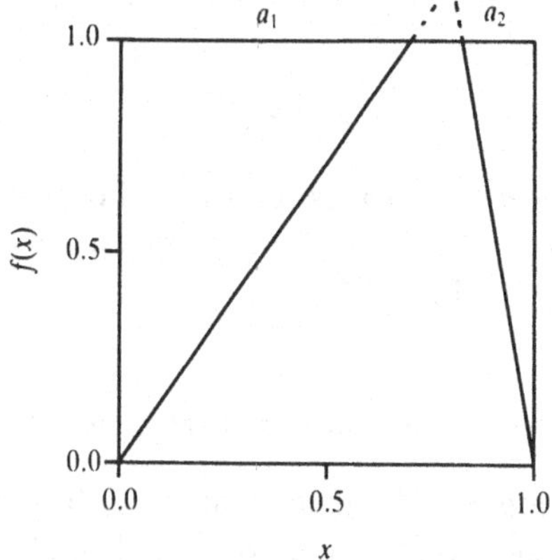

orbit, or it may leave the phase space. This phenomenon of incidentally chaotic motion for a long but finite time is called *transient chaos*. The theoretical explanation for transient chaos is provided by the existence of a repeller in the phase space: trajectories starting close to the repeller will stay in its neighbourhood for quite a while. During this time they exhibit all the typical properties of chaotic motion. Sooner or later, however, they will escape from the vicinity of the repeller, and this implies the end of the chaotic phase (a realization concerning Hamiltonian systems is the phenomenon of irregular scattering; see, for example, Eckhardt (1988) and Smilansky (1990) for reviews). Of course, if we just look at a single trajectory, the length of the transient phase very strongly fluctuates with the initial condition. For a statistical description, a reasonable idea is to consider an ensemble of initial values and to ask for the fraction of trajectories that have *not* escaped after a certain time.

Let us cover the phase space $X$ uniformly with a sufficiently large number $S_0$ of initial points. Iterating $f$, many of these points will leave $X$ after a certain number of iterations. After $N$ iteration steps, let $S_N$ be the number of survivors, i.e., the number of the trajectories staying in $X$. Typically, for large $N$ one observes an exponential decay of $S_N$:

$$\frac{S_N}{S_0} \sim \exp(-\kappa N), \qquad (N \to \infty). \tag{18.2.1}$$

The corresponding rate of decay $\kappa$ is called *escape rate*.

It is clear that in the one-dimensional case the number $S_N$ of survivors of an initially uniform distribution is proportional to the total length of the $N$-cylinders. Hence we have

$$S_N \sim \sum_j l_j^{(N)} \sim \exp(-\kappa N). \tag{18.2.2}$$

This means

$$\kappa = -\mathscr{P}(1) \tag{18.2.3}$$

which relates the escape rate to a special value of the topological pressure.

For the map of eq. (18.1.2) with $\mu = 2.05$ the escape rate is numerically estimated as $\kappa = 0.0711$ (Tél 1990).

## *18.3 The Feigenbaum attractor as a repeller

As described in section 1.3, the logistic map exhibits a period doubling scenario. Let $\mu_\infty$ be the critical value of period doubling accumulation.

For $\mu = \mu_\infty$ the orbits of the map approach the Feigenbaum attractor. We would like to analyse this attractor by means of the thermodynamic formalism and the transfer operator method. However, in full mathematical rigour the applicability of the formalism is proved for special classes of maps only, namely for hyperbolic maps, and the logistic map is not hyperbolic, since $f'(0) = 0$.

But this difficulty can be avoided by a simple trick: without going into details we mention that it is possible to construct a hyperbolic map $F(x)$ whose repeller is identical to the Feigenbaum attractor. This construction was suggested in Ledrappier and Misiurewicz (1985), Collet, Lebowitz and Porzio (1987) and Feigenbaum (1988). It should be stressed that the dynamics generated by $F$ and by $f$ are totally different. For example, the Liapunov exponent of $F$ is positive, whereas the one of $f$ is zero. This, however, does not matter as long as one is interested in purely geometrical properties of the Feigenbaum attractor. The graph of $F$ is plotted in fig. 18.4. The function $F(x)$ consists of two branches: a straight line on the left, and a slightly curved branch on the right. $\alpha = -2.50291\ldots$ is the Feigenbaum constant; the function $g$ is Feigenbaum's universal function (Feigenbaum 1978, 1979) satisfying

$$\alpha g(g(x/\alpha)) = g(x). \tag{18.3.1}$$

A more detailed introduction to eq. (18.3.1) and its solution $g$ will be given in section 21.7, where we shall deal with renormalization group methods in nonlinear dynamics. For the moment it is sufficient to know that $g(x)$ is just a special function with the series expansion

$$g(x) = 1 - 1.528x^2 + 0.105x^4 + 0.027x^6 - \cdots. \tag{18.3.2}$$

Fig. 18.4 The map $F$ whose repeller is the Feigenbaum attractor.

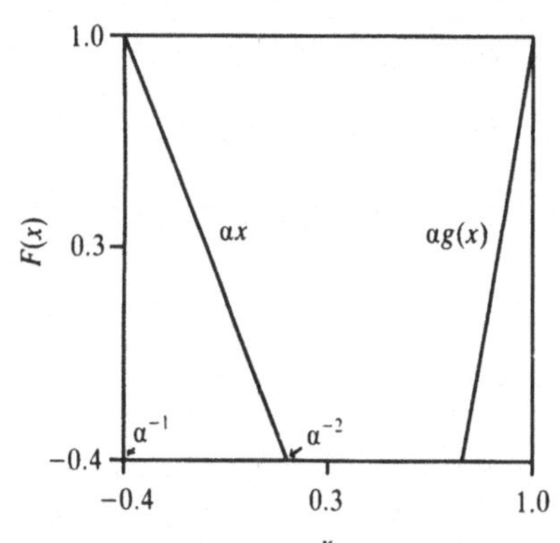

Combining this repeller approach with the generalized Perron–Frobenius operator method described in section 17.4, Kovács (1989) was able to determine the Rényi dimensions of the Feigenbaum attractor with a precision of ten digits, whereas a usual box counting algorithm yields at most two or three digits. This illustrates the usefulness of thermodynamic tools. In the next chapter we shall explain how it is possible to proceed from the topological pressure to the Rényi dimensions for hyperbolic maps.

*Remark*: It should be clear that there are several different numerical methods to determine the Rényi dimensions of the Feigenbaum attractor with quite a high precision. The conceptually and numerically simplest method described in Grassberger (1985) indeed yields very precise results. Another method is based on the zeta-function approach (Christiansen, Paladin, and Rugh 1990), which we shall explain in section 20.4.

Part V · Advanced thermodynamics

# 19
# Thermodynamics of expanding maps

Most of the considerations in this chapter are only valid for special classes of chaotic maps $f$, namely for either expanding or hyperbolic maps (for the definitions see section 15.6). We shall first derive a very important variational principle for the topological pressure. The notion of Gibbs measures and SRB measures (Sinai–Ruelle–Bowen measures) will be introduced. We shall then show that for one-dimensional expanding systems one type of free energy is sufficient: all interesting quantities such as the (dynamical) Rényi entropies, the generalized Liapunov exponents, and the Rényi dimensions can be derived from the topological pressure. A disadvantage is that, due to the expansion condition, we consider quite a restricted class of systems.

## 19.1 A variational principle for the topological pressure

Similar to the principle of minimum free energy in conventional statistical mechanics (see section 6.3) there is also a variational principle for the topological pressure. This variational principle allows us to distinguish the natural invariant measure of an expanding (or hyperbolic) map from other, less important invariant measures. In fact, the 'physical meaning' of the variational principle could be formulated as follows. Among all possible invariant measures of a map one is distinguished in the sense that it is the smoothest one along the unstable manifold. Just as the Gibbs measure (the canonical distribution) in statistical mechanics is the distribution that assumes the least structure consistent with the conservation laws (see section 6.1), the natural invariant measure of a map is the measure that assumes the least structure along the unstable manifold, consistent with the invariance of the measure. Indeed, any such 'structure' is expected to smoothe out due to the expanding property of the map.

For reasons of concreteness we shall formulate the variational principle for $d$-dimensional expanding maps, but an analogous formulation for hyperbolic maps is also possible. According to a general theorem of Bowen and Ruelle, for expanding maps a Markov partition exists (Ruelle 1989). It possesses all the properties that we described in section 3.6. Remember that for a fixed $N$ we labelled each $N$-cylinder by the index $j$. The cylinder was denoted by $J_j^{(N)}$, and its length by $l_j^{(N)}$. The topological pressure with respect to a test function $\phi(x)$ is defined as

$$\mathscr{P}([\phi]) = \lim_{N\to\infty} \frac{1}{N} \ln \sum_j \exp[S_N\phi(x_0^{(j)})]. \tag{19.1.1}$$

We have chosen a general test function $\phi(x)$, as explained in section 16.5. The operator $S_N$ stands for the expression

$$S_N\phi(x) = \phi(x) + \phi(f(x)) + \cdots + \phi(f^{N-1}(x)). \tag{19.1.2}$$

The point $x_0^{(j)}$ is situated in the $j$th $N$-cylinder. We may interpret

$$E_j^{(N)} = -\beta^{-1} S_N\phi(x_0^{(j)}) \tag{19.1.3}$$

as the 'energy' of this cylinder. The topological pressure

$$\mathscr{P}([\phi]) = \lim_{N\to\infty} \frac{1}{N} \ln \sum_j \exp(-\beta E_j^{(N)}) \tag{19.1.4}$$

is the negative generalized free energy $\Psi$ of all $N$-cylinders divided by $N$ in the limit $N \to \infty$.

The principle of minimum free energy can be applied to any partition function, either in conventional thermodynamics or in nonlinear dynamics. The topological pressure is a negative free energy, hence the principle of minimum free energy will now become a *principle of maximum topological pressure.* Let us attribute a normalized probability distribution $p_j^{(N)}$ to the $N$-cylinders. A distinguished role is played by those probability distributions that minimize the function

$$\Psi_N([\phi], p) = \sum_j [p_j^{(N)}\beta E_j^{(N)} + p_j^{(N)} \ln p_j^{(N)}], \tag{19.1.5}$$

which is the analogue of the free energy of eq. (6.3.5). The minimum is achieved for the canonical distribution

$$P_j^{(N)} = \exp[\Psi_N + S_N\phi(x_0^{(j)})]. \tag{19.1.6}$$

The corresponding free energy is

$$\Psi_N([\phi]) = -\ln \sum_j \exp[S_N\phi(x_0^{(j)})] = \min_p \Psi_N([\phi], p). \quad (19.1.7)$$

It is distinguished by the fact that it is the minimum value that expression (19.1.5) can take on with respect to all possible probability distributions $p$. We may also interpret this fact in the sense that in the space of probability distributions the canonical distribution is stable with respect to small variations.

To obtain well defined finite quantities in the limit $N \to \infty$, we divide eq. (19.1.5) by $-1/N$:

$$-\frac{1}{N}\Psi_N([\phi], p) = \frac{1}{N}\sum_j p_j^{(N)} S_N\phi(x_0^{(j)}) - \frac{1}{N}\sum_j p_j^{(N)} \ln p_j^{(N)}. \quad (19.1.8)$$

According to eq. (19.1.1) and (19.1.7), the left hand side converges to the topological pressure provided we take the supremum over all possible probability distributions $p$. The probabilities $p_j^{(N)}$ are related to the $N$-cylinders $J_j^{(N)}$ and the measure $\sigma$ by

$$p_j^{(N)} = \int_{J_j^{(N)}} \mathrm{d}\sigma(x) \quad (19.1.9)$$

(see section 3.6). Hence we may write

$$\mathscr{P}([\phi]) = \sup_\sigma \lim_{N\to\infty} \left[ -\frac{1}{N}\Psi_N([\phi], p) \right]. \quad (19.1.10)$$

The second term on the right hand side of eq. (19.1.8) obviously converges to the KS entropy with respect to the measure $\sigma$ (see section 14.1):

$$\lim_{N\to\infty} \left( -\frac{1}{N}\sum_j p_j^{(N)} \ln p_j^{(N)} \right) = h(\sigma) \quad (19.1.11)$$

(the supremum over all partitions in the definition (14.1.9) of $h(\sigma)$ can be omitted for a generating partition).

Finally, let us deal with the first term on the right hand side of eq. (19.1.8). This term can be further simplified if we do not consider arbitrary measures $\sigma$ but only invariant measures $\mu$ of $f$ (for the definition of an invariant measure see section 2.2). The invariance property implies that for any test function $g(x)$ in the limit $N \to \infty$

$$\sum_j p_j^{(N)} g(f(x_0^{(j)})) = \sum_j p_j^{(N)} g(x_0^{(j)}). \quad (19.1.12)$$

In our case this yields

$$\frac{1}{N}\sum_j p_j^{(N)} S_N \phi(x_0^{(j)}) = \sum_j p_j^{(N)} \phi(x_0^{(j)}). \tag{19.1.13}$$

The limiting value of this expression is the mean value $\langle\phi\rangle$ of $\phi(x)$.

Hence we can formulate the following *variational principle for the topological pressure*: by comparison of all invariant measures $\mu$, the topological pressure with respect to the test function $\phi(x)$ is given by the supremum

$$\mathscr{P}([\phi]) = \sup_\mu [h(\mu) + \langle\phi\rangle]. \tag{19.1.14}$$

This variational principle is sometimes used as an alternative definition of the topological pressure.

It is clear that in the thermodynamic interpretation – up to trivial factors – the topological pressure corresponds to the free energy, $h$ to the entropy, and $\langle\phi\rangle$ to the mean energy. The variational principle distinguishes the invariant measure for which the free energy takes on its minimum value.

Unfortunately, the variational principle can only be rigorously formulated for quite a restricted class of maps. First of all, we need the existence of proper $N$-cylinders. Secondly, the definition of the topological pressure $\mathscr{P}([\phi])$ only makes sense if it does not depend on the precise choice of $x_0^{(j)}$ inside the corresponding cylinder, respectively on the choice of the ensemble $H_N$ (see section 16.1). The necessary assumptions are satisfied for expanding or hyperbolic maps. For the precise mathematical conditions see Ruelle (1989) and Sinai (1986, 1993).

## 19.2 Gibbs measures and SRB measures

For each test function $\phi$ there is an invariant measure $\mu_\phi$ that maximizes the variational principle (19.1.14). This measure is called *Gibbs measure* (Sinai 1972, Ruelle 1969, 1978a; Haydn 1987; Baladi and Keller 1990; Baladi 1991). It has the form of a canonical distribution

$$P_j^{(N)} = \mu_\phi(J_j^{(N)}) = c_j^{(N)} \exp[-N\mathscr{P}[\phi] + S_N\phi(x_0^{(j)})]. \tag{19.2.1}$$

However, we have to generalize slightly in the sense that now there are some constants $c_j^{(N)}$ in front of the exponential function. The reason that we have to introduce the constants is that in eq. (19.1.14) we take a supremum with respect to all *invariant* probability measures. The constants $c_j^{(N)}$ have to be adjusted in such a way that the measure is invariant under the dynamics (this will become clearer at the end of this section). Only *invariant* measures are of interest in nonlinear dynamics. For expanding maps it can be shown (Ruelle 1989) that:

(1) the variational principle has a unique solution of the form (19.2.1);
(2) the constants $c_j^{(N)}$ are bounded between $N$-independent positive numbers, say

$$c_1 \leqslant c_j^{(N)} \leqslant c_2. \tag{19.2.2}$$

As the constants $c_j^{(N)}$ are bounded, we may neglect them in the thermodynamic limit $N \to \infty$. This does not have any effect on the topological pressure. The proof presented in section 16.2 applies again.

Let us now restrict ourselves to one-dimensional maps and the special test function

$$\phi(x) = -\beta \ln|f'(x)|. \tag{19.2.3}$$

For each $\beta$ there is an invariant Gibbs measure $\mu_\beta$ that minimizes the free energy. The special measure with $\beta = 1$ is called the *SRB measure* (Sinai–Ruelle–Bowen measure). Further below we shall show that for expanding maps the SRB measure is identical with the natural invariant measure, which was introduced in section 2.2. This means that the variational principle for the topological pressure allows us to distinguish the *natural* invariant measure from other, less important invariant measures.

As we have chosen the special test function (19.2.3), the length scale interpretation applies (see section 16.2). Neglecting all constants of order 1, we may write according to eqs. (19.1.6), (19.1.7), and (16.2.7)

$$P_j^{(N)} = \frac{\exp[S_N\phi(x_0^{(j)})]}{\sum_{j'} \exp[S_N\phi(x_0^{(j')})]} = \frac{(l_j^{(N)})^\beta}{\sum_{j'} (l_{j'}^{(N)})^\beta}. \tag{19.2.4}$$

That means the Gibbs measure $\mu_\beta$ attributes to each cylinder a probability proportional to the cylinder size taken to the power $\beta$. The probabilities (19.2.4), of course, can be regarded as escort distributions. In contrast to the escort distributions occurring in section 11.1, here we form escort distributions with respect to cylinder lengths, rather than with respect to the probabilities contained in boxes of equal size. Notice that the SRB measure obtained for $\beta = 1$ is a *smooth* measure: there are no singularities, as all probabilities are proportional to the cylinder lengths, and the proportionality constants are bounded.

Let us now do a somewhat more careful analysis and consider the constants $c_j^{(N)}$ occurring in eq. (19.2.1) as well. In particular, we want to prove our previous statement, namely the fact that the Gibbs measure for $\beta = 1$, the SRB measure, is identical with the natural invariant measure of the expanding map $f$.

As mentioned before, the expansiveness condition guarantees the existence of a Markov partition (see section 3.6). An essential property of a Markov partition is the fact that the preimage of an arbitrary $N$-cylinder $J_j^{(N)}$ can be expressed as a union of an appropriate set of $(N + 1)$-cylinders (see also section 17.5). This fact can actually be used to *define* a Markov partition. Let the $N$-cylinder $J_j^{(N)}$ have the cylinders $J_{k_1}{}^{(N+1)}$, $J_{k_2}{}^{(N+1)}, \ldots,$ $J_{k_l}{}^{(N+1)}$ as preimages. The condition that a measure $\mu$ is invariant then reads

$$\mu(J_j^{(N)}) = \mu(J_{k_1}{}^{(N+1)}) + \cdots + \mu(J_{k_l}{}^{(N+1)}) = \sum_{\tau=1}^{l} \mu(J_{k_\tau}{}^{(N+1)}). \quad (19.2.5)$$

For the Gibbs measure $\mu_\phi$ given by eq. (19.2.1) this equation takes on the form

$$c_j{}^{(N)} \exp[S_N\phi(x_0{}^{(j)})] = \exp(-\mathscr{P}[\phi]) \sum_{\tau=1}^{l} c_{k_\tau}{}^{(N+1)} \exp\{S_{N+1}\phi(x_0{}^{(k_\tau)})\}. \quad (19.2.6)$$

For the special test function $\phi(x) = -\beta \ln|f'(x)|$ we have

$$\exp[S_N\phi(x_0{}^{(j)})] = \exp\left[-\beta \sum_{n=0}^{N-1} \ln|f'(x_n{}^{(j)})|\right] = |f^{N\prime}(x_0{}^{(j)})|^{-\beta} \quad (19.2.7)$$

and

$$\exp[S_{N+1}\phi(x_0{}^{(k_\tau)})] = |f^{N+1\prime}(x_0{}^{(k_\tau)})|^{-\beta}. \quad (19.2.8)$$

We now consider the limit $N \to \infty$. In this limit the cylinder $J_j^{(N)}$ shrinks to a point $y$ and the cylinders $J_{k_\tau}{}^{(N+1)}$ to points $x_\tau$ satisfying $f(x_\tau) = y$. In other words, the $x_\tau$ are the preimages of $y$ under $f$. For the limiting values of the constants we write

$$\lim_{N\to\infty} c_j{}^{(N)} = \rho(y) \quad (19.2.9)$$

and

$$\lim_{N\to\infty} c_{k_\tau}{}^{(N+1)} = \rho(x_\tau). \quad (19.2.10)$$

Thus for $N \to \infty$ eq. (19.2.6) approaches

$$\rho(y)|f^{N\prime}(y)|^{-\beta} = \{\exp[-\mathscr{P}(\beta)]\} \sum_{\tau} \rho(x_\tau)|f^{N+1\prime}(x_\tau)|^{-\beta}. \quad (19.2.11)$$

As

$$f^{N+1\prime}(x_\tau) = f'(x_\tau)f^{N\prime}(y), \tag{19.2.12}$$

we arrive at

$$\rho(y) = \{\exp[-\mathscr{P}(\beta)]\} \sum_\tau \rho(x_\tau)|f'(x_\tau)|^{-\beta}. \tag{19.2.13}$$

Notice that this is a rederivation of the generalized Perron–Frobenius equation of section 17.4, in particular of relation (17.4.5). But now we are in a position to learn further facts, since we started from the Gibbs measures of order $\beta$: for the choice $\beta = 1$ and for the case that there is no escape ($\kappa = -\mathscr{P}(1) = 0$) we just obtain the usual Perron–Frobenius fixed point equation of section 17.2. That means the SRB measure indeed corresponds to the natural invariant measure for expanding systems. It is identical with the Gibbs measure for $\beta = 1$, and it fulfils the variational principle (19.1.14) for the choice $\phi(x) = -\ln|f'(x)|$. For general $\beta$, we now find an interpretation of the eigenfunction belonging to the largest eigenvalue of the generalized Perron–Frobenius operator. According to eq. (19.2.9), it is the density function of the Gibbs measure of order $\beta$.

As a by-product we have also found the generalization of the Perron–Frobenius operator for systems with escape (Szépfalusy and Tél 1986; Tél 1986, 1987; Feigenbaum 1988). In this case the so called conditionally invariant density $\rho$ obeys the functional equation

$$\rho(y) = (\exp\kappa) \sum_\tau \rho(x_\tau)|f'(x_\tau)|^{-1}, \tag{19.2.14}$$

where $\kappa$ is the escape rate. The conditionally invariant density can be regarded as an invariant density obtained by 'pumping' new points into the system just according to the escape rate $\kappa$, in order to compensate the loss of escaping trajectories. For expanding maps with escape, the Gibbs measure of order $\beta = 1$ is again a kind of natural invariant measure on the repeller. It is smooth, too.

We would like to emphasize once more that the equality between the SRB measure and natural invariant measure holds for expanding (or hyperbolic) systems only. The condition $|f'(x)| > 1$ guarantees that the constants $c_j^{(N)}$ are bounded. This means that the natural invariant density does not diverge, it is a smooth function.

A simple example is the *binary shift map* (see section 16.3). For this map the SRB measure, that is to say the natural invariant measure, is the Lebesgue measure. The lengths $l_j^{(N)}$ are indeed proportional to the natural invariant measure. The factors $c_j^{(N)}$ are 1 in this simple case.

An example where the expansiveness conditions fails is the *Ulam map*

$$f(x) = 1 - 2x^2. \tag{19.2.15}$$

This map is not expanding, since $|f'(x)| = 0$ for $x = 0$. The natural invariant density is

$$\rho(x) = 1/\pi(1 - x^2)^{1/2} \tag{19.2.16}$$

and the natural invariant measure of an $N$-cylinder

$$\mu(J_j^{(N)}) = \int_{J_j^{(N)}} \mathrm{d}x\, \rho(x) \tag{19.2.17}$$

scales as $(l_j^{(N)})^{1/2}$ at the edges of the phase space $X = [-1, 1]$, whereas it scales as $l_j^{(N)}$ in the interior of the phase space. Hence $\mu(J_j^{(N)})$ is not an SRB measure, as it is not proportional to the lengths of the cylinders in the entire phase space.

## 19.3 Relations between topological pressure and the Rényi entropies

We shall now show that for *one-dimensional expanding maps* one type of free energy, namely the topological pressure, is sufficient to determine the other ones, namely the dynamical Rényi entropies, the generalized Liapunov exponents, and the Rényi dimensions. In this section we restrict ourselves to the Rényi entropies. The generalized Liapunov exponents and the Rényi dimensions will be considered in sections 19.4 and 19.5, respectively.

First of all, we remind the reader that we have introduced the Rényi entropies $K(\beta)$ with respect to the *natural* invariant measure (see section 14.2). As we restrict ourselves to expanding systems, the natural invariant measure coincides with the SRB measure. We may, however, also generalize our concept and consider Rényi entropies for arbitrary Gibbs measures $\mu_q$ parametrized by a temperature $1/q$. These are denoted by $K(\beta, q)$. Notice that $K(1, q) = h([\mu_q])$ and $K(\beta, 1) = K(\beta)$.

The Rényi entropies are defined by the asymptotic scaling relation

$$\sum_j (P_j^{(N)})^\beta \sim \exp[N(1-\beta)K(\beta, q)], \qquad N \to \infty. \tag{19.3.1}$$

The $P_j^{(N)}$ are related to the Gibbs measure $\mu_q$ by

$$P_j^{(N)} = \int_{J_j^{(N)}} \mathrm{d}\mu_q(x). \tag{19.3.2}$$

They scale as

$$P_j^{(N)} \sim (1/Z_N^{\text{top}})(l_j^{(N)})^q, \tag{19.3.3}$$

where according to eqs. (16.2.2) and (16.2.7) with $Z_N^{\text{top}} \approx \tilde{Z}_N^{\text{top}}$

$$Z_N^{\text{top}} \sim \sum_j (l_j^{(N)})^q \sim \exp[N\mathscr{P}(q)], \qquad N \to \infty. \tag{19.3.4}$$

Hence

$$\begin{aligned}\sum_j (P_j^{(N)})^\beta &\sim \exp[-N\beta\mathscr{P}(q)] \sum_j (l_j^{(N)})^{q\beta} \\ &\sim \exp[-N\beta\mathscr{P}(q)] \exp[N\mathscr{P}(q\beta)] \\ &\sim \exp[N(1-\beta)K(\beta, q)].\end{aligned} \tag{19.3.5}$$

We obtain

$$K(\beta, q) = \frac{1}{1-\beta}[\mathscr{P}(\beta q) - \beta\mathscr{P}(q)]. \tag{19.3.6}$$

In particular for the case $q = 1$:

$$K(\beta) = \frac{1}{1-\beta}[\mathscr{P}(\beta) - \beta\mathscr{P}(1)]. \tag{19.3.7}$$

According to eq. (18.2.3), $-\mathscr{P}(1)$ is the escape rate $\kappa$. Thus we obtain the following relation between Rényi entropy, topological pressure, and escape rate

$$K(\beta) = \frac{1}{1-\beta}[\mathscr{P}(\beta) + \kappa\beta]. \tag{19.3.8}$$

An important special case is $K(0) = \mathscr{P}(0)$, which relates the topological entropy to the topological pressure. To obtain a relation for the KS entropy $K(1)$, we assume that $\mathscr{P}(\beta)$ is 'smooth', i.e., that there is the following series expansion

$$\mathscr{P}(\beta) = \mathscr{P}(1) + (\beta - 1)\mathscr{P}'(1) + O((\beta - 1)^2). \tag{19.3.9}$$

Inserting this into eq. (19.3.7) and taking the limit $\beta \to 1$, we obtain

$$K(1) = \mathscr{P}(1) - \mathscr{P}'(1). \tag{19.3.10}$$

$\mathscr{P}(1)$ is the negative escape rate, but what is the 'physical' meaning of $\mathscr{P}'(1)$? According to eqs. (16.1.3) and (16.1.6), the topological pressure is given by

$$\mathscr{P}(q) = \lim_{N\to\infty} \frac{1}{N} \ln \sum_j \exp[-qNE_N(x_0^{(j)})]. \tag{19.3.11}$$

Differentiating eq. (19.3.11) we obtain

$$\mathscr{P}'(q) = -\lim_{N\to\infty} \sum_j P_j^{(N)} E_N(x_0^{(j)}) = -\langle E\rangle(q). \tag{19.3.12}$$

Here $E_N(x_0^{(j)})$ is the local expansion rate of the $j$th $N$-cylinder (remember that for expanding systems for large $N$ the length $l_j^{(N)}$ is very small, and the local expansion rate is practically independent of the precise position of $x_0^{(j)}$ in the cylinder $J_j^{(N)}$). The quantity $\langle E\rangle(q)$ is the mean expansion rate with respect to the Gibbs measure $\mu_q$. In the one-dimensional case $\langle E\rangle(1)$ is simply the Liapunov exponent $\lambda$. Hence, eq. (19.3.10) becomes

$$K(1) = \lambda - \kappa. \tag{19.3.13}$$

This equation relates the KS entropy, the Liapunov exponent, and the escape rate of a one-dimensional expanding map. Among physicists it became well known from a paper of Kantz and Grassberger (1986), see also Kadanoff and Tang (1984). We recognize that the KS entropy and the Liapunov exponent of a one-dimensional system only coincide if there is no escape ($\kappa = 0$).

The extension of this formula to higher-dimensional expanding maps goes as follows. One simply replaces $\lambda$ by the mean volume expansion rate

$$\langle E\rangle(1) = \lambda_1 + \cdots + \lambda_d, \tag{19.3.14}$$

where $\lambda_1, \ldots, \lambda_d$ are the positive Liapunov exponents of the system. However, this derivation applies only to the case that the $d$-dimensional map $f$ expands into all directions (see section 15.6 for the definition of the expanding property). In particular, all Liapunov exponents must be positive. In general, if some of the $\lambda_\alpha$ are negative as well, the correct formula is

$$K(1) = \sum_{\lambda_\alpha > 0} \lambda_\alpha - \kappa \tag{19.3.15}$$

provided Pesin's identity, obtained for $\kappa = 0$ (no escape), is valid (see section 15.5). Again our previous remark applies: in full rigour, eq. (19.3.15) or (19.3.13) can be proved for appropriate classes of maps only.

What is the 'physical meaning' of the so far rather abstractly introduced Gibbs measure $\mu_q$? Remember that, analogous to the spectrum $f(\alpha)$ of local dimensions or the spectrum $g(\gamma)$ of dynamical bit-numbers, we may introduce a spectrum $S(E)$ of local cylinder expansion rates $E$

$$S(E) = \mathscr{P}(q) + qE \tag{19.3.16}$$

by a Legendre transformation of the topological pressure (see eq. (16.2.10)). $E = -\mathscr{P}'(q)$ is just a shorthand notation for $\langle E \rangle(q)$ in eq. (19.3.12). According to the variational principle of section 19.1, with

$$\langle \phi \rangle = -q \langle E \rangle(q), \tag{19.3.17}$$

the right hand side of eq. (19.3.16) is identical with the KS entropy with respect to the Gibbs measure $\mu_q$:

$$S(E) = h(\mu_q) \tag{19.3.18}$$

(remember the definition of $h$ for an arbitrary measure $\sigma$ by eq. (14.1.9)). Therefore the spectrum $S(E)$ of local cylinder expansion rates is sometimes called the 'entropy function' (Bohr and Rand 1987; Rand 1989). It should be clear that this relates to the entropy of the Gibbs measure $\mu_q$. These Gibbs measures are just constructions that yield the spectrum of local cylinder expansion rates in the case of expanding maps.

## 19.4 Relations between topological pressure and the generalized Liapunov exponents

The generalized Liapunov exponents $\Lambda(\beta)$ of order $\beta$ were introduced in section 15.4. They are essentially defined by the asymptotic scaling behaviour of a partition function $Z_N{}^{\text{Lia}}(\beta)$ that contains local expansion rates weighted with the natural invariant measure $\mu$:

$$Z_N{}^{\text{Lia}}(\beta) = \int d\mu(x_0) \exp[\beta N E_N(x_0)] \sim \exp[\beta N \Lambda(\beta)], \qquad N \to \infty. \tag{19.4.1}$$

When we have a set of $N$-cylinders $J_j^{(N)}$, we may discretize and write the above integral for large $N$ as a sum

$$Z_N{}^{\text{Lia}}(\beta) \approx \sum_j p_j^{(N)} \exp[\beta N E_N(x_0^{(j)})] \sim \sum_j p_j^{(N)} (l_j^{(N)})^{-\beta}, \tag{19.4.2}$$

where

$$p_j^{(N)} = \int_{J_j^{(N)}} d\mu(x_0). \tag{19.4.3}$$

For the Gibbs measure of order $q$ we have according to eqs. (19.3.3) and (19.3.4)

$$p_j^{(N)} = P_j^{(N)} \sim (l_j^{(N)})^q \exp[-N\mathscr{P}(q)]. \tag{19.4.4}$$

Thus in the limit $N \to \infty$

$$\begin{aligned} Z_N^{\text{Lia}}(\beta) &\sim \sum_j (l_j^{(N)})^{q-\beta} \exp[-N\mathscr{P}(q)] \\ &\sim \exp[-N\mathscr{P}(q) + N\mathscr{P}(q-\beta)] \\ &\sim \exp[\beta N \Lambda(\beta, q)]. \end{aligned} \tag{19.4.5}$$

Here $\Lambda(\beta, q)$ denotes the generalized Liapunov exponent taken with respect to the Gibbs measure of order $q$. We obtain

$$\Lambda(\beta, q) = \beta^{-1}[\mathscr{P}(q-\beta) - \mathscr{P}(q)]. \tag{19.4.6}$$

Of physical relevance is the case $q = 1$, since for expanding maps the Gibbs measure of order 1 (the SRB measure) coincides with the natural invariant measure. Using $\kappa = -\mathscr{P}(1)$, relation (19.4.6) reduces to

$$\Lambda(\beta) = \beta^{-1}[\mathscr{P}(1-\beta) + \kappa]. \tag{19.4.7}$$

Hence, the topological pressure can be used to calculate the generalized Liapunov exponents. However, once again it should be stressed that this simple relation can be proved in the case of expanding (or hyperbolic) maps only. For the nonexpanding case see chapter 20.

Assuming that the topological pressure is a smooth function, we may expand $\mathscr{P}(1-\beta)$ in eq. (19.4.7) as

$$\mathscr{P}(1-\beta) = \mathscr{P}(1) - \beta\mathscr{P}'(1) + O(\beta^2). \tag{19.4.8}$$

As a consequence, eq. (19.4.7) yields in the limit $\beta \to 0$

$$\Lambda(0) = -\mathscr{P}'(1) = \lambda. \tag{19.4.9}$$

This is a rederivation of our result of section 15.4, namely that the generalized Liapunov exponent of order 0 coincides with the usual Liapunov exponent.

## 19.5 Relations between topological presssure and the Rényi dimensions

Recall the definition of the Rényi dimensions for cells of variable size $l_i$ given in section 11.4. One has

$$D(\beta) = \frac{\tau}{\beta - 1}, \tag{19.5.1}$$

where $\tau$ is the value for which

$$Z(\beta, \tau) = \begin{cases} \inf_{\{\sigma\}} \sum_j \frac{p_j^\beta}{l_j^\tau} & \text{for } \beta \leqslant 1, \tau \leqslant 0 \\ \sup_{\{\sigma\}} \sum_j \frac{p_j^\beta}{l_j^\tau} & \text{for } \beta > 1, \tau > 0 \end{cases} \tag{19.5.2}$$

neither diverges nor goes to zero in the limit $l \to \infty$. Of course, the Rényi dimensions depend on the measure given by the set of $p_j$.

Now let us assume that we have a generating partition and an expanding one-dimensional map. Let us take for the cells $N$-cylinders $J_j^{(N)}$ of length $l_j^{(N)}$ and for $p_j$ the Gibbs measure $P_j^{(N)}$ corresponding to the inverse temperature $q$. Due to the generating partition, the infimum, respectively supremum, has already been reached. For large $N$ we have

$$P_j^{(N)} \sim [l_j^{(N)}]^q \exp[-N\mathscr{P}(q)]. \tag{19.5.3}$$

Inserting this relation into eq. (19.5.2), the following expression remains finite (i.e., of order 1) in the limit of $N \to \infty$:

$$\exp[-N\beta\mathscr{P}(q)] \sum_j [l_j^{(N)}]^{q\beta - \tau} = O(1). \tag{19.5.4}$$

This means

$$\exp[-N\beta\mathscr{P}(q) + N\mathscr{P}(q\beta - \tau)] = O(1), \qquad N \to \infty. \tag{19.5.5}$$

Hence, the Rényi dimensions with respect to the Gibbs measure of eq. (19.5.3) are determined by the value $\tau$ for which

$$\mathscr{P}(q\beta - \tau) - \beta\mathscr{P}(q) = 0. \tag{19.5.6}$$

Of 'physical interest' is the case $q = 1$. We obtain

$$\mathscr{P}[\beta - (\beta - 1)D(\beta)] + \beta\kappa = 0 \tag{19.5.7}$$

as an implicit relation between the topological pressure, the Rényi dimensions, and the escape rate. An important special case is the choice $\beta = 0$. In this case eq. (19.5.7) reduces to

$$\mathscr{P}[D(0)] = 0. \tag{19.5.8}$$

This means, that the fractal dimension (Hausdorff dimension) is given by the value of $q$ for which the topological pressure $\mathscr{P}(q)$ vanishes. Eq. (19.5.8) is known as the Bowen–Ruelle formula. Remember that we have already recognized the validity of this formula for a special example, namely the classical Cantor set discussed in section 18.1.

For $\beta = 1$ eq. (19.5.7) reduces to the known statement that $\mathscr{P}(1) = -\kappa$. Another important relation is obtained by differentiating eq. (19.5.7) with respect to $\beta$:

$$\mathscr{P}'[\beta - (\beta - 1)D(\beta)][1 - D(\beta) - \beta D'(\beta) + D'(\beta)] + \kappa = 0. \quad (19.5.9)$$

For the special choice $\beta = 1$ we obtain

$$\mathscr{P}'(1)[1 - D(1)] + \kappa = 0, \quad (19.5.10)$$

that is to say,

$$\lambda[1 - D(1)] = \kappa. \quad (19.5.11)$$

This relation, first conjectured by Kantz and Grassberger (1985), connects the escape rate with the Liapunov exponent and the information dimension.

A disadvantage of eq. (19.5.7) is the fact that it is an implicit relation. It turns out that if we perform a Legendre transformation, the corresponding spectra $S(E)$ and $f(\alpha)$ satisfy a simpler, explicit relation. This we shall derive in the following.

Remember the definition of a crowding index $\alpha$ by eqs. (11.1.1) and (11.1.2):

$$p_i(\varepsilon) \sim \varepsilon^\alpha, \qquad \varepsilon \to 0. \quad (19.5.12)$$

Here $\varepsilon$ denotes the box size. Let us restrict ourselves to the subset of cylinders that have a certain expansion rate $E$. Their size $l_j^{(N)}$ can be regarded as a fixed local cell size $\varepsilon$, it is given by

$$l_j^{(N)} = \varepsilon \sim \exp[-NE_N(x_0^{(j)})] = \exp(-NE). \quad (19.5.13)$$

Now for the Gibbs measure we obtain

$$\begin{aligned} P_j^{(N)} &\sim [l_j^{(N)}]^q \exp[-N\mathscr{P}(q)] \sim \exp[-N(\mathscr{P}(q) + qE)] \\ &\sim \exp[-EN(q + \mathscr{P}(q)/E)]. \end{aligned} \quad (19.5.14)$$

On the other hand

$$P_j^{(N)} \sim \varepsilon^\alpha \sim \exp(-N\alpha E), \quad (19.5.15)$$

which implies

$$E = \frac{\mathscr{P}(q)}{\alpha - q}. \quad (19.5.16)$$

We notice that a fixed crowding index $\alpha$ corresponds to a fixed expansion rate $E$ and vice versa.

The number $M$ of cells with a certain crowding index $\alpha$ scales as

$$M \sim \exp[-f_q(\alpha) \ln \varepsilon] = \exp[f_q(\alpha)NE]. \quad (19.5.17)$$

The index $q$ indicates that $f_q(\alpha)$ is the spectrum $f(\alpha)$ with respect to the Gibbs measure of temperature $1/q$. On the other hand, the number $\tilde{N}$ of cells with a certain expansion rate $E$ scales as

$$\tilde{N} \sim \exp[NS(E)] \tag{19.5.18}$$

(this is just the 'physical meaning' of the functions $f_q(\alpha)$ and $S(E)$). For the choice (19.5.16) we have $M = \tilde{N}$, which implies

$$f_q(\alpha) = S(E)/E \qquad \text{with } E = \mathscr{P}(q)/(\alpha - q). \tag{19.5.19}$$

The $f_q(\alpha)$-spectrum is equal to $f(\alpha)$ for $q = 1$. This yields

$$f(\alpha) = S(E)/E \qquad \text{with } E = \kappa/(1 - \alpha). \tag{19.5.20}$$

This equation is equivalent to eq. (19.5.7). Sometimes, however, it is easier to handle, because it is an *explicit* relation between $f(\alpha)$ and $S(E)$.

**Higher-dimensional systems** A generalization of the results of sections 19.3–19.5 to higher-dimensional systems is possible for two-dimensional hyperbolic maps (see, for example, Tél (1990)) or for systems with isotropic expansion. In the higher-dimensional case it is advantageous to introduce so called 'partial dimensions', which describe the fractal structure along the stable and unstable manifolds. All relations between topological pressure, generalized Liapunov exponents, and Rényi dimensions remain valid for two-dimensional hyperbolic maps, provided we relate all these quantities for the unstable manifold. This means from the topological pressure, defined by a projection onto the unstable manifold, we can obtain statements on the largest generalized Liapunov exponent and on the partial dimension along the unstable manifold. In general, however, this pressure does not contain information about the stable manifold (for example, on negative Liapunov exponents and on partial dimensions along the stable manifold). Exceptions are maps with constant Jacobi determinant, where the topological pressure yields all the necessary information on the contracting directions as well. This is rather clear, since for arbitrary points $x$ in the phase space the eigenvalues of the Jacobi matrix $Df(x)$ satisfy $\eta_1 \eta_2 = \text{const}$. Thus $\eta_1$ can be used to determine $\eta_2$.

*Remark*: Some of the results of this chapter seem to be valid for more general cases than just expanding or hyperbolic maps, but a rigorous proof cannot be given in these cases.

# 20
# Thermodynamics with several intensive parameters

Whereas for one-dimensional expanding (or hyperbolic) systems one type of free energy (the topological pressure) is sufficient to derive the other ones (Rényi dimensions, Rényi entropies, and generalized Liapunov exponents), for nonhyperbolic systems all these free energies are independent quantities. There is no simple general relation between them. In this chapter a generalization of the previously described formalism will be developed that is of relevance for nonhyperbolic systems. The partition function will now depend on several different intensity parameters. This corresponds to the introduction of a pressure ensemble or a grand canonical ensemble in conventional thermodynamics. This generalization is advantageous for nonhyperbolic systems, since it allows us to unify the concepts of Rényi dimensions, Rényi entropies, topological pressure, and generalized Liapunov exponents in one theory. We shall describe several possibilities for introducing grand canonical partition functions, and also discuss the zeta-function approach in this context.

## 20.1 The pressure ensemble

Let us consider a more general type of thermodynamics of chaotic sytems (Kohmoto 1987; Tél 1988). The important point is that now there will be a partition function that contains two fluctuating quantities, namely both fluctuating length scales $l_j^{(N)}$ and fluctuating probabilities $p_j^{(N)}$ attributed to the $N$-cylinders. Of course, this only yields more information if the probabilities $p_j^{(N)}$ do not depend on the length scales $l_j^{(N)}$ in a simple $j$-independent way. Notice that in the previous chapter we dealt with Gibbs measures, which were essentially defined by the condition

$$p_j^{(N)} \sim (l_j^{(N)})^q, \tag{20.1.1}$$

i.e., the probability of a cylinder behaves as a power $q$ of the length scale, where $q$ is independent of the index $j$. For these types of measures, of utmost interest for hyperbolic dynamical systems, the extended thermodynamics that we are going to describe in this chapter is not necessary, as all the information about the $p_j^{(N)}$ is already contained in the $l_j^{(N)}$. However, in general the natural invariant measure of a nonhyperbolic map does not satisfy condition (20.1.1). As a counter example, we have already mentioned the Ulam map at the end of section 19.2. For all these nonhyperbolic cases, the thermodynamics with two fluctuating quantities $p_j^{(N)}$, $l_j^{(N)}$ is relevant.

Let us assume that a generating partition exists (otherwise an extremum has to be taken over all possible partitions). For a given probability measure on the phase space we define a partition function depending on two intensive parameters $\beta$ and $q$ as follows:

$$Z_N(\beta, q) = \sum_j (p_j^{(N)})^\beta (l_j^{(N)})^q. \tag{20.1.2}$$

As before, $l_j^{(N)}$ denotes the length (i.e., the Lebesgue measure) of the $N$-cylinder $J_j^{(N)}$, and $p_j^{(N)}$ is the probability attributed to it. To simplify the notation, in the following we shall suppress the index $N$. We have already considered this partition function in section 11.4, where we introduced the definition of the Rényi dimensions for cells of variable size. There $q = q(\beta)$ was chosen in such a way that $Z_N(\beta, q)$ remained finite in the limit $N \to \infty$. Now, however, we regard $q$ as an independent parameter.

Again it is useful to introduce escort distributions

$$w_j = (p_j^\beta l_j^q)/Z_N(\beta, q). \tag{20.1.3}$$

We may regard these probabilities $w_j$ as the probabilities of a generalized canonical ensemble, for example, a pressure ensemble (see section 6.2). To demonstrate this, we may introduce a fluctuating 'energy' $E_j$ and a fluctuating 'volume' $V_j$ by writing

$$p_j^\beta = \exp(-\beta E_j) \tag{20.1.4}$$

$$l_j^q = \exp(-q V_j). \tag{20.1.5}$$

Moreover, we define a 'pressure' $\Pi$ (not to be confused with the topological pressure) by the quotient

$$\Pi = q/\beta. \tag{20.1.6}$$

With this notation we may write the escort distribution as

$$w_j = \exp[\beta(G - E_j - \Pi V_j)], \tag{20.1.7}$$

where the Gibbs free energy $G$ is defined as

$$G(\beta, \Pi) = -\beta^{-1} \ln Z_N(\beta, q). \tag{20.1.8}$$

For large $N$ we expect the Gibbs free energy to grow in proportion to $N$. Again it appears plausible to call $N$ the 'particle number'. We may then define in the thermodynamic limit $N \to \infty$ the Gibbs free energy per particle (i.e., per iteration step) as

$$g(\beta, \Pi) = \lim_{N\to\infty} \frac{1}{N} G(\beta, \Pi) = -\lim_{N\to\infty} \frac{1}{\beta N} \ln Z_N(\beta, q). \tag{20.1.9}$$

This function provides us with quite a general characterization of a chaotic dynamical system. The Rényi entropies $K(\beta)$ and the topological pressure $\mathscr{P}(q)$ are given as special values of the more general thermodynamic potential $g(\beta, \Pi)$:

$$K(\beta) = \frac{\beta}{\beta - 1} g(\beta, 0), \tag{20.1.10}$$

$$\mathscr{P}(q) = \lim_{N\to\infty} \frac{1}{N} \ln Z_N(0, q) \tag{20.1.11}$$

(we always take for the probability distribution the natural invariant measure of the map). Moreover, according to their definition in section 11.4, the Rényi dimensions $D(\beta)$ are determined by the implicit relation

$$g\left(\beta, \frac{1-\beta}{\beta} D(\beta)\right) = 0. \tag{20.1.12}$$

Finally, the generalized Liapunov exponent defined in section 15.4 is given by

$$\Lambda(q) = -\frac{1}{q} g(1, -q) = -\frac{1}{\Pi} g(1, -\Pi). \tag{20.1.13}$$

Here we have used the first two equations of section 19.4.

Summarizing: The function $g(\beta, \Pi)$ unifies the concepts of Rényi dimensions, Rényi entropies, generalized Liapunov exponents, and topological pressure for nonhyperbolic systems. Further, it yields additional information about the dynamical system.

*Remark 1*: Kovács and Tél (1992) have introduced a powerful method for calculating Gibbs free energies numerically. $g(\beta, \Pi)$ is obtained from the largest eigenvalue of a bivariate extension of the transfer operator of chapter 17.
*Remark 2*: By a Legendre transformation of $g(\beta, \Pi)$ one can again proceed to an entropy function $S$, the bivariate analogue of eq. (19.3.16) (Stoop and Parisi 1991; Stoop, Peinke and Parisi 1991).

## *20.2 Equation of state of a chaotic system

For the pressure ensemble, both the energies

$$E_j^{(N)} = -\ln p_j^{(N)} \tag{20.2.1}$$

as well as the volumes

$$V_j^{(N)} = -\ln l_j^{(N)} \tag{20.2.2}$$

of the microstates fluctuate, whereas the 'particle number' $N$ has a sharp value. For large $N$ the partition function scales as

$$Z_N(\beta, q) = \sum_j (p_j^{(N)})^\beta (l_j^{(N)})^q \sim \exp[-\beta N g(\beta, \Pi)], \qquad N \to \infty, \tag{20.2.3}$$

where the pressure $\Pi$ is defined by eq. (20.1.6).

The volumes $V_j^{(N)}$ fluctuate since the lengths of the $N$-cylinders of the generating partition are usually not constant. On the other hand, in many cases we are interested in a uniform partition of the phase space into small boxes of equal size $\varepsilon$, rather than cells of variable size $l_j^{(N)}$. For a uniform partition the volume $V = -\ln \varepsilon$ has a sharp value, and only the energies associated with the probabilities still fluctuate. Let us denote the probability associated with the $j$th box of a uniform partition by $\bar{p}_j$. Formally, we can again evaluate the partition function of the pressure ensemble for a uniform partition, replacing $p_j^{(N)}$ by $\bar{p}_j$ and $l_j^{(N)}$ by $\varepsilon$. For the corresponding partition function we write $Z_V^{\text{uni}}(\beta, q)$, since it is characterized by a uniform partition of fixed volume $V$, rather than a nonuniform partition of fixed particle number $N$. We obtain

$$\begin{aligned} Z_V^{\text{uni}}(\beta, q) &= \sum_j (\bar{p}_j)^\beta \varepsilon^q \sim \varepsilon^{(\beta-1)D(\beta)+q} \\ &= \exp\{[(1-\beta)D(\beta) - q]V\}, \qquad V \to \infty. \end{aligned} \tag{20.2.4}$$

Here $D(\beta)$ are the Rényi dimensions, as defined for boxes of equal size in section 11.2. The thermodynamic limit means $N \to \infty$ for the partition function $Z_N(\beta, q)$ and $V \to \infty$ for the partition function $Z_V^{\text{uni}}(\beta, q)$. Comparing the two different partition functions, a useful quantity is the ratio

$$\sigma = N/V. \tag{20.2.5}$$

As this is the ratio of particle number and volume, we shall call it the 'particle density' (not to be confused with the natural invariant density of a map). The particle density describes the relative size of temporal ($N$) versus spatial ($V$) degrees of freedom.

It is quite clear that we prefer to have a unique description of the chaotic system that does not depend on the partition chosen (either generating or uniform). We want to have the same generalized free energy in both cases. Thus we may equate the exponents in eqs. (20.2.3) and (20.2.4). Using $\Pi = q/\beta$, we obtain in the thermodynamic limit

$$\sigma(\beta, \Pi) = \frac{N}{V} = \frac{(\beta - 1)\beta^{-1}D(\beta) + \Pi}{g(\beta, \Pi)}. \tag{20.2.6}$$

This equation describes the way in which the coupled limit $N \to \infty$, $V \to \infty$ has to be performed in order to have equivalence between a description with cells of variable size and boxes of equal size, i.e., between the generalized free energies of the partition functions $Z_N$ and $Z_V{}^{\text{uni}}$. We may interpret eq. (20.2.6) as an equation of state, as it yields the particle density $\sigma(\beta, \Pi)$ as a function of the 'inverse temperature' $\beta$ and the 'pressure' $\Pi$.

An important consistency condition is immediately seen if we write eq. (20.2.6) as

$$g(\beta, \Pi) = [(\beta - 1)\beta^{-1}D(\beta) + \Pi]\sigma^{-1}. \tag{20.2.7}$$

Provided $\sigma \neq 0$, we notice that the function $g(\beta, \Pi)$ vanishes for the special choice $\Pi = -(\beta - 1)\beta^{-1}D(\beta)$. This is just eq. (20.1.12). But now we see that indeed the assumption $\sigma \neq 0$ implies that the two different definitions of the Rényi dimensions coincide, i.e., the definition for boxes of equal size (see section 11.2) and that for cells of variable size (see section 11.4). Namely, in eq. (20.2.7) $D(\beta)$ is defined for boxes of equal size, whereas in eq. (20.1.12) it is defined for cells of variable size.

We may interpret eq. (20.2.7) in a direct thermodynamic way. Remember that in chapter 11 we interpreted $(\beta - 1)\beta^{-1}D(\beta)$ as a Helmholtz free energy $F$ per volume $V$. Moreover, $g(\beta, \Pi)$ is the Gibbs free energy $G$ per particle number $N$. Hence we can write eq. (20.2.7) as

$$\lim_{N \to \infty} \frac{G}{N} = \lim_{N \to \infty} \left(\frac{F}{V} + \Pi\right)\frac{V}{N} = \lim_{N \to \infty} \frac{1}{N}(F + \Pi V). \tag{20.2.8}$$

We recognize the relation

$$G = F + \Pi V \tag{20.2.9}$$

of conventional thermodynamics (see section 6.2). The situation in nonlinear dynamics, however, is slightly different from that in conventional thermodynamics. In the latter, eq. (20.2.9) is always valid, i.e., for arbitrary finite particle densities $\sigma(\beta, \Pi)$. It just describes the fact that in the thermodynamic limit $G$ and $F$ are Legendre transforms of each other. In nonlinear dynamics, eq. (20.2.9) is valid only for a certain specific particle density $\sigma(\beta, \Pi) = N/V$. In other words, the validity of eq. (20.2.9) *defines* a particle density. That we have this modification compared to conventional thermodynamics is rather clear if we look at the information contained in the different partition functions. $Z_V^{\mathrm{uni}}(\beta, q)$ just contains information about the static free energy. $Z_N(\beta, q)$ contains much more information, namely on the probabilities $p_j^{(N)}$ as well as on the cylinder lengths $l_j^{(N)}$. Hence, for nonhyperbolic systems, the corresponding free energies $F$ and $G$ cannot be related by a simple Legendre transformation. They are not equivalent, since $G$ contains more information than $F$. Nevertheless, the equation of state helps us to circumvent this difficulty: if we perform the coupled limits $N \to \infty$, $V \to \infty$ such that $N/V$ is given by the right hand side of eq. (20.2.6), then eq. (20.2.9) is valid in the thermodynamic limit.

The equation of state of an ideal gas is

$$\Pi V = \tfrac{3}{2} N k_B T. \tag{20.2.10}$$

With $\sigma = N/V$, the Boltzmann constant $k_B$ set equal to 1, and $\beta = 1/T$ this can be written as

$$\sigma(\beta, \Pi) = \tfrac{2}{3} \Pi \beta. \tag{20.2.11}$$

What, however, is the equation of state of a simple chaotic system? To answer this question, one has to calculate both the Rényi dimensions and the Gibbs free energy $g$, obtaining $\sigma(\beta, \Pi)$ from eq. (20.2.6).

Let us consider a very simple example, namely the *binary shift map*. We know that the Rényi dimensions have the trivial value $D(\beta) = 1$, as the natural invariant density is the uniform distribution on [0, 1]. For the partition function we obtain

$$\begin{aligned} Z_N(\beta, q) &= \sum_j (p_j^{(N)})^\beta (l_j^{(N)})^q \\ &= 2^N (\tfrac{1}{2})^{N\beta} (\tfrac{1}{2})^{Nq} \\ &= \exp[-N(-1 + \beta + q) \ln 2]. \end{aligned} \tag{20.2.12}$$

Hence

$$g(\beta, \Pi) = \lim_{N\to\infty} [-(\beta N)^{-1} \ln Z_N(\beta, q)]$$

$$= \left(-\frac{1}{\beta} + 1 + \frac{q}{\beta}\right) \ln 2$$

$$= \left(\frac{\beta - 1}{\beta} + \Pi\right) \ln 2 \tag{20.2.13}$$

and

$$\sigma(\beta, \Pi) = 1/\ln 2 = \text{const.} \tag{20.2.14}$$

For this trivial example the particle density $\sigma$ is independent of the inverse temperature and the pressure. The equation of state is that of (idealized) condensed matter.

In general, however, $\sigma$ will depend on both $\beta$ and $\Pi$. As an example we shall treat the Ulam map in section 21.4. Another nontrivial example is the two-scale Cantor set with a multiplicative two-scale measure (see section 11.4), which we shall treat now. As in section 11.4, we denote the shrinking ratios of the probabilities $p_j^{(N)}$ by $w_1$, $w_2$ and those of the length scales $l_j^{(N)}$ by $a_1$, $a_2$. We consider the special situation that $a_2 = a_1^2$. First let us determine the Rényi dimensions $D(\beta)$. According to eq. (11.4.7) the function $\tau(\beta) = (\beta - 1)D(\beta)$ is determined by the condition

$$\frac{w_1^\beta}{a_1^\tau} + \frac{w_2^\beta}{a_1^{2\tau}} = 1. \tag{20.2.15}$$

Setting $x = a_1^\tau$, eq. (20.2.15) is equivalent to the quadratic equation

$$x^2 - w_1^\beta x - w_2^\beta = 0 \tag{20.2.16}$$

with the solution

$$x_{1/2} = \frac{w_1^\beta}{2} \pm \left(\frac{w_1^{2\beta}}{4} + w_2^\beta\right)^{1/2} = a_1^\tau. \tag{20.2.17}$$

This yields

$$\tau(\beta) = (\beta - 1)D(\beta) = \frac{1}{\ln a_1} \ln\left[\frac{w_1^\beta}{2} + \left(\frac{w_1^{2\beta}}{4} + w_2^\beta\right)^{1/2}\right]. \tag{20.2.18}$$

On the other hand, according to eq. (11.4.6) with $q = -\tau$, the partition function $Z_N(\beta, q)$ is given by

$$Z_N(\beta, q) = (w_1^\beta a_1^q + w_2^\beta a_1^{2q})^N. \tag{20.2.19}$$

Thus

$$g(\beta, \Pi) = -\lim_{N\to\infty} \frac{1}{\beta N} \ln Z_N(\beta, q)$$

$$= -\frac{1}{\beta} \ln(w_1{}^\beta a_1{}^{\Pi\beta} + w_2{}^\beta a_1{}^{2\Pi\beta}), \tag{20.2.20}$$

and the equation of state is

$$\sigma(\beta, \Pi) = -\frac{\dfrac{1}{\ln a_1} \ln\left[\dfrac{w_1{}^\beta}{2} + \left(\dfrac{w_1{}^{2\beta}}{4} + w_2\right)^{1/2}\right] + \Pi\beta}{\ln(w_1{}^\beta + w_2{}^\beta a_1{}^{\Pi\beta}) + \Pi\beta \ln a_1}. \tag{20.2.21}$$

## 20.3 The grand canonical ensemble

Instead of considering a fluctuating volume $V_j^{(N)} = -\ln l_j^{(N)}$ of the cylinder $J_j^{(N)}$ we may also present a formulation where the cells are boxes of constant size $\varepsilon$, but the iteration number $N$ fluctuates. Indeed, we may incorporate dynamical information into the partition function by considering a fluctuating 'particle number' in the following way. To each box $i$ of constant size we attribute the probability

$$\bar{p}_i = \int_{\text{box } i} \mathrm{d}x\, \rho(x), \tag{20.3.1}$$

where $\rho(x)$ is the natural invariant density. Now for each box we may ask the following question: How often do we have to iterate the map such that the box $i$ is mapped onto the unit interval (or, in general, onto a region with size of the order of the whole phase space)? Let $N_i$ denote the necessary number of iterations for box $i$. We then define for a given volume $V = -\ln \varepsilon$ a partition function with two parameters $\beta, q$ as

$$Z_V(\beta, q) = \sum_i (\bar{p}_i)^\beta \exp(qN_i). \tag{20.3.2}$$

To distinguish the corresponding escort distributions from the previously introduced pressure ensemble, we identify them with the distribution of a grand canonical ensemble with constant volume

$$P_i = \exp[\beta(\Omega - E_i + \mu N_i)] \tag{20.3.3}$$

(see eq. (6.2.7)). This is achieved by putting

$$q = \beta\mu, \tag{20.3.4}$$

$$(\bar{p}_i)^\beta = \exp(-\beta E_i), \tag{20.3.5}$$

$$-\frac{1}{\beta} \ln Z_V(\beta, q) = \Omega(\beta, \mu). \tag{20.3.6}$$

The fluctuating iteration number $N_i$ is now the analogue of the particle number in a microstate $i$ of conventional statistical mechanics. $\mu$ is the corresponding 'chemical potential'.

In particular, we are interested in the quantity

$$\omega(\beta, \mu) = \lim_{V \to \infty} \frac{1}{V} \Omega(\beta, \mu). \tag{20.3.7}$$

It is obvious that the Rényi dimensions $D(\beta)$ are related to $\omega$ by

$$\omega(\beta, 0) = \frac{\beta - 1}{\beta} D(\beta). \tag{20.3.8}$$

The other values of $\omega(\beta, \mu)$ provide us with further information about the chaotic system. For example, for $\beta = 0$ we obtain a quantity similar to the topological pressure. It describes fluctuations of expansion rates, where each box has the same *a priori* probability.

The above consideration is just one of many possibilities to introduce a grand canonical ensemble in nonlinear dynamics. In fact, it turns out that practically all of the previously introduced partition functions $Z_N(\beta)$ (such as those corresponding to the Rényi entropies, the generalized Liapunov exponents, and the topological pressure) can be extended to grand canonical partition functions by a method that we are going to describe now. The advantage is that the generalized free energy associated with these grand canonical partition functions does not only contain information on the limit case $N \to \infty$ but also on finite $N$.

Let us first return to the grand canonical ensemble in conventional thermodynamics. We consider a system in thermodynamic equilibrium with fluctuating energy $E_i$ and fluctuating particle number $N_i$, where the volume $V$ and the temperature $T = 1/\beta$ are assumed to be fixed. The partition function of the system is

$$\tilde{Z}(\beta, \mu) = \sum_i \exp(-\beta E_i + \beta\mu N_i). \tag{20.3.9}$$

Here $\mu$ denotes the chemical potential. The summation runs over all microstates $i$ of the system. $\tilde{Z}(\beta, \mu)$ is related to the grand canonical potential $\Omega(\beta, \mu)$ by

$$\Omega(\beta, \mu) = -\frac{1}{\beta} \ln \tilde{Z}(\beta, \mu). \tag{20.3.10}$$

In conventional thermodynamics it is often advantageous to perform a partial summation in eq. (20.3.9), namely a summation over all those states that correspond to a certain fixed particle number $N$. Let us agree on $N \geqslant 1$. We may then write eq. (20.3.9) in the form

$$\tilde{Z}(\beta, \mu) = \sum_{N=1}^{\infty} Z_N(\beta) \exp(\beta\mu N), \tag{20.3.11}$$

where

$$Z_N(\beta) = \sum_i \exp(-\beta E_i^{(N)}) \tag{20.3.12}$$

is a canonical partition function corresponding to all energy states with a certain $N$. The quantity

$$s = \exp(\beta\mu) \tag{20.3.13}$$

is called the *fugacity*.

In the previous chapters we defined several partition functions $Z_N(\beta)$ that are of interest for the dynamical analysis of chaotic systems. Among them were the partition functions for the Rényi entropies (section 14.2), for the generalized Liapunov exponents (section 15.4), and for the topological pressure (section 16.1). It is quite clear that for all these partition functions, depending on a single intensive parameter $\beta$, we can proceed to a grand canonical partition function $\tilde{Z}$, depending on two intensive parameters $\beta$, $\mu$, via eq. (20.3.11). This has the advantage that $\tilde{Z}$ does not only contain information on the limit $N \to \infty$ but also on finite $N$ properties of the system. The chemical potential $\mu$ determines the relative weight of the various states for finite $N$: the larger $\mu$, the larger is the relative weight of the large $N$ states, provided $\beta > 0$. In principle, we can also define a grand canonical partition function for the static canonical ensemble used in section 11.1 if we replace $N$ by $V = -\ln \varepsilon$. In this case we can define

$$\tilde{Z}(\beta, \mu) = \sum_{V=1}^{\infty} Z_V(\beta) \exp(\beta\mu V), \tag{20.3.14}$$

where

$$Z_V(\beta) = \sum_i [p_i(\varepsilon)]^{\beta} \tag{20.3.15}$$

is the static partition function for boxes of equal size $\varepsilon$.

The generalized free energy associated with the grand canonical partition function is $\Omega(\beta, \mu)$ (see eq. (20.3.10)). It exists (i.e., it does not diverge) for a certain limited range of the chemical potential $\mu$. To see that we have to remember that the partition function $Z_N(\beta)$ scales as

$$Z_N(\beta) \sim \exp[\Psi(\beta)N], \qquad N \to \infty, \tag{20.3.16}$$

where $\Psi(\beta)$ is the generalized free energy associated with the partition function $Z_N(\beta)$. For example, for the Rényi entropies $K(\beta)$ we have

$$\Psi(\beta) = (1 - \beta)K(\beta), \tag{20.3.17}$$

for the generalized Liapunov exponents $\Lambda(\beta)$

$$\Psi(\beta) = \beta\Lambda(\beta), \tag{20.3.18}$$

and for the topological pressure $\mathscr{P}(\beta)$

$$\Psi(\beta) = \mathscr{P}(\beta). \tag{20.3.19}$$

Summing up the various $Z_N$ we obtain

$$\tilde{Z}(\beta, \mu) = \sum_{N=1}^{\infty} Z_N(\beta) \exp(\beta\mu N) \approx \sum_{N=1}^{\infty} \exp[(\beta\mu + \Psi)N]. \tag{20.3.20}$$

The last term on the right hand side is a geometrical series that converges for

$$\beta\mu + \Psi < a < 0. \tag{20.3.21}$$

The *critical point*, i.e., the radius of convergence, is determined by the condition

$$\mu_c = -\Psi/\beta. \tag{20.3.22}$$

For $\mu < \mu_c$ and $\beta > 0$, respectively $\mu > \mu_c$ and $\beta < 0$, the partition function $\tilde{Z}(\beta, \mu)$ exists. A variation of $\mu$ then corresponds to a different weighting of the finite $N$ partition function $Z_N(\beta)$. For $\mu \to \mu_c$ we formally attribute the maximum weight to the case $N = \infty$. An analogous consideration applies to the partition function (20.3.14), replacing $N$ by $V$.

A grand partition function can certainly also be defined if we take for $Z_N$ the partition function $Z_N(\beta, q)$ of the pressure ensemble (20.1.2). The grand canonical partition function $Z$ then depends on three intensive parameters:

$$\tilde{Z}(\beta, q, \mu) = \sum_{N=1}^{\infty} Z_N(\beta, q) \exp(\beta\mu N). \tag{20.3.23}$$

This construction corresponds to the introduction of a grand canonical ensemble with fluctuating energy $E_i$, fluctuating volume $V_i$, and fluctuating particle number $N_i$ (compare eqs. (20.1.4) and (20.1.5)). Eq. (20.3.23) appears to be one of the most general partition functions that can be defined for a chaotic system.

## 20.4 Zeta functions

We now deal with an effective method for calculating the topological pressure of a given dynamical system. This method is based on the so called 'zeta-function' approach (Ruelle 1978a, 1989; Cvitanović 1988; Artuso, Aurell and Cvitanović 1990; Tél 1990; Bedford, Keane and Series 1991; Campbell 1992). It is closely related to the introduction of a grand canonical ensemble, where we use in eq. (20.3.11) the partition function $Z_N^{\text{top}}(\beta)$ corresponding to the topological pressure. The zeta-function approach has turned out to be a very precise method for calculating the topological pressure, at least for hyperbolic dynamical systems (for nonhyperbolic systems the convergence can be rather slow, see Grassberger *et al.* (1989)). In general, the theory of zeta functions is a very interesting and beautiful branch of mathematics (see, for example, Iyanaga and Kawada (1977)). Here we just restrict ourselves to the use of zeta functions within the theory of chaotic systems. A complete description is beyond the scope of this book, we just sketch the main idea.

Remember that the partition function for the topological pressure $\mathscr{P}(\beta)$ is

$$Z_N^{\text{top}}(\beta) = \sum_j \exp[-\beta N E_N(x_0^{(j)})] \sim \exp[N\mathscr{P}(\beta)], \qquad N \to \infty. \quad (20.4.1)$$

The quantities $E_N(x_0^{(j)})$ are the local expansion rates corresponding to the initial values $x_0^{(j)}$:

$$E_N(x_0^{(j)}) = \frac{1}{N} \sum_{n=0}^{N-1} \ln|f'(x_n^{(j)})|. \quad (20.4.2)$$

$x_n^{(j)}$ again denotes the $n$th iterate of $x_0^{(j)}$.

For the following considerations it is advantageous to take for the set $H_N$ of initial values $x_0^{(j)}$ the second choice described in section 16.1, namely the set of all fixed points of $f^N$. In other words, the $x_0^{(j)}$ are all solutions of the equation

$$f^N(x_0) = x_0. \quad (20.4.3)$$

As described in the previous section, we may introduce a grand canonical partition function $\tilde{Z}(s, \beta)$ for the canonical partition function (20.4.1) as follows (Ruelle 1978a, Feigenbaum 1987)

$$\tilde{Z}(s, \beta) = \sum_{N=1}^{\infty} s^N Z_N^{\text{top}}(\beta). \tag{20.4.4}$$

Here we regard $\tilde{Z}$ as a function of the fugacity $s = \exp(\beta\mu)$ rather than the chemical potential $\mu$. Of course, this does not make any difference, but is just a matter of convenient notation. $s$ is regarded as an arbitrarily chosen variable that may even take on complex values. Notice that for the special choice $s_c = \exp[-\mathscr{P}(\beta)]$ the partition function $Z(s_c, \beta)$ diverges, whereas for $|s| < s_c$ it is expected to converge.

Let us now analyse the solutions of eq. (20.4.3) in more detail. As already mentioned in section 1.1, the fixed points of $f^N$ can also be regarded as periodic orbits of the map $f$. However, it should be clear that the length $n$ of a periodic orbit corresponding to a fixed point of $f^N$ does not necessarily coincide with $N$: for example, a fixed point of $f^6$ can be a periodic orbit of length $n = 1, 2, 3$ or 6. To emphasize the difference, the periodic orbits of $f$ are often called *primitive cycles.* This elucidates that we mean by a periodic orbit a cycle that cannot be decomposed into smaller periodic cycles.

In general, a map possesses a certain set of primitive cycles, which we label by the index $p$ (not to be confused with the notation $p$ for a probability distribution). Its length is denoted by $n_p$. An arbitrary fixed point $x_0^{(j)}$ of $f^N$ always corresponds to a primitive cycle of length $n_p$ with $r$ repetitions:

$$N = rn_p. \tag{20.4.5}$$

The corresponding expansion rate satisfies, according to eq. (20.4.2),

$$NE_N(x_0^{(j)}) = rn_p E_{n_p}(x_0^{(j)}). \tag{20.4.6}$$

We now rearrange the sum over $N$ occurring in eq. (20.4.4) in an appropriate way. Namely, we replace it by a sum over all primitive cycles $p$ and a sum over repetitions $r$. Using eqs. (20.4.5) and (20.4.6) we obtain

$$\tilde{Z}(s, \beta) = \sum_p n_p \sum_{r=1}^{\infty} \{s^{n_p} \exp[-\beta n_p E_{n_p}(x_0^{(p)})]\}^r. \tag{20.4.7}$$

Here $x_0^{(p)}$ is an arbitrary orbit element of the primitive cycle $p$. The number $n_p$ multiplies the sum over $r$ because for each primitive cycle of length $n_p$ we take just one initial value $x_0^{(p)}$ (in total the cycle yields the same contribution $n_p$ times). To abbreviate the notation, let us introduce the shorthand notation

$$H_p = n_p E_{n_p}(x_0^{(p)}). \tag{20.4.8}$$

We regard it as the 'energy contribution' of the $p$th primitive cycle. The second sum occurring in eq. (20.4.7) is a geometric series. Summing up, we obtain

$$\begin{aligned}\tilde{Z}(s, \beta) &= \sum_p n_p \sum_{r=1}^{\infty} [s^{n_p} \exp(-\beta H_p)]^r \\ &= \sum_p \frac{n_p s^{n_p} \exp(-\beta H_p)}{1 - s^{n_p} \exp(-\beta H_p)}. \end{aligned} \tag{20.4.9}$$

Let us now define a zeta function $\zeta_\beta(s)$ by

$$\zeta_\beta(s) = \prod_p [1 - s^{n_p} \exp(-\beta H_p)]. \tag{20.4.10}$$

The product is taken over all primitive cycles of the map $f$. Taking the logarithm we obtain

$$\ln \zeta_\beta(s) = \sum_p \ln[1 - s^{n_p} \exp(-\beta H_p)] \tag{20.4.11}$$

and

$$\frac{\partial}{\partial s} \ln \zeta_\beta(s) = \sum_p \frac{-n_p s^{n_p - 1} \exp(-\beta H_p)}{1 - s^{n_p} \exp(-\beta H_p)} = -\frac{1}{s} \tilde{Z}(s, \beta) \tag{20.4.12}$$

according to eq. (20.4.9). Hence the logarithmic derivative of the zeta function is identical with the grand canonical partition function $\tilde{Z}$ (up to a factor $-1/s$). This provides us with a powerful method for calculating $\tilde{Z}(s, \beta)$, because often the expression (20.4.10) for the zeta function converges quite rapidly. Typically it is sufficient to take a product over the first few primitive cycles of the map $f$, already obtaining quite a good approximation (Cvitanović 1988). In particular, this method allows us to determine the topological pressure. Suppose we have found a root $s_0$ of the zeta function:

$$\zeta_\beta(s_0) = 0, \tag{20.4.13}$$

then, of course, both $\ln \zeta_\beta(s_0)$ and its derivative $(\partial/\partial s_0) \ln \zeta_\beta(s_0)$ diverge. Thus also $\tilde{Z}(s_0, \beta)$ diverges. According to the remark made at the beginning of this section, the topological pressure $\mathscr{P}(\beta)$ is determined by the smallest positive value $s_c$ of $s$ that makes the function $\tilde{Z}(s, \beta)$ diverge. It is identical with the smallest positive root of the zeta function:

$$s_c = \min_{s_0 > 0} s_0 = \exp[-\mathscr{P}(\beta)]. \tag{20.4.14}$$

This provides us with a simple method of determining $\mathscr{P}(\beta)$.

Let us interpret eq. (20.4.14) in a thermodynamic way. The fugacity $s_c$ is related to the chemical potential $\mu_c$ by $s_c = \exp(\beta\mu_c)$. Hence we can write eq. (20.4.14) as

$$-\frac{1}{\beta}\mathscr{P}(\beta) = \mu_c. \qquad (20.4.15)$$

Notice that the left hand side is just the *topological expansion free energy* per particle. According to the length scale interpretation of the topological pressure, using the notation of eqs. (20.1.3) and (20.1.11), we may regard $-\beta^{-1}\mathscr{P}(\beta)$ as the Gibbs free energy per particle of a pressure ensemble with fluctuating volume $V_j$ in the limit $N \to \infty$:

$$-\frac{1}{\beta}\mathscr{P}(\beta) = \lim_{N\to\infty}\frac{G}{N}. \qquad (20.4.16)$$

Thus eq. (20.4.15) is the analogue of the Gibbs–Duhem relation of conventional thermodynamics

$$G = \mu N \qquad (20.4.17)$$

relating the Gibbs free energy to the chemical potential $\mu$ (see section 7.3). In conventional thermodynamics this relation is valid for homogeneous systems. Similarly, eq. (20.4.14) can only be rigorously proved for expanding or hyperbolic dynamical systems.

## *20.5 Partition functions with conditional probabilities

As we have seen, several types of partition function can be defined for chaotic systems. It depends on the question under investigation which one is the best for the particular problem. An alternative approach to introduce a thermodynamics with several intensive parameters relates to partition functions with conditional probabilities.

Suppose we have an arbitrary partition of the phase space consisting of $R$ cells. The cells may have equal or variable size. Again we denote by $p(i_0, \ldots, i_{N-1})$ the probability of observing a symbol sequence $i_0, \ldots, i_{N-1}$ (see section 3.6). In particular, we shall denote by $p(i)$ the probability of the single event $i$. The symbols $i_n$ take values in $\{1, \ldots, R\}$. For the initial distribution we again choose the natural invariant measure of the map.

As the $N$ events $i_n$ are not statistically independent, the probabilities $p(i_0, \ldots, i_{N-1})$ do not factorize with respect to the single events $i_n$. But we can always decompose them using conditional probabilities. Again using the notation of previous chapters, the conditional probability $p(B\,|\,A)$ denotes the probability of event $B$ under the condition that event $A$ has occurred. For example, we can decompose $p(i_0, \ldots, i_{N-1})$ in the following different ways:

$$p(i_0, \ldots, i_{N-1}) = p(i_0)p(i_1, \ldots, i_{N-1} | i_0) \tag{20.5.1}$$

or

$$p(i_0, \ldots, i_{N-1}) = p(i_1, \ldots, i_{N-1})p(i_0 | i_1, \ldots, i_{N-1}) \tag{20.5.2}$$

or

$$p(i_0, \ldots, i_{N-1}) = p(i_0)p(i_1 | i_0)p(i_2, \ldots, i_{N-1} | i_0, i_1) \tag{20.5.3}$$

etc. As, in general, a dynamical system does not generate a Markov process, the conditional probabilities depend on the entire symbol sequences contained in the events $A$ and $B$. Suppose we have chosen a certain decomposition of the probability $p(i_0, \ldots, i_{N-1})$ into a product of a probability $p^{(0)}$ and a product of $L$ conditional probabilities $p^{(1)}, \ldots, p^{(L)}$:

$$p = p^{(0)}p^{(1)} \cdots p^{(L)}. \tag{20.5.4}$$

We may then attribute to each probability $p^{(k)}$ an inverse temperature $\beta_k$ and define a partition function as

$$Z_N(\beta_0, \ldots, \beta_L) = \sum_{i_0, \ldots, i_{N-1}} (p^{(0)})^{\beta_0}(p^{(1)})^{\beta_1} \cdots (p^{(L)})^{\beta_L}, \tag{20.5.5}$$

where the sum runs over all allowed sequences $i_0, \ldots, i_{N-1}$. In particular, we are interested in the *free energy per particle* in the limit $N \to \infty$:

$$J(\beta_0, \ldots, \beta_L) = -\lim_{N \to \infty} \frac{1}{N\beta_0} \ln Z_N(\beta_0, \ldots, \beta_L). \tag{20.5.6}$$

The number $L$ can be chosen arbitrarily, as long as $L \leqslant N - 1$. One might even couple both $L$ and $N$ by setting $L = N - 1$, obtaining for large $N$ a partition function similar to a path integral (Schulman 1981; Roepstorff 1991) with the factorization

$$p(i_0, \ldots, i_{N-1}) = p(i_0)p(i_1 | i_0)p(i_2 | i_0, i_1) \cdots p(i_{N-1} | i_0, \ldots, i_{N-2}). \tag{20.5.7}$$

In the following, let us consider a simple example with $L = 1$. We choose a grid of boxes of equal size $\varepsilon$ and the factorization given by eq. (20.5.1). Let us denote the two different 'inverse temperatures' for the probability $p(i_0)$ and the conditional probabilities $p(i_1, \ldots, i_{N-1} | i_0)$ of eq. (20.5.1) by $\beta$ and $q$. The partition function is

$$Z_N(\beta, q, \sigma) = \sum_{i_0, \ldots, i_{N-1}} p(i_0)^\beta p(i_1, \ldots, i_{N-1} | i_0)^q. \tag{20.5.8}$$

The sum runs over all allowed symbol sequences $i_0, \ldots, i_{N-1}$. Besides the two inverse temperatures $\beta$ and $q$, the partition function $Z_N$ certainly depends on the 'particle density' $\sigma = N/V$, where again $V$ is related to the box size $\varepsilon$ by $V = -\ln \varepsilon$, and $N$ is the length of the symbol sequence. In contrast to section 20.2, $\sigma$ is now an arbitrarily chosen parameter. It describes the way in which the coupled limit $N \to \infty$, $V \to \infty$ is performed. In other words, it describes the relative size of temporal versus spatial degrees of freedom. A thermodynamic potential $J$ may then be defined as

$$J(\beta, q, \sigma) = -\lim_{N\to\infty} \frac{1}{N\beta} \ln Z_N(\beta, q, \sigma). \tag{20.5.9}$$

To obtain the Rényi entropies from $J(\beta, q, \sigma)$ we have to set $\beta = q$. Moreover, we remember that in the definition (14.2.10) of the Rényi entropies we first had to let $N \to \infty$, keeping the box size $\varepsilon$ finite. This corresponds to a particle density $\sigma \to \infty$. Hence

$$\frac{\beta}{\beta - 1} J(\beta, \beta, \infty) = K(\beta). \tag{20.5.10}$$

In a similar way $J(\beta, 0, 0)$ yields information about the Rényi dimensions, and $J(0, q, \infty)$ is a quantity similar to the topological pressure. The other values $J(\beta, q, \sigma)$ provide us with quite a general characterization of the scaling properties of a chaotic system, both with respect to static and dynamical aspects. The advantage of this concept is that it can be applied quite generally to any dynamical system, as the phase space is simply divided into boxes of equal size. The disadvantage is that there seems to be no constructive method of calculating $J(\beta, q, \sigma)$, and numerical calculations may converge very slowly.

# 21
# Phase transitions

An interesting phenomenon concerning the thermodynamics of chaotic systems is the occurrence of 'phase transitions' (Ott *et al.* 1984; Cvitanović 1987; Szépfalusy and Tél 1987; Szépfalusy *et al.* 1987; Katzen and Procaccia 1987; Badii and Politi 1987; Bohr and Jensen 1987). These are singularities of the free energy (or more generally, of some other characteristic quantity) at certain critical parameter values. Phase transitions are a typical phenomenon for nonhyperbolic dynamical systems. We shall present a short introduction to the subject and classify the various types of phase transitions. Our emphasis lies on simple illustrative examples rather than a complete description of this field. Remember that in the previous chapters we discussed free energies of various types. We called $\beta^{-1}(\beta - 1)D(\beta)$ 'static', $\beta^{-1}(\beta - 1)K(\beta)$ 'dynamic', and $-\beta^{-1}\mathscr{P}(\beta)$ and $-\Lambda(\beta)$ 'expansion' free energies. It is a natural question to investigate whether these free energies depend on $\beta$ in a nonanalytical way for certain dynamical systems. Depending on which kind of free energy is exhibiting the nonanalytical behaviour, we shall distinguish between static, dynamical, or expansion phase transitions. Moreover, there are phase transitions with respect to the 'volume' and with respect to external control parameters. In general, one type of phase transition does not imply another.

## 21.1 Static phase transitions

In section 13.3 we dealt with the Ulam map

$$f(x) = 1 - 2x^2 \qquad x \in [-1, 1]. \tag{21.1.1}$$

We saw that the Rényi dimensions $D(\beta)$ are not differentiable with respect to $\beta$ at the 'critical point' $\beta_c = 2$ (see eq. (13.3.7)). The reason is that the natural invariant density

$$\rho(x) = 1/[\pi(1 - x^2)^{1/2}] \tag{21.1.2}$$

possesses a power law singularity at the endpoints $x = \pm 1$ of the phase space $X = [-1, 1]$.

Let us investigate this phenomenon more generally. We consider a probability density $\rho(x)$ that is singular at a finite number of points $x_\sigma$ of a $d$-dimensional phase space in such a way that for $x \to x_\sigma$ it scales as

$$\rho(x) \sim |x - x_\sigma|^\gamma, \qquad \gamma \neq 0. \tag{21.1.3}$$

For all other points the density is assumed to be finite and nonzero.

Standard examples of maps where the natural invariant density has two singular points are one-dimensional maps of the type

$$f(x) = 1 - 2|x|^z, \qquad x \in [-1, 1], \quad z \geqslant 1. \tag{21.1.4}$$

The corresponding Perron–Frobenius fixed point equation is

$$\rho(x) = \frac{1}{2z}\left(\frac{2}{1-x}\right)^{(z-1)/z}\left\{\rho\left[\left(\frac{1-x}{2}\right)^{1/z}\right] + \rho\left[-\left(\frac{1-x}{2}\right)^{1/z}\right]\right\} \tag{21.1.5}$$

(the special case $z = 2$ has already been treated in section 17.2). Eq. (21.1.5) indicates that the density $\rho(x)$ diverges at $x = 1$ with the exponent $\gamma = (1 - z)/z$, and, in fact, for these types of maps divergences of the density occur at the edges $\pm 1$ only.

Now let us determine the Rényi dimensions for densities of type (21.1.3). If we partition the phase space into small boxes of equal size $\varepsilon$ (with volume $\varepsilon^d$), there is just a finite number of boxes $i$ containing $x_\sigma$ where the probabilities

$$p_i = \int_{\text{box } i} \mathrm{d}x\, \rho(x) \tag{21.1.6}$$

scale as $\varepsilon^{\gamma+1}\varepsilon^{d-1} = \varepsilon^{\gamma+d}$. All other probabilities $p_i$ scale as $\varepsilon^d$. The number of boxes with $p_i \sim \varepsilon^d$ is of the order $\varepsilon^{-d}$. Hence for $\varepsilon \to 0$

$$\sum_i (p_i)^\beta \sim \varepsilon^{(\gamma+d)\beta} + \varepsilon^{-d}\varepsilon^{\beta d} \sim \varepsilon^{(\beta-1)D(\beta)}. \tag{21.1.7}$$

This implies

$$(\beta - 1)D(\beta) = \min[(\gamma + d)\beta, (\beta - 1)d]. \tag{21.1.8}$$

The critical point $\beta_c$ where the minimum value switches, is given by

$$(\gamma + d)\beta_c = (\beta_c - 1)d. \tag{21.1.9}$$

Hence $\beta_c = -d/\gamma$. For the static Helmholtz free energy per volume $\varphi(\beta) = \beta^{-1}(\beta - 1)D(\beta)$ we obtain for $\gamma < 0$

$$\varphi(\beta) = \begin{cases} [1 - (1/\beta)]d & \text{for } \beta \leqslant \beta_c \\ d + \gamma & \text{for } \beta \geqslant \beta_c \end{cases} \tag{21.1.10}$$

and for $\gamma > 0$:

$$\varphi(\beta) = \begin{cases} d + \gamma & \text{for } \beta \leqslant \beta_c \\ [1 - (1/\beta)]d & \text{for } \beta \geqslant \beta_c. \end{cases} \tag{21.1.11}$$

Notice that $\varphi(\beta)$ is continuous at $\beta_c = -d/\gamma$. However, the derivative $\varphi'(\beta)$ does not exist at the critical point $\beta_c$ for $\gamma \neq 0$. Hence the corresponding phase transition is a phase transition of first order.

It is easy to construct probability densities with a more complicated phase diagram. One possibility is to take a density that scales at some finite points $x_\sigma$ in the phase space as $|x - x_\sigma|^{\gamma_\sigma}$, where the $\gamma_\sigma$ are different. An example is the following probability density on the interval $[0, 1]$:

$$\rho(x) = \frac{x^{\gamma_1}(1 - x)^{\gamma_2}}{B(\gamma_1 + 1, \gamma_2 + 1)}. \tag{21.1.12}$$

Here $B(x, y) = \Gamma(x)\Gamma(y)[\Gamma(x + y)]^{-1}$ denotes the beta function. $\rho(x)$ is normalized (Gradshteyn and Ryzhik 1965). We notice that $\rho(x)$ scales at $x = 0$ with the exponent $\gamma_1$ and at $x = 1$ with the exponent $\gamma_2$. Let us assume that $\gamma_1 \in (0, \infty)$ and $\gamma_2 \in (-1, 0)$. For this case a typical graph of $\rho(x)$ is plotted in fig. 21.1. The Rényi dimensions can be calculated by

Fig. 21.1 Example of a probability density exhibiting a double phase transition ($\gamma_1 = \frac{3}{4}$, $\gamma_2 = -\frac{1}{2}$).

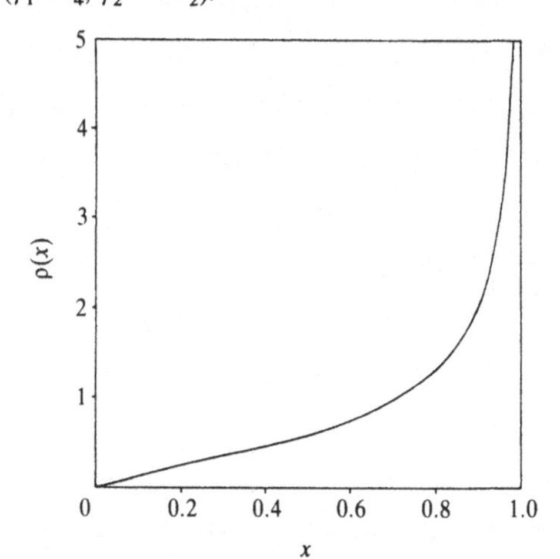

using just the same arguments as before. In the $d$-dimensional case we obtain the free energy

$$\varphi(\beta) = \begin{cases} d + \gamma_1 & \text{for } \beta \leqslant -d/\gamma_1 \\ [1 - (1/\beta)]d & \text{for } -d/\gamma_1 \leqslant \beta \leqslant -d/\gamma_2 \\ d + \gamma_2 & \text{for } \beta \geqslant -d/\gamma_2. \end{cases} \quad (21.1.13)$$

The example of eq. (21.1.12) corresponds to the case $d = 1$. The corresponding Rényi dimensions $D(\beta) = \beta(\beta - 1)^{-1}\varphi(\beta)$ are plotted in Fig. 21.2. The system exhibits a double first order phase transition; one for negative inverse temperature $\beta$, one for positive $\beta$ (Beck 1990a).

What is the 'physical meaning' of these phase transitions? As we have already emphasized in previous chapters, we can 'scan' the influence of different subregions of the phase space by a variation of $\beta$. At a phase transition point, the dominant contribution to the free energy suddenly comes from a different subregion of the phase space. For instance, for the Ulam map the density $\rho(x)$ given by eq. (21.1.2) has two singularities at the edges of the phase space. For $\beta < 2$, all boxes in the interior of this interval are dominant in the sum $\sum_i (p_i)^\beta$. For $\beta > 2$, however, the two boxes at the edges prevail, which scale with a different power of $\varepsilon$.

The examples treated so far were analytically solvable. But static phase transitions occur quite frequently for all kinds of nonhyperbolic maps. More complicated examples are the Hénon map (Grassberger *et al.* 1988) or maps of Kaplan–Yorke type (Beck 1991c). For these systems one

Fig. 21.2 Rényi dimensions of the probability density plotted in fig. 21.1.

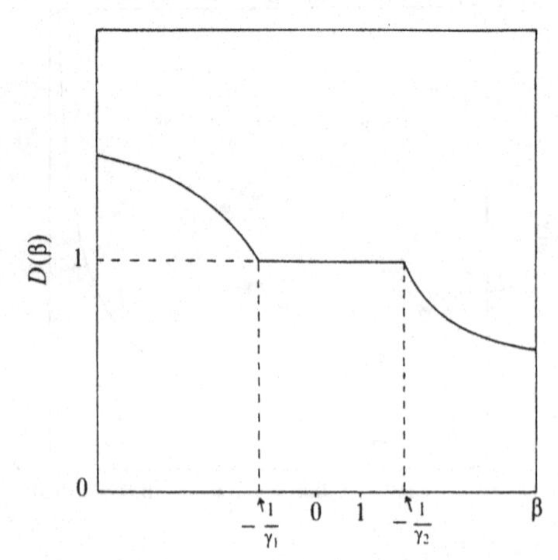

observes homoclinic tangencies, i.e., the stable and unstable manifolds touch each other in a tangential way (see section 15.6). At these points typically the invariant density scales in a nontrivial way, and the same consideration as before applies against: changing $\beta$, either the boxes containing the homoclinic tangencies or those containing the rest of the attractor dominate the static partition function.

*Remark*: For certain exceptional probability measures there are even static phase transitions with a stronger degree of nonanalyticity: in this case the function $D(\beta)$ can be discontinuous at a critical point. This strange behaviour is caused by the fact that the probability $p_i$ of at least one box does not scale as $\varepsilon^\alpha$, but in an anomalous way, for example as $(-\ln \varepsilon)^{-1}$, or as a constant. The latter case arises if the probability measure contains distribution-like ingredients such as Dirac's delta function (see Csordás and Szépfalusy (1989a) and Beck (1990a) for more details).

**$f(\alpha)$ spectra for systems with static phase transitions** In the case of a first order phase transition, the function $D(\beta)$ is no longer differentiable. Thus, to calculate the Legendre transform of $(\beta - 1)D(\beta)$, which is the $f(\alpha)$ spectrum, we have to use the more general definition of a Legendre transform that is also applicable to nondifferentiable functions. This concept was described in section 7.1. Hence, in the case of phase transitions we define $f(\alpha)$ by the variational principle

$$\tau(\beta) = \min_\alpha [\beta\alpha - f(\alpha)], \tag{21.1.14}$$

$$f(\alpha) = \min_\beta [\beta\alpha - \tau(\beta)], \tag{21.1.15}$$

where $\tau(\beta) = (\beta - 1)D(\beta)$. If $\tau(\beta)$ and $f(\alpha)$ are differentiable, this definition reduces to the usual one

$$\tau(\beta) = \beta\alpha(\beta) - f(\alpha(\beta)), \qquad \tau'(\beta) = \alpha(\beta), \qquad f'(\alpha) = \beta \tag{21.1.16}$$

of section 11.3.

On the basis of this definition it is easy to evaluate $f(\alpha)$ for the examples presented above. One simply inserts the known function $\tau(\beta) = \beta\varphi(\beta)$ into eq. (21.1.15) and determines the minimum. If the probability density $\rho(x)$ scales with the constant exponent $\gamma$ at a finite number of points $x_\sigma$ of the $d$-dimensional phase space, the result is

$$f(\alpha) = -\frac{\alpha - d}{\gamma} d + d, \qquad \alpha \in I \tag{21.1.17}$$

where the interval $I$ is given by

$$I = \begin{cases} [\gamma + d, d] & \text{for } \gamma < 0 \\ [d, \gamma + d] & \text{for } \gamma > 0. \end{cases} \tag{21.1.18}$$

Notice that $f(\alpha)$ is defined for a finite range of scaling indices $\alpha$ only. For the example of the density $\rho(x)$ with two scaling indices $\gamma_1 > 0$ and $\gamma_2 < 0$ we obtain

$$f(\alpha) = \begin{cases} -\dfrac{\alpha - d}{\gamma_2} d + d & \text{for } \alpha \in [\gamma_2 + d, d] \\[2ex] -\dfrac{\alpha - d}{\gamma_1} d + d & \text{for } \alpha \in [d, \gamma_1 + d] \end{cases} \tag{21.1.19}$$

(see fig. 21.3). All the spectra have a triangular shape and are not differentiable at $\alpha = d$.

## 21.2 Dynamical phase transitions

Not only the Rényi dimensions $D(\beta)$, or the static free energy, but also the Rényi entropies $K(\beta)$, corresponding to the dynamical free energy may exhibit nonanalytic behaviour at critical values $\beta_c$. Meanwhile, several examples are known where this phenomenon occurs (although dynamical phase transitions seem to be less common than static phase transitions). There are basically two different mechanisms that lead to dynamical phase transitions: either probabilities for certain symbol sequences decrease *more slowly* than exponentially with increasing length

Fig. 21.3 $f(\alpha)$ spectrum for the probability density plotted in fig. 21.1.

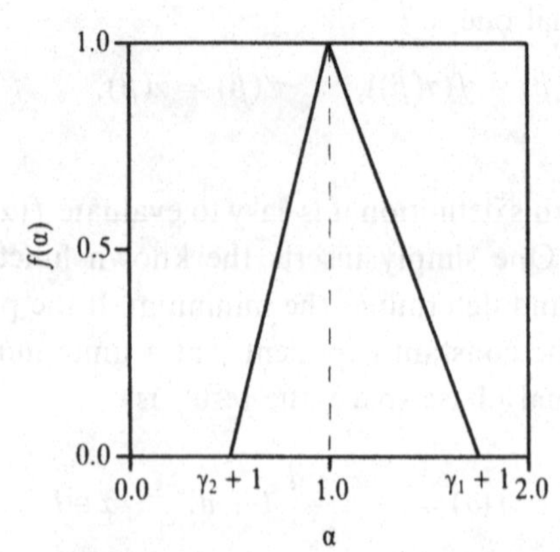

of the sequence, or they decrease *more rapidly*. This will be worked out in the following.

Typically for a hyperbolic system the probability of observing a certain symbol sequence $i_0, \ldots, i_{N-1}$ of length $N$ decreases exponentially with increasing $N$. For example, for the binary shift may we have

$$p(i_0, \ldots, i_{N-1}) = 2^{-N} = \exp(-N \ln 2). \qquad (21.2.1)$$

For a nonhyperbolic system, on the other hand, there may be symbol sequence probabilities that decrease much more slowly with increasing $N$. A concrete example is provided by the intermittency phenomenon (see section 1.3). For an intermittent system the orbit stays for a long time in a certain subregion of the phase space. It becomes almost periodic for a long time, and this results in long-time correlations of the symbols. Typically in such cases there is at least one symbol sequence $\tilde{i}_0, \tilde{i}_1, \ldots, \tilde{i}_{N-1}$ (corresponding to the laminar phase) for which the probability decreases more slowly than exponentially with increasing $N$, let us say in a polynomial way:

$$p(\tilde{i}_0, \ldots, \tilde{i}_{N-1}) = CN^{-\alpha}. \qquad (21.2.2)$$

Here $\alpha$ and $C$ are positive constants. For $\beta \to \infty$ this largest probability $p(\tilde{i}_0, \ldots, \tilde{i}_{N-1})$ dominates the dynamical partition function, and we obtain

$$\begin{aligned} K(\beta) &= \lim_{N\to\infty} \frac{1}{N}\frac{1}{1-\beta} \ln \sum_{i_0,\ldots,i_{N-1}} p^\beta(i_0, \ldots, i_{N-1}) \\ &\approx \lim_{\beta\to\infty} \lim_{N\to\infty} \frac{1}{N}\frac{1}{1-\beta} \ln p^\beta(\tilde{i}_0, \ldots, \tilde{i}_{N-1}) \\ &= \lim_{N\to\infty} \frac{1}{N}\frac{\beta}{1-\beta}(\ln C - \alpha \ln N) = 0. \end{aligned} \qquad (21.2.3)$$

Thus $K(\infty) = 0$. But an even stronger statement is true. Since, in general, $K(\beta) \geqslant 0$ and $K(\beta) \leqslant (\beta - 1)^{-1}\beta K(\infty)$ for $\beta > 1$ (see section 11.2), we must have

$$K(\beta) = 0, \qquad \beta > 1, \qquad (21.2.4)$$

i.e., not only $K(\infty)$, but all $K(\beta)$ with $\beta > 1$ vanish. On the other hand, if the system exhibits chaotic behaviour, then by definition it has a positive KS entropy:

$$K(1) = h > 0. \qquad (21.2.5)$$

Eqs. (21.2.5) and (21.2.4) imply that at the critical parameter value $\beta_c = 1$ the function $K(\beta)$ is discontinuous. This is a direct consequence of the fact that there is at least one symbol sequence whose probability decreases more slowly than exponentially. Nonanalytic behaviour of this type has, in fact, been observed for various maps, all of them exhibiting intermittent behaviour (Szépfalusy *et al.* 1987; Bene, Szépfalusy and Fülöp 1989; Kaufmann and Szépfalusy 1989; Szépfalusy, Tél and Vattay 1991).

The second possible reason for the occurrence of a dynamical phase transition is an extremely unstable dynamics or, in other words, the existence of at least one point in the phase space with an infinite local Liapunov exponent. In this case the corresponding symbol sequence probability $p(\tilde{i}_0, \ldots, \tilde{i}_{N-1})$ decreases more rapidly than exponentially with increasing symbol sequence length $N$. For $\beta \to -\infty$ this smallest probability dominates the dynamical partition function:

$$\begin{aligned} K(\beta) &\approx \lim_{\beta \to -\infty} \lim_{N \to \infty} \frac{1}{N} \frac{1}{1-\beta} \ln p^{\beta}(\tilde{i}_0, \ldots, \tilde{i}_{N-1}) \\ &= \infty . \end{aligned} \tag{21.2.6}$$

Thus $K(-\infty) = \infty$. However, in general, the inequality

$$K(\beta) \geqslant \frac{\beta}{\beta - 1} K(-\infty), \qquad \beta < 0 \tag{21.2.7}$$

is valid (see section 11.2). Thus we actually have

$$K(\beta) = \infty \qquad \text{for all } \beta < 0. \tag{21.2.8}$$

On the other hand, in the case of a generating partition consisting of a finite number $R$ of cells the topological entropy $K(0)$ is finite: it describes the growth rate of the number of allowed symbol sequences (see section 14.2). Thus, in this case, the function $K(\beta)$ must be discontinuous at the critical parameter value $\beta_c = 0$. The situation is somewhat different if the generating partition contains infinitely many cells. An example is the continued fraction map

$$f(x) = \frac{1}{x} - \left\lfloor \frac{1}{x} \right\rfloor, \qquad x \in [0, 1]. \tag{21.2.9}$$

Here the generating partition indeed consists of infinitely many cells, namely the infinite number of preimages of the interval [0, 1] (see section 17.2). For this map $K(0) = \infty$, and the transition occurs later, namely at $\beta_c = \frac{1}{2}$ (Csordás and Szépfalusy 1989b). Notice that at $x = 0$ the derivative $f'(x)$ diverges, i.e., there is, indeed, an infinite local Liapunov exponent, which causes the dynamical phase transition.

Another example where dynamical phase transitions occur is provided by the following class of piecewise parabolic maps

$$f(x) = \frac{1}{2\mu}\{\mu + 1 - [(\mu - 1)^2 + 4\mu|1 - 2x|]^{1/2}\}. \quad (21.2.10)$$

The parameter $\mu$ takes on values in the range $[-1, 1]$; the phase space is the interval $X = [0, 1]$. Szépfalusy *et al.* (Csordás and Szépfalusy 1989a) found phase transitions for both the Rényi dimensions and the Rényi entropies. Depending on the parameter $\mu$, the behaviour of $K(\beta)$ is qualitatively different. For example, for $\mu = -1$ there is a critical point $\beta_c = 0$ where $K(\beta)$ jumps from infinity to a finite value. For $\mu = 1$ there is another critical point $\beta_c' = 1$ where $K(\beta)$ jumps from a finite value to 0. The above mentioned authors have used the notions 'chaotic chaos phase', 'stochastic chaos phase', and 'regular chaos phase' for the regions of the inverse temperature $\beta$ where $K(\beta)$ is finite, infinite, or zero, respectively.

## 21.3 Expansion phase transitions

After static and dynamical phase transitions, let us now deal with phase transitions with respect to the topological pressure $\mathscr{P}(\beta)$. We call them *topological expansion phase transitions.* To explain the occurrence of such a phase transition, we may use similar arguments to those used before. The partition function $Z_N^{\text{top}}(\beta)$ consists of a sum of contributions of the $N$-cylinders (or, in general, of the members of the ensemble $H_N$) raised to the power $\beta$. At a critical temperature the dominant contribution to the partition function switches from one subset of cylinders to another one. Moreover, in the case that some cylinder lengths decrease more slowly or more rapidly than exponentially with increasing $N$, we expect similar phenomena to those for the Rényi entropies. We shall deal with two important examples in this section, namely with the Ulam map and the Julia set of the complex logistic map $f(x) = x^2 + C$ for $C = \frac{1}{4}$.

**The Ulam map** In section 16.3 we derived that the topological pressure of this map is

$$\mathscr{P}(\beta) = \begin{cases} -2\beta \ln 2 & \text{for } \beta \leqslant -1 \\ (1 - \beta) \ln 2 & \text{for } \beta \geqslant -1. \end{cases} \quad (21.3.1)$$

We notice that $\mathscr{P}(\beta)$ is not differentiable at the critical inverse temperature $\beta_{\text{crit}} = -1$. Hence the above map is a simple example of a dynamical system that exhibits a topological expansion phase transition. We have also seen that the same system exhibits a static phase transition at $\beta_c = 2$, because the Rényi dimensions of the natural invariant density of this map satisfy

$$D(\beta) = \begin{cases} 1 & \text{for } \beta \leqslant 2 \\ \dfrac{\beta}{2(\beta - 1)} & \text{for } \beta \geqslant 2. \end{cases} \tag{21.3.2}$$

Both phase transitions, however, are related to free energies of different types. *A priori*, they are independent of each other and occur at different critical points $\beta_{crit}$ and $\beta_c$. The reason for the expansion phase transitions is the fact that locally, in the vicinity of the unstable fixed point $-1$, i.e., at the edges of the phase space, a small interval expands with the local rate $2 \ln 2$, whereas almost all other small subintervals of the phase space expand with the local rate $\ln 2$. If $\beta$ passes the critical value $-1$, the main contribution to the global expansion rate switches from the local contribution $2 \ln 2$ to the average contribution $\ln 2$. In contrast to this, the static phase transition is caused by the fact that the natural invariant density diverges at the edges of the phase space with a power law singularity. The expansion phase transition is caused by the behaviour of the interval lengths, the static phase transition by the behaviour of the probability density. For nonhyperbolic systems, these phenomena need not have anything to do with each other.

The question arises of whether the Ulam map also exhibits a dynamical phase transition, besides the static and expansion phase transitions. The answer is *no*: for all $\beta$ the Rényi entropies are given by

$$K(\beta) = \ln 2. \tag{21.3.3}$$

To understand this, remember that the natural invariant measure of an $N$-cylinder is

$$p_j^{(N)} = \mu(J_j^{(N)}) = \int_{J_j^{(N)}} dx\, \rho(x), \qquad j = 1, \ldots, 2^N \tag{21.3.4}$$

(see chapter 3). In the interior of the phase space $X = [-1, 1]$, the length $l_j^{(N)}$ of a cylinder scales like $2^{-N}$, as explained in the context of eq. (16.3.10). On the other hand, the length $l_1^{(N)}$ at the endpoint of the interval scales like $l_1^{(N)} \sim 2^{-2N}$. At the endpoint $\rho(x)$ diverges in proportion to $\Delta^{-1/2}$ if $\Delta$ is the distance from the endpoint. Hence, we obtain

$$p_1^{(N)} \sim (l_1^{(N)})^{1/2} = 2^{-N}. \tag{21.3.5}$$

These effects, the scaling of the length and the relative scaling of the measure $\mu$ with the length, compensate each other, resulting in the fact that for all cylinders we have the same scaling behaviour $\mu(J_j^{(N)}) \sim 2^{-N}$. This leads to constant Rényi entropies $K(\beta) = \ln 2$, and no dynamical phase transition occurs.

The Ulam map is a somewhat special example in the sense that the static and the expansion effects cancel each other if we pass to the Rényi entropies. We learn from this simple example that topological pressure, dynamical Rényi entropies, and Rényi dimensions may exhibit different phase diagrams. We need all three quantities for a sufficient description of a nonhyperbolic dynamical system.

**Julia set of the logistic map for $C = \frac{1}{4}$** In section 16.3 we dealt with the topological pressure $\mathscr{P}(\beta)$ of the Julia set of the complex logistic map

$$x_{n+1} = f(x_n) = x_n^2 + C. \tag{21.3.6}$$

Barnsley, Geronimo, and Harrington (1983) have proved that for $|C| < \frac{1}{4}$ the topological pressure is a smooth function, i.e., there is no phase transition. The situation changes if we consider the parameter value $C = \frac{1}{4}$. That something special might happen at this parameter value can already be guessed if we look at the stability properties of the fixed point

$$x^* = \tfrac{1}{2} + (\tfrac{1}{4} - C)^{1/2}. \tag{21.3.7}$$

For $C = \frac{1}{4}$ we have

$$|f'(x^*)| = 1, \tag{21.3.8}$$

i.e., the system is no longer hyperbolic. In fact, numerically a phase transition of $\mathscr{P}(\beta)$ at the critical value $\beta_c = D_H$ is observed (Katzen and Procaccia 1987). Here $D_H$ denotes the Hausdorff dimension of the Julia set. The theoretical explanation for this phase transition is similar to that for the dynamical phase transition in the previous section. For $C = \frac{1}{4}$ it is observed that there are some length scales (namely $l^{(N)}(1, 1, \ldots, 1)$ and $l^{(N)}(-1, -1, \ldots, -1)$) that decrease with increasing $N$ as $1/N$, rather than in an exponential way. This leads us to $\mathscr{P}(\beta) = 0$ for $\beta \to \infty$ (same arguments as in the previous section). From eq. (5.3.14) we know that $\mathscr{P}(\beta)$ is monotonically decreasing with increasing $\beta$. Moreover, the definition of the Hausdorff dimension (see section 10.3 and eq. (19.5.8)) implies that

$$\mathscr{P}(\beta) \geqslant 0 \qquad \text{for } \beta < D_H \tag{21.3.9}$$

and

$$\mathscr{P}(\beta) \leqslant 0 \qquad \text{for } \beta > D_H. \tag{21.3.10}$$

Hence $\mathscr{P}(\infty) = 0$ actually implies the stronger statement

$$\mathscr{P}(\beta) = 0 \qquad \text{for all } \beta > D_H. \tag{21.3.11}$$

On the other hand, for an object with nontrivial Hausdorff dimension we expect that $\mathscr{P}(\beta)$ does not vanish for $\beta < D_H$, otherwise the definition of the Hausdorff dimension would not work. This makes it plausible that indeed the critical point of the phase transition is given by $\beta_c = D_H$.

> *Remark*: Phase transitions of the topological pressure can also be studied with the help of the transfer operator method. In this approach the phase transition manifests itself in a crossing of the two largest eigenvalues of the transfer operator (Feigenbaum, Procaccia and Tél 1989).

**Metric expansion phase transitions** The generalized Liapunov exponents $\Lambda(\beta)$, defined by

$$Z_N^{\text{Lia}}(\beta) = \sum_j p_j^{(N)} (l_j^{(N)})^{-\beta} \sim \exp[\beta N \Lambda(\beta)], \qquad N \to \infty \tag{21.3.12}$$

can also exhibit nonanalytic behaviour for nonhyperbolic dynamical systems. We call this a *metric* expansion phase transition.

For the somewhat special example of the Ulam map, there is no difference between the metric and the topological phase transition behaviour. As we attribute to each cylinder the *constant* probability $p_j^{(N)} = 2^{-N}$, we have

$$Z_N^{\text{Lia}}(\beta) = 2^{-N} \sum_j (l_j^{(N)})^{-\beta} \sim \exp\{-N \ln 2 + N\mathscr{P}(-\beta)\}. \tag{21.3.13}$$

Hence, in this special case the generalized Liapunov exponent $\Lambda(\beta)$ is directly given by the topological pressure, up to trivial constants:

$$\Lambda(\beta) = \frac{\mathscr{P}(-\beta) - \ln 2}{\beta} = \begin{cases} \ln 2 & \text{for } \beta \leqslant 1 \\ [(2\beta - 1)/\beta] \ln 2 & \text{for } \beta \geqslant 1. \end{cases} \tag{21.3.14}$$

This yields the $N \to \infty$ asymptotics of eqs. (15.4.19)–(15.4.23) for all those values of $\beta$ where the integral exists. In general, however, the probabilities $p_j^{(N)}$ are fluctuating quantities. This can lead to differences between metric and topological expansion free energies. Thus the phase diagrams may differ.

## 21.4 Bivariate phase transitions

In the previous sections we have seen that very simple nonhyperbolic dynamical systems such as the Ulam map may exhibit several types of phase transitions. It is a natural question to ask if it is possible to embed these different phase transitions in a unified thermodynamic description. An appropriate tool is, of course, the bivariate thermodynamics of chapter 20. We may investigate the Gibbs free energy $g(\beta, \Pi)$, or alternatively the equation of state, for a sudden change of behaviour at critical parameter values. As there are two intensive parameters now, in general we expect critical lines $\Pi_c(\beta)$ instead of critical points, where the nonanalytic behaviour takes place.

In general, such a bivariate phase diagram may be quite complicated and accessible by numerical experiments only (Kovács and Tél 1992). However, for the Ulam map an analytic treatment is still possible using the same arguments as in the previous sections. As it illustrates the most important features, we shall work out this example in detail.

According to eq. (20.1.2), we have to consider the bivariate partition function

$$Z_N(\beta, q) = \sum_j (p_j^{(N)})^\beta (l_j^{(N)})^q. \tag{21.4.1}$$

Using

$$p_j^{(N)} \sim (\tfrac{1}{2})^N \tag{21.4.2}$$

and

$$l_j^{(N)} \sim \begin{cases} (\frac{1}{4})^N & \text{for } j = 1 \text{ and } j = 2^N \\ (\frac{1}{2})^N & \text{otherwise} \end{cases} \tag{21.4.3}$$

we obtain the following result:

$$\begin{aligned} Z_N(\beta, q) &\sim 2^N(\tfrac{1}{2})^{N\beta}(\tfrac{1}{2})^{Nq} + 2(\tfrac{1}{2})^{N\beta}(\tfrac{1}{4})^{Nq} \\ &= \exp[-N \ln 2(-1 + \beta + q)] + 2\exp[-N \ln 2(\beta + 2q)] \\ &\sim \exp[-N\beta g(\beta, \Pi)] \end{aligned} \tag{21.4.4}$$

with $\Pi = q/\beta$. This yields

$$\beta g(\beta, \Pi) = \min(\beta - 1 + q, \beta + 2q) \ln 2 \tag{21.4.5}$$

or

$$g(\beta, \Pi) = \begin{cases} (1 + 2\Pi) \ln 2 & \text{for } q \leqslant -1 \\ [\beta^{-1}(\beta - 1) + \Pi] \ln 2 & \text{for } q > -1. \end{cases} \tag{21.4.6}$$

$g(\beta, \Pi)$ mainly contains information about the expansion phase transition at $q = -1$. Therefore it is useful to proceed to the 'equation of state'

$$\sigma(\beta, \Pi) = \frac{\beta^{-1}(\beta - 1)D(\beta) + \Pi}{g(\beta, \Pi)} \tag{21.4.7}$$

(see section 20.2). This has the advantage that the quantity $\sigma(\beta, \Pi)$ also contains information on the static phase transition. As

$$\frac{\beta-1}{\beta}D(\beta)=\begin{cases}\dfrac{\beta-1}{\beta} & \text{for } \beta\leqslant 2\\ \tfrac{1}{2} & \text{for } \beta>2,\end{cases} \tag{21.4.8}$$

we obtain

$$\sigma(\beta,\Pi)=\frac{1}{\ln 2}\begin{cases}\dfrac{\beta+\beta\Pi-1}{\beta+2\beta\Pi} & \text{for I:}\quad \beta\leqslant 2,\ \beta\Pi\leqslant -1\\ 1 & \text{for II:}\ \ \beta\leqslant 2,\ \beta\Pi>-1\\ \tfrac{1}{2} & \text{for III: } \beta>2,\ \beta\Pi\leqslant -1\\ \dfrac{\beta+2\beta\Pi}{2(\beta+\beta\Pi-1)} & \text{for IV: } \beta>2,\ \beta\Pi>-1.\end{cases} \tag{21.4.9}$$

We recognize that there are four different phases I–IV with different equations of state. Regions II and III correspond to condensed matter (of different density), whereas in regions I and IV there is an explicit dependence of the particle density $\sigma$ on both the pressure and the temperature. The phase diagram is plotted in fig. 21.4. There are three critical lines, namely $\Pi_c=-1/\beta$, $\beta_c=0$, and $\beta_c=2$.

## *21.5 Phase transitions with respect to the volume

We would like to broaden our view of phase transitions in nonlinear dynamics, in the sense that we may also look at a sudden change of

Fig. 21.4 Phase diagram of the Ulam map.

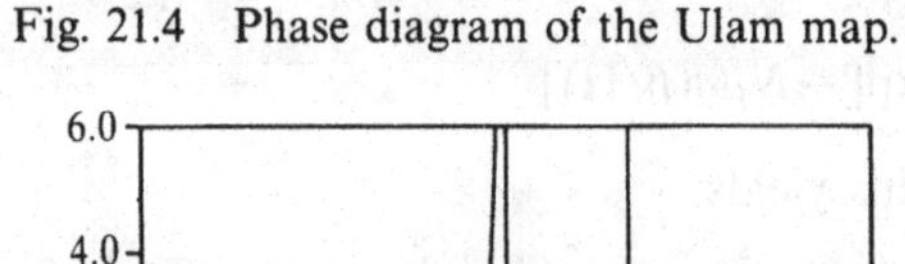
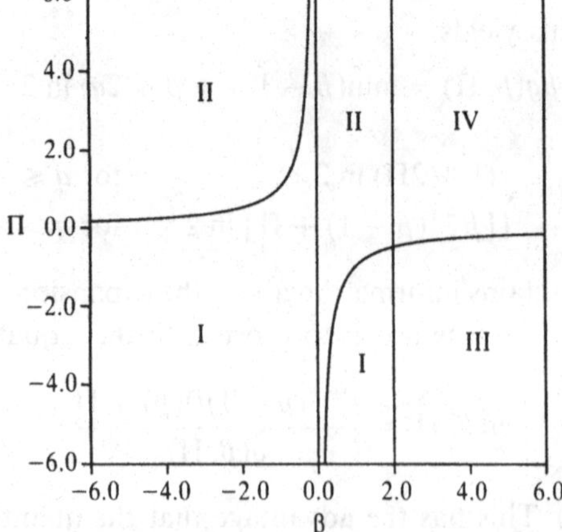

behaviour of the free energy if the box size $\varepsilon$, respectively the value of the 'volume' $V = -\ln \varepsilon$, is varied (notice that the volume of a box in the phase space is $\varepsilon^d$). A decrease of $\varepsilon$ means an increase of the resolving ability of the observation. Hence, $V$ may be regarded as a measure of the resolving ability. If by increasing $V$ a quantity such as the free energy suddenly changes its behaviour, this gives information about a new fine structure of the system. Thus it is interesting to study such changes. Generally it is known that phase transitions can only occur in the thermodynamic limit $V \to \infty$ and $N \to \infty$. In fact, it turns out that the phase transitions that we are going to describe here are more precisely phase transitions with respect to the 'particle density' $\sigma = N/V$ occurring in the limit $N \to \infty$.

Let us consider the *logistic map of order z*

$$f(x) = 1 - \mu|x|^z \tag{21.5.1}$$

where $z > 1$ and $\mu \in [0, 2]$. If we iterate the map for a parameter value $\mu$ that is slightly smaller than the critical value $\mu_\infty$ of the period doubling accumulation, the map possesses a stable periodic orbit with a rather large period length $2^N$. Now let us determine the static free energy for finite volume

$$\tilde{F}(\beta, V) = -\frac{1}{\beta} \ln \sum_i p_i^\beta \tag{21.5.2}$$

(see section 13.2). If $|\mu - \mu_\infty|$ is very small and if the volume $V$ is not too large, the result for $\tilde{F}(\beta, V)$ will be essentially the same as that for the order $z$ Feigenbaum attractor ($\mu = \mu_\infty$). That is, we obtain

$$\tilde{F}(\beta, V) = \frac{\beta - 1}{\beta} [D(\beta)V + C(\beta)], \tag{21.5.3}$$

where $D(\beta)$ is the Rényi dimension and $C(\beta)$ the reduced Rényi information of the Feigenbaum attractor. However, the picture drastically changes if we further increase $V$ keeping $\mu - \mu_\infty$ constant. Then there is a critical value $V_c$ of $V$ where the resolution is large enough to recognize that there is a stable periodic orbit rather than the Feigenbaum attractor. This means that we obtain the picture sketched in fig. 21.5. Above the critical value $V_c$ the free energy no longer depends on $V$. The reason is that for $V > V_c$ all orbit elements have a distance larger than the box size $\varepsilon$. Hence, the sum in eq. (21.5.2) no longer changes if the box size $\varepsilon$ is further decreased.

What is our theoretical prediction for $V_c$? In the vicinity of 0 the distance between neighbouring orbit elements scales as $|\alpha|^{-N}$, where $\alpha(z)$ is the Feigenbaum constant (see sections 1.3 and 11.4). As this distance is mapped onto the vicinity of 1 by a map with a maximum of order $z$, there the distance between orbit elements scales as $|\alpha|^{-zN}$, and this is indeed the minimum distance of neighbouring orbit elements of the entire period $2^N$:

$$d_{\min} \sim |\alpha|^{-zN}. \tag{21.5.4}$$

Moreover, according to the definition of the Feigenbaum constant $\delta(z)$, the parameter value $\mu$ for a stable orbit of length $2^N$ scales as

$$|\mu - \mu_\infty| \sim \delta^{-N} \tag{21.5.5}$$

for $N \to \infty$.

By elimination of $N$, we obtain for the critical box size $\varepsilon_c$ where the phase transition occurs

$$\varepsilon_c \sim d_{\min} \sim |\mu - \mu_\infty|^\kappa \tag{21.5.6}$$

with

$$\kappa = z\frac{\ln|\alpha|}{\ln \delta}. \tag{21.5.7}$$

The corresponding critical volume is

$$V_c = -\ln \varepsilon_c = -\kappa \ln|\mu - \mu_\infty|. \tag{21.5.8}$$

This consideration has been presented in Beck and Roepstorff (1987b) in a slightly different context, namely in considering the effects that round-off errors have on nonlinear mappings. Actually, all the above scaling relations are valid in the limit $N \to \infty$ only. In this limit the quantities occurring in eq. (21.5.8) diverge. So we may better write eq. (21.5.8) with the help of eq. (21.5.5) in the equivalent, but finite form

$$N/V_c = \sigma_c = 1/(z \ln|\alpha|), \tag{21.5.9}$$

Fig. 21.5 A phase transition with respect to the volume.

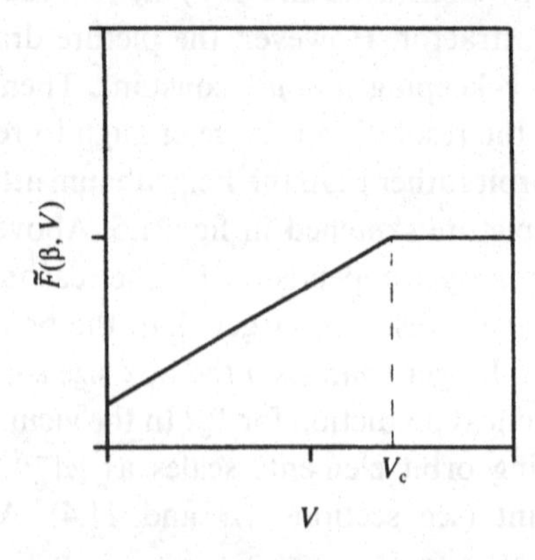

Table 21.1. *Critical density $\sigma_c$ of the Feigenbaum attractor for various values of the universality class z.*

| $z$ | $\sigma_c$ |
|---|---|
| 1 | 0 |
| 1.02 | 0.313954 |
| 1.04 | 0.3627452 |
| 1.06 | 0.3949330 |
| 1.08 | 0.4190062 |
| 1.1 | 0.4380739 |
| 1.2 | 0.49557318 |
| 1.4 | 0.53850221 |
| 1.6 | 0.54975803 |
| 1.6922 | 0.55060896 |
| 1.8 | 0.54973143 |
| 2 | 0.54498692 |
| 3 | 0.50788009 |
| 4 | 0.47627414 |
| 5 | 0.4525179 |
| 6 | 0.4343383 |
| 8 | 0.4084623 |
| 10 | 0.3909036 |
| 15 | 0.364442 |
| 20 | 0.349538 |
| 25 | 0.33990 |
| 30 | 0.33312 |
| 35 | 0.32807 |
| 40 | 0.32416 |
| 45 | 0.32100 |
| 50 | 0.31845 |
| ⋮ | ⋮ |
| $\infty$ | 0.294 |

where we again have introduced the particle density $\sigma = N/V$. Thus the phase transition with respect to the volume is more precisely a phase transition with respect to the density $\sigma$: at a finite critical value $\sigma_c = (z \ln|\alpha|)^{-1}$ the static free energy changes the behaviour in the thermodynamic limit $N \to \infty$.

Table 21.1 shows numerical results for $(z \ln|\alpha(z)|)^{-1}$ as obtained in Beck and Roepstorff (1987b). Surprisingly enough, the critical density $\sigma_c$ of the Feigenbaum attractor cannot exceed a certain maximum value $\sigma_{max}$ for arbitrary values of the universality class $z$. Indeed, the critical density $\sigma_c(z)$ is a monotonically increasing function in the interval $[1, z^*]$ and monotonically decreasing in the interval $[z^*, \infty]$. At $z^* = 1.6922\ldots$ it takes on the maximum value $\sigma_{max} = \sigma_c(z^*) = 0.55061\ldots$. We notice that for reasons of principle the critical density of the order-$z$ Feigenbaum attractor cannot exceed a certain threshold. In other words, there is a basic nontrivial bound on resolving an orbit of length $2^N$ on the Feigenbaum attractor of order $z$ with a certain resolving ability $V$. This bound is independent of the universality class.

If the parameter $\mu$ is chosen to be slightly larger than $\mu_\infty$, that means if we enter into the chaotic regime, another transition occurs. In this case the attractor of the logistic map consists of $2^N$ separated bands with chaotic behaviour inside (see section 1.3). The natural invariant density scales at the edges of the bands with the same power law singularity as for the case $\mu = 2$. Hence, as there is only a finite number of bands, the Rényi dimensions are again given by

$$D(\beta) = \begin{cases} 1 & \text{for } \beta \leqslant 2 \\ \dfrac{1}{2}\dfrac{\beta}{\beta - 1} & \text{for } \beta \geqslant 2. \end{cases} \tag{21.5.10}$$

Now, if $V$ is not too large and if $|\mu - \mu_\infty|$ is small, the free energy first follows the Feigenbaum laws. Only if the box size exceeds a critical value, is the chaotic behaviour, i.e., eq. (21.5.10), revealed. At the critical volume $V_c$ the free energy switches from the Feigenbaum behaviour to the asymptotic chaotic behaviour. As an estimate, eq. (21.5.8) for $V_c$, respectively eq. (21.5.9) for $\sigma_c$, is again valid.

## 21.6 External phase transitions of second order

A map usually depends on a control parameter or on a set of such parameters. An example is the parameter $\mu$ of the logistic map or the parameter $\lambda$ of the Kaplan–Yorke map. If we incorporate this dependence into the thermodynamic formalism, these control parameters enter into the description as further thermal variables. The various free energies then become functions of such parameters as well. It is not surprising that there are many systems for which the free energy exhibits nonanalytical behaviour not only with respect to $\beta$ or $V$ but also with respect to the control parameter. We call these phase transitions *external phase transitions*, because they are related to an external parameter of the system. On the other hand, we call all previously described phase transitions (sections 21.1–21.5) *internal phase transitions*, because they arise from the intrinsic thermodynamic treatment of a dynamical system, keeping the control parameter fixed. In this section we shall deal with external phase transitions of *second order*. Just as in conventional statistical mechanics, these phase transitions are characterized by the occurrence of *critical exponents*.

A simple example is again the *logistic map* near the accumulation point $\mu_\infty$ of period doubling. It is well known that the Liapunov exponent $\lambda = \lambda(\mu)$ scales near $\mu_\infty$ as

$$\lambda \sim |\mu - \mu_\infty|^\gamma, \qquad \gamma = \ln 2/\ln \delta \tag{21.6.1}$$

where $\delta$ is the Feigenbaum constant (see section 1.3). In section 14.5 we showed that the Liapunov exponent $\Lambda(\beta = 0) = \lambda$ can be regarded as the metric expansion free energy of a system with inverse temperature $\beta = 0$. Thus the above scaling behaviour can be regarded as an external metric expansion phase transition with respect to $\mu$, occurring at $\beta = 0$. It is a phase transition of second order, since there is a critical exponent $\gamma$. This exponent is essentially given by the Feigenbaum constant $\delta$.

The Liapunov exponent is just one possible observable that can be used to describe the behaviour near $\mu_\infty$. Other useful 'order parameters' are period lengths $L(\mu)$ or correlation functions. All these quantities scale near $\mu_\infty$ in a nontrivial way, and the critical exponents can be related to the Feigenbaum constants (see, for example, Schuster (1984) and Schlögl (1987)).

Another external phase transition of second order is observed at the critical point $\mu_T$ of tangent bifurcation (see section 1.3). Approaching $\mu_T$ from below, the average length $\bar{L}$ of the laminar phase of intermittency scales with the parameter $\mu$ as

$$\bar{L} \sim |\mu - \mu_T|^\xi. \tag{21.6.2}$$

The critical exponent can be related to the order $z$ of the map $f(x) = 1 - \mu|x|^z$ (see Hu and Rudnick (1982) and Hirsch, Nauenberg and Scalapino (1982)):

$$\xi = 1/z - 1. \tag{21.6.3}$$

Critical exponents are also observed for dynamical systems with escape: near a critical parameter value $a_c$ where transient chaotic behaviour sets in (so called crisis points) the escape rate $\kappa$ scales as

$$\kappa \sim |a - a_c|^\eta \tag{21.6.4}$$

(see Grebogi, Ott and Yorke (1986) and Grebogi, Ott, Romeiras and Yorke (1987) for details). As $\kappa = -\mathscr{P}(1)$, this behaviour can be interpreted as a topological expansion phase transition with respect to the external control parameter $a$, occurring for the inverse temperature $\beta = 1$.

A large number of further scaling laws, for different maps and different 'order parameters' are known in nonlinear dynamics. Reporting on all of them would by far exceed the scope of this book. There is, however, one important feature shared by all of these phenomena: usually the critical exponents do not depend on the details of the map. For example, the Feigenbaum constants only depend on the order of the maximum of the map and nothing else. A similar generality is observed for the other critical exponents as well. Thus, as in conventional statistical mechanics, the critical exponents are *universal.* Just as in conventional statistical mechanics an explanation of this universality is provided by a powerful mathematical tool, namely the renormalization group approach. This method also yields numerical values of the exponents.

## *21.7 Renormalization group approach

We shall sketch the renormalization group method of nonlinear dynamics for just one example that plays a central role in this book, namely the Feigenbaum attractor (Feigenbaum 1978, 1979; Coullet and Tresser 1978; Lanford 1982). It should, however, be clear that renormalization group methods have been successfully applied to a variety of other maps as well (see, for example, Feigenbaum, Kadanoff and Shenker (1982), Rand *et al.* (1982), Hirsch *et al.* (1982), Hu and Rudnick (1982), MacKay (1983), Baesens (1991)).

The key observation is that for $\mu \approx 1.4$ the twice iterated logistic map $f^2(x)$ has a certain similarity with the original map $f(x)$, provided we put the graph of $f^2$ upside down, rescale it in an appropriate way and restrict ourselves to the vicinity of $x = 0$. Of course, this similarity is just fulfilled in an approximate way, but one can ask the following question: does a function $g(x)$ exist such that the rescaled twice iterated function $g^2(x)$ is *exactly* equal to the original function $g(x)$? This is expressed by the equation

$$\alpha g(g(x/\alpha)) = g(x), \tag{21.7.1}$$

where $\alpha$ denotes the 'rescaling factor'. *A priori*, this equation can have many solutions, but we are interested in a solution that is an analytic function in $x^2$, i.e., of the form

$$g(x) = \sum_{k=0}^{\infty} c_k x^{2k}, \tag{21.7.2}$$

where the $c_k$ are appropriate coefficients. If we choose $c_0 = 1$, the ansatz (21.7.2) guarantees that $g(0) = 1$ and $g'(0) = 0$, i.e., in the vicinity of $x = 0$ the function $g$ has the same properties as the logistic map $f$.

With the ansatz (21.7.2) the renormalization group equation (21.7.1) can be solved numerically. The first few coefficients $c_k$ are obtained as

$$c_k = 1, \; -1.5276330, \; +0.1048152, \; +0.0267057, \; -0.0035274, \; +0.0000816$$
$$(k = 0, \quad 1, \quad 2, \quad 3, \quad 4, \quad 5, \quad ).$$

Eq. (21.7.1) yields for the special choice $x = 0$ the rescaling factor $\alpha = 1/g(1) = -2.5029079$.

The graphs of $g$, $g^2$, $g^4$, $g^8$, $g^{16}$, and $g^{32}$ are plotted in fig. 21.6. The

Fig. 21.6 Feigenbaum's universal function $g(x)$ and the iterates $g^{2^k}(x)$, (*a*) $k = 0$, (*b*) 1.

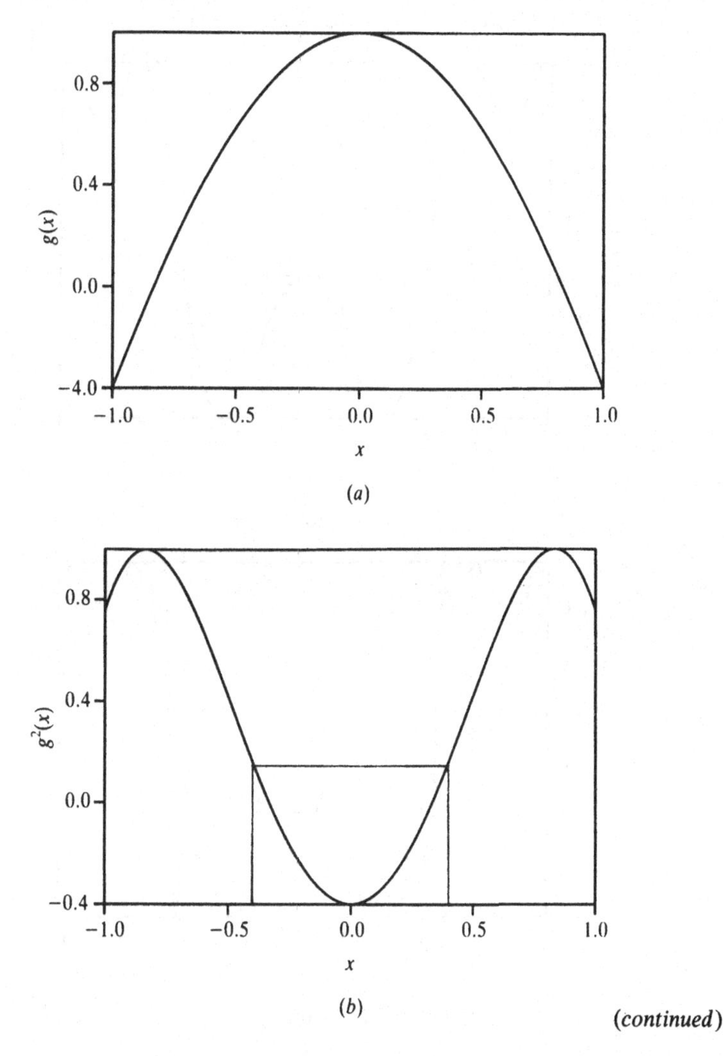


(*continued*)

self-similarity in the marked boxes is evident. According to eq. (21.7.1), we can interpret $g$ as the fixed point of an operator $R$, the so called *doubling operator* defined by

$$Rf(x) = \alpha f(f(x/\alpha)) = \alpha f^2(x/\alpha). \tag{21.7.3}$$

It should be clear that this fixed point is an entire fixed point function, it should not be confused with the fixed points of a map in the phase space, which were mentioned in section 1.1. Applying the doubling operator

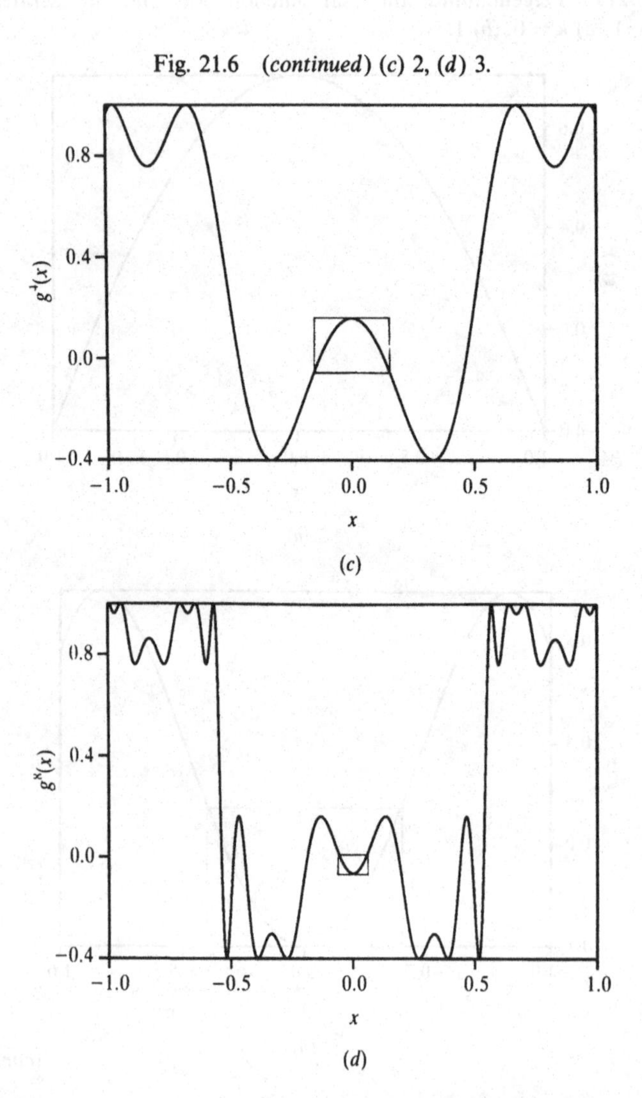


Fig. 21.6 (*continued*) (*c*) 2, (*d*) 3.

$k$-times to some function $f(x)$, we obtain

$$R^k f(x) = \alpha^k f^{2^k}(x/\alpha^k). \tag{21.7.4}$$

The fixed point $g$ satisfies

$$R^k g(x) = \alpha^k g^{2^k}(x/\alpha^k) = g(x). \tag{21.7.5}$$

Thus in the boxes marked in fig. 21.6 we see exactly the graph of the original function $g$ rescaled by a factor $\alpha^k$. As $\alpha$ is negative, for odd $k$ we have to turn the graph upside down.

Fig. 21.6 (*continued*) (*e*) 4, (*f*) 5.

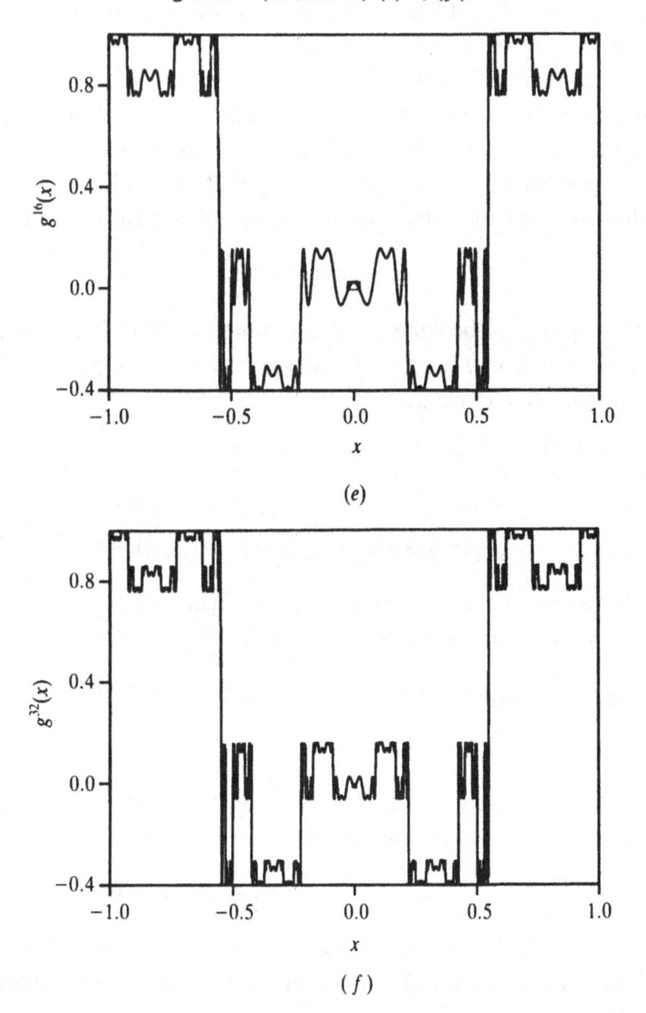

The idea behind universality is that at the critical point $\mu_\infty$ of period doubling entire classes of single humped maps $f$ with a quadratic maximum converge to the function $g$ under successive iteration and rescaling:

$$\lim_{k\to\infty} R^k f(x) = g(x), \qquad \mu = \mu_\infty. \tag{21.7.6}$$

Here $f$ can be either the logistic map, or any other map with a quadratic maximum, with the parameter value $\mu = \mu_\infty$ chosen such that one is at the critical point of period doubling accumulation. The fixed point $g$, obtained in the limit $k \to \infty$, as well as the Feigenbaum constant $\alpha$ are universal in the sense that they do not depend on further details of the map $f$. The critical value $\mu_\infty$ of the parameter $\mu$, however, is nonuniversal and depends on the precise form of the map.

Not only is the behaviour at the critical point $\mu_\infty$ of period doubling universal, but also the behaviour in the vicinity of $\mu_\infty$. Let us consider some map $f$ with a parameter value $\mu = \tilde{\mu}_k$ chosen in such a way that there is a superstable periodic orbit of length $2^k$. If $k$ is large, $f$ will only slightly deviate from the universal function $g$. Hence we may write

$$f(x) = g(x) + \varepsilon h(x), \tag{21.7.7}$$

where $\varepsilon$ is supposed to be small, and $h$ is some function. Iterating $f$ twice we obtain the function $f^2$, which possesses a superstable period of length $2^{k-1}$. From eq. (21.7.7) we get

$$\begin{aligned} f(f(x)) &= f(g(x) + \varepsilon h(x)) \\ &= g(g(x) + \varepsilon h(x)) + \varepsilon h(g(x) + \varepsilon h(x)) \\ &= g(g(x)) + \varepsilon h(x) g'(g(x)) + \varepsilon h(g(x)) + O(\varepsilon^2). \end{aligned} \tag{21.7.8}$$

Here we have used Taylor's formula, and the symbol $O(\varepsilon^2)$ denotes terms of order $\varepsilon^2$, which can be neglected. Thus, from eq. (21.7.3) we obtain

$$\begin{aligned} Rf(x) &= \alpha g(g(x/\alpha)) + \varepsilon\{\alpha h(x/\alpha) g'(g(x/\alpha)) + \alpha h(g(x/\alpha))\} \\ &= g(x) + \varepsilon DRh(x). \end{aligned} \tag{21.7.9}$$

In the last step we used eq. (21.7.1); moreover, we defined the operator $DR$, the so called 'linearized doubling operator', by

$$DRh(x) = \alpha h(x/\alpha) g'(g(x/\alpha)) + \alpha h(g(x/\alpha)). \tag{21.7.10}$$

So far the function $h$ has been arbitrary. An important simplification, however, arises if we can find a function $h$ that is an eigenfunction of the linearized doubling operator $DR$:

$$DRh = \lambda h. \tag{21.7.11}$$

Here $\lambda$ is the corresponding eigenvalue of the operator. Feigenbaum has solved eq. (21.7.11) numerically. In fact, there are several eigenfunctions; however, there is only one eigenfunction with an eigenvalue $\lambda$ that satisfies $|\lambda| > 1$. This largest eigenvalue is given by

$$\lambda_{\max} = \delta = 4.6692011\ldots. \tag{21.7.12}$$

Thus we can interpret the Feigenbaum constant $\delta$ as the largest eigenvalue of the linearized doubling operator. This largest eigenvalue determines the universal behaviour in the vicinity of the period doubling accumulation point: as soon as we move from a superstable period $2^k$ to a superstable period $2^{k-1}$, the 'distance' from the limit point $\mu_\infty$ of period doubling accumulation (the distance from the universal function $g$) increases by a factor $\delta$:

$$|\mu_\infty - \tilde{\mu}_{k-1}| = \delta|\mu_\infty - \tilde{\mu}_k|, \qquad k \to \infty. \tag{21.7.13}$$

This equation is equivalent to eq. (1.3.7)

*Remark*: It is very important to specify the class of functions for which the renormalization group equations (21.7.1) and (21.7.11) will be valid. For example, one can easily verify that the function

$$\hat{g}(x) = (x^{1-z} - b)^{1/(1-z)}, \qquad z > 1, b = \text{const} \tag{21.7.14}$$

also satisfies eq. (21.7.1) with $\hat{\alpha} = 2^{1/(z-1)}$. This solution, however, is not relevant for the period doubling scenario, because it does not belong to the space of functions in which we are interested, namely single humped functions with a maximum of order $z = 2$. Nevertheless, there are applications for the function $\hat{g}$ as well, namely for the intermittency phenomenon. In fact, both renormalization group equations can be solved analytically in the case of intermittency (Hirsch *et al.* 1982; Hu and Rudnick 1982). Several useful results can be derived in this way, for example eq. (21.6.3).

## 21.6 External phase transitions of first order

There are also *first order* phase transitions with respect to an external control parameter. In this case the generalized free energy depends on the control parameter in a continuous, but nondifferentiable way. Certainly we do *not* have critical exponents in that case. A somewhat trivial, but illustrative example is the following two-dimensional map on the unit square:

$$f(x, y): \begin{cases} x_{n+1} = 2x_n \bmod 1 \\ y_{n+1} = (by_n + x_n) \bmod 1, \qquad b > 0. \end{cases} \tag{21.8.1}$$

This possesses the constant Jacobi matrix

$$Df(x, y) = \begin{pmatrix} 2 & 0 \\ 1 & b \end{pmatrix}. \tag{21.8.2}$$

The eigenvalues are

$$\eta_1 = 2, \qquad \eta_2 = b. \tag{21.8.3}$$

Thus the Liapunov exponents are

$$\lambda_1 = \ln 2, \qquad \lambda_2 = \ln b. \tag{21.8.4}$$

According to Pesin's identity (see section 15.5), which is valid in this trivial case, the KS entropy $h \equiv K(1)$ is given by the sum of the *positive* Liapunov exponents of the system. Thus

$$h(b) = \begin{cases} \ln 2 & \text{for } b \leqslant 1 \\ \ln 2 + \ln b & \text{for } b > 1. \end{cases} \tag{21.8.5}$$

At the critical point $b_c = 1$, the KS entropy is not differentiable with respect to $b$. As the Rényi entropies $K(\beta)$ are a kind of dynamical free energy, this phenomenon can be interpreted as an external dynamical phase transition of first order, occurring for the inverse temperature $\beta = 1$.

Our next example is an external *static* phase transition of first order, i.e., a phase transition concerning the Rényi dimension rather than the Rényi entropies. This is a good example since it turns out to have a distinguished physical meaning: it describes a generic transition scenario from complex chaotic to Gaussian random behaviour. Let us consider the two-dimensional map $f$ of Kaplan–Yorke type

$$f: \begin{cases} x_{n+1} = T(x_n) = 1 - 2x_n{}^2 \\ y_{n+1} = \lambda y_n + x_{n+1} \end{cases} \tag{21.8.6}$$

(see section 1.4 for a physical motivation to study this map). Fig. 21.7 shows the attractor of this map for various values $\lambda$. A qualitative change is observed for the critical parameter value $\lambda_c = \frac{1}{2}$. Here the attractor changes from a fractal object with noninteger dimension to an object with dimension 2. In fact, the information dimension of the attractor is given by

$$D(1, \lambda) = \begin{cases} 1 + \dfrac{\ln 2}{|\ln \lambda|} & \text{for } \lambda \leqslant \frac{1}{2} \\ 2 & \text{for } \lambda > \frac{1}{2}. \end{cases} \quad (21.8.7)$$

This is a consequence of the Kaplan–Yorke conjecture (Kaplan and Yorke 1979), which quite generally provides us with a relation between the Liapunov exponents of a map and the information dimension of the

Fig. 21.7 Attractor of the map (21.8.6) for (*a*) $\lambda = 0.3$, (*b*) 0.5, (*c*) 0.7, (*d*) 0.9.

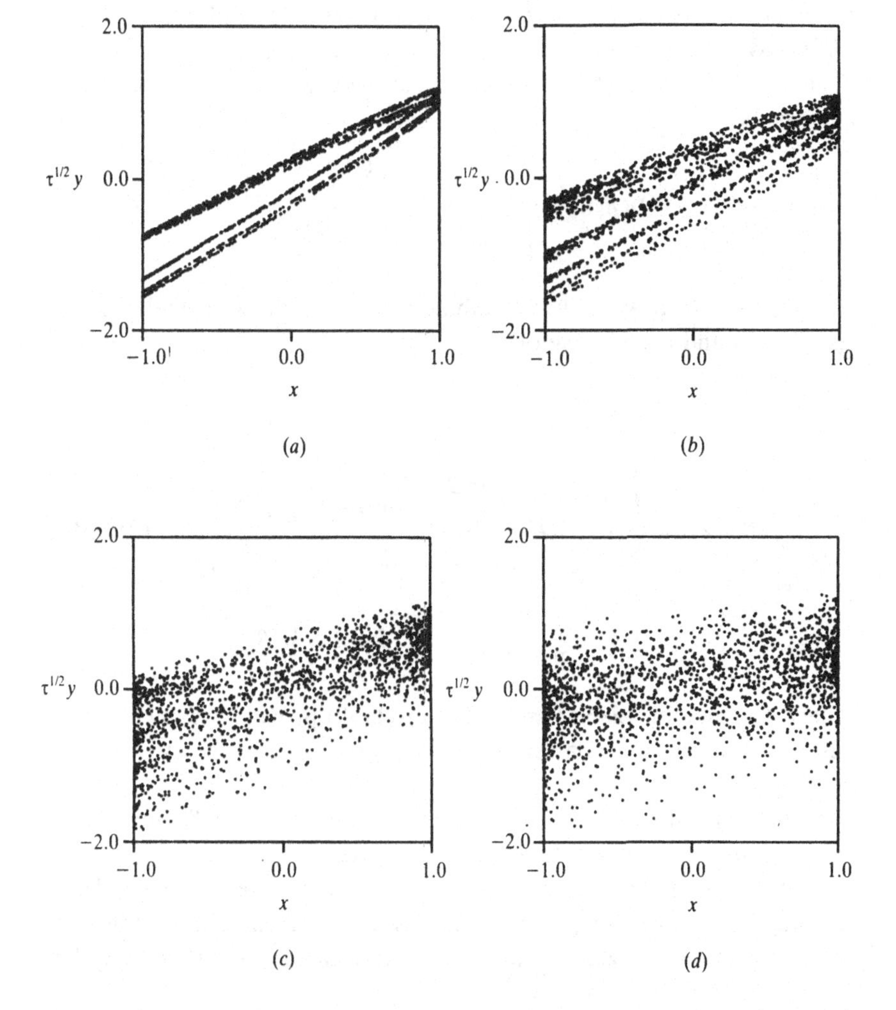

attractor. For general two-dimensional maps of type (1.4.1) it reads

$$D(1,\lambda) = \begin{cases} 1 + \dfrac{\chi}{|\ln \lambda|} & \text{for } \lambda \leqslant \exp(-\chi) \\ 2 & \text{for } \lambda > \exp(-\chi). \end{cases} \tag{21.8.8}$$

where $\chi$ is the (positive) Liapunov exponent of the dynamics $T$. The Kaplan–Yorke conjecture has been proved by Young (1982) under quite general assumptions for a two-dimensional diffeomorphism. Although maps of Kaplan–Yorke type are not diffeomorphisms, eq. (21.8.8) is believed to be true in this case as well, and this is well supported by numerical experiments. Hence, in general, the critical point $\lambda_c$ where the attractor changes from a fractal to a two-dimensional object is given by $\lambda_c = \exp(-\chi)$. At this point $D(1, \lambda)$ is not differentiable with respect to $\lambda$.

We may even determine the Rényi dimensions $D(\beta, \lambda)$ of the attractor of the map $f$. For $\lambda = 0$ this map reduces to the Ulam map with the Rényi dimensions

$$D(\beta, 0) = \begin{cases} 1 & \text{for } \beta \leqslant 2 \\ \dfrac{\beta}{2(\beta - 1)} & \text{for } \beta \geqslant 2 \end{cases} \tag{21.8.9}$$

(see eq. (13.3.7)). If we naively combine the $\beta$- and the $\lambda$-dependences of eqs. (21.8.7) and (21.8.9), we obtain

$$D(\beta, \lambda) = \begin{cases} 1 + \dfrac{\ln 2}{|\ln \lambda|} & \text{for } \lambda \leqslant \frac{1}{2},\ \beta \leqslant 2 \\ \dfrac{\beta}{2(\beta - 1)} + \dfrac{\ln 2}{|\ln \lambda|} & \text{for } \lambda \leqslant \frac{1}{2},\ \beta \geqslant 2 \\ 2 & \text{for } \lambda \geqslant \frac{1}{2},\ \beta \leqslant 2 \\ 1 + \dfrac{\beta}{2(\beta - 1)} & \text{for } \lambda \geqslant \frac{1}{2},\ \beta \geqslant 2. \end{cases} \tag{21.8.10}$$

In fact, Chennaoui, Liebler and Schuster (1990) provide arguments that this is the correct formula for the Rényi dimensions of the attractor of the map $f$.

At the critical point $\lambda_c = \frac{1}{2}$ the Rényi dimensions $D(\beta, \lambda)$ are not differentiable with respect to $\lambda$ for all $\beta$. Moreover, at $\beta_c = 2$, $D(\beta, \lambda)$ is not differentiable with respect to $\beta$ for all $\lambda$. Hence there are *critical lines* defining first order static phase transitions (one external and one internal phase transition). In general, for more complicated maps there may be much more complicated phase diagrams. According to the Kaplan–Yorke conjecture, external phase transitions of the above kind with respect to $\lambda$ are generic and quite a typical phenomenon (Kaplan and Yorke 1979; Badii *et al.* 1988).

*Remark*: The notion 'critical line' just means that there is non-analytical behaviour. Following the analogy with conventional thermodynamic systems, we should call such a line merely a 'transition line', the crossing of which is connected with a phase transition. In the gas–liquid system for instance, that point in the phase diagram where the coexistence of two different phases ceases is called the 'critical point'. At this point the transition line, i.e., the coexistence line of gas and liquid ends.

Let us now come to the physical interpretation of the external phase transition (Beck 1991c). As described in section 1.4, the $y$-variable of the Kaplan–Yorke map can be interpreted as the stroboscopic velocity of a kicked damped particle. The map $T$ determines the dynamics of the kicks. The parameter $\lambda$ is related to the time difference $\tau$ between kicks by

$$\lambda = \exp(-\gamma\tau), \tag{21.8.11}$$

where $\gamma$ is the viscosity of the damping medium. It has been proved that for $\tau \to 0$ the rescaled process $\tau^{1/2} y_n$ converges to a Gaussian stochastic process, the so called Ornstein–Uhlenbeck process or dynamical Brownian motion, provided the map $T$ satisfies certain strong mixing properties (Beck and Roepstorff 1987a; Beck 1991b). Thus decreasing $\tau$ further and further ($\lambda \to 1$), we have a transition scenario changing from complex chaotic to Gaussian stochastic behaviour. We recognize that this transition scenario does not take place in a 'smooth' way, but there is a critical time scale

$$\tau_c = -\frac{1}{\gamma} \ln \lambda_c = \frac{\chi}{\gamma}, \tag{21.8.12}$$

where the corresponding strange attractor loses its fractal structure. Of course, this loss of fractal properties is necessary in order to approach a smooth Gaussian distribution in the limit $\tau \to 0$. What is surprising is the fact that the loss takes place at a finite time scale $\tau_c$ rather than in the limit $\tau \to 0$. According to the Kaplan–Yorke conjecture, this appears to be a generic phenomenon for an arbitrary chaotic dynamics $T$ of the kicks.

Besides the Rényi dimensions of the attractor, there are further interesting quantities that exhibit nonanalytic behaviour with respect to the parameter $\lambda$ (or $\tau$). One of them is the relaxation time $v$, which in general describes the decay rate of correlations of the iterates of a map. It is related to the second largest eigenvalue $\eta_1$ of the Perron–Frobenius operator by $|\eta_1| = \exp(-1/\nu)$ (see section 17.3). In Beck (1990c) it was proved that under quite general assumptions there is a critical parameter value $\lambda_{crit}$ (respectively $\tau_{crit}$), where the relaxation time $\nu(\lambda)$ of the map $f$ is not differentiable with respect to the control parameter $\lambda$. This is given by

$$\lambda_{crit} = \exp(-\gamma\tau_{crit}) = \exp(-1/\xi). \tag{21.8.13}$$

Here $\xi$ denotes the relaxation time of the map $T$. In other words, the critical time scale where the relaxation time $\nu$ exhibits a kind of phase transition behaviour is given by

$$\tau_{crit} = 1/(\xi\gamma). \tag{21.8.14}$$

In general, the phase transition point $\lambda_c$ of the Rényi dimensions does not coincide with the phase transition point $\lambda_{crit}$ of the relaxation time, although there are several examples where this is the case. For instance, $\lambda_{crit} = \lambda_c$ for $T(x) = 1 - 2x^2$. However, $\lambda_c \neq \lambda_{crit}$ for the continuous fraction map (Beck 1991c).

# References


R. Adler, B. Weiss, *Proc. Nat. Acad. Sci. USA* **57**, 1573 (1967).

V. M. Alekseev, M. V. Yakobson, *Phys. Rep.* **75**, 287 (1981).

A. Arneodo, M. Holschneider, *J. Stat. Phys.* **50**, 995 (1988).

V. I. Arnol'd, A. Avez, *Ergodic Problems of Classical Mechanics*, Benjamin, New York (1968).

R. Artuso, E. Aurell, P. Cvitanović, *Nonlinearity* **3**, 325, 361 (1990).

R. Badii, A. Politi, *Phys. Rev.* **35A**, 1288 (1987).

R. Badii, G. Broggi, B. Derighetti, M. Ravani, S. Ciliberto, A. Politi, M. A. Rubio, *Phys. Rev. Lett.* **60**, 979 (1988).

R. Badii, *Nuovo Cimento* **10**, 819 (1988).

C. Baesens, *Physica* **53D**, 319 (1991).

V. Baladi, J.-P. Eckmann, D. Ruelle, *Nonlinearity* **2**, 119 (1989).

V. Baladi, G. Keller, *Commun. Math. Phys.* **127**, 459 (1990).

V. Baladi, *J. Stat. Phys.* **62**, 239 (1991).

M. Barnsley, J. S. Geronimo, A. N. Harrington, *Commun. Math. Phys.* **88**, 479 (1983).

M. Barnsley, *Fractals Everywhere*, Academic Press, Boston (1988).

R. J. Baxter, *Exactly Solved Models in Statistical Mechanics*, Academic Press, London (1982).

C. Beck, G. Roepstorff, *Physica* **145A**, 1 (1987a).

C. Beck, G. Roepstorff, *Physica* **25D**, 173 (1987b).

C. Beck, *Phys. Lett.* **136A**, 121 (1989).

C. Beck, *Physica* **41D**, 67 (1990a).

C. Beck, *Physica* **169A**, 324 (1990b).

C. Beck, *Commun. Math. Phys.* **130**, 51 (1990c).

C. Beck, G. Roepstorff, *Physica* **165A**, 270 (1990).

C. Beck, *Physica* **50D**, 1 (1991a).

C. Beck, *Nonlinearity* **4**, 1131 (1991b).

C. Beck, Transitions from Chaotic to Brownian Motion Behaviour, in *Solitons and Chaos*, ed. I. Antoniou, F. Lambert, Springer, Berlin (1991c).

C. Beck, Digital machines iterating chaotic maps: Roundoff induced periodicity, in *Treizième Colloque sur le Traitement du Signal et des Images*, GRETSI, Juan les Pins (1991d).

C. Beck, G. Roepstorff, C. Schroer, *Driven Chaotic Motion in Single- and Double-Well Potentials*, preprint, RWTH Aachen (1992).

C. Beck, D. Graudenz, *Phys. Rev.* **46A**, 6265 (1992).

T. Bedford, M. Keane, C. Series (eds.), *Ergodic Theory, Symbolic Dynamics and Hyperbolic Spaces*, Oxford University Press, Oxford (1991).

J. Bene, P. Szépfalusy, A. Fülöp, *Phys. Rev.* **40A**, 6719 (1989).

R. Benzi, G. Paladin, G. Parisi, A. Vulpiani, *J. Phys.* **17A**, 3521 (1984).

D. Bessis, G. Paladin, G. Turchetti, S. Vaienti, *J. Stat. Phys.* **51**, 109 (1988).

P. Billingsley, *Convergence of Probability Measures*, Wiley, New York (1968).

T. Bohr, D. Rand, *Physica* **25D**, 387 (1987).

T. Bohr, M. H. Jensen, *Phys. Rev.* **36A**, 4904 (1987).

T. Bohr, T. Tél, The Thermodynamics of Fractals, in *Directions in Chaos* **2**, ed. Hao Bai-Lin, World Scientific, Singapore (1988).

R. Bowen, *Amer. J. Math.* **92**, 725 (1970).

R. Bowen, *Equilibrium States and the Ergodic Theory of Anosov Diffeomorphisms*, Lecture Notes in Mathematics **470**, Springer, New York (1975).

R. Bowen, *On Axiom A diffeomorphisms*, in CBMS Conference Series **35**, AMS (1978).

D. K. Campbell (ed.), *Chaos* **2**, number 1, *Chaos focus issue on periodic orbit theory* (1992).

A. Chennaoui, J. Liebler, H. G. Schuster, *J. Stat. Phys.* **59**, 1311 (1990).

A. B. Chhabra, C. Meneveau, R. V. Jensen, K. R. Sreenivasan, *Phys. Rev.* **40A**, 5284 (1989).

F. Christiansen, G. Paladin, H. H. Rugh, *Phys. Rev. Lett.* **65**, 2087 (1990).

P. Collet, J.-P. Eckmann, *Iterated Maps on the Interval as Dynamical Systems*, Birkhäuser, Basel (1980).

P. Collet, J. Lebowitz, A. Porzio, *J. Stat. Phys.* **47**, 609 (1987).

P. Coullet, J. Tresser, *C.R. Hebd. Séances Acad. Sci. Series* **A 287**, 577 (1978a).
P. Coullet, J. Tresser, *J. Phys.* **C5**, 25 (1978b).
I. P. Cornfeld, S. V. Fomin, Ya. G. Sinai, *Ergodic Theory*, Springer, New York (1982).
I. Csiszár, *Magyar Tud. Akad. Mat. Kutato Int. Közl.* **8**, 35 (1963).
I. Csiszár, *Stud. Sci. Math. Hung.* **2**, 299 (1967).
A. Csordás, P. Szépfalusy, *Phys. Rev.* **39A**, 4767 (1989a).
A. Csordás, P. Szépfalusy, *Phys. Rev.* **40A**, 2221 (1989b).
C. D. Cutler, *J. Stat. Phys.* **62**, 651 (1991).
P. Cvitanović (ed.), *Universality in Chaos*, Adam Hilger, Bristol (1984).
P. Cvitanović, in: *XV. Int. Coll. on Group Theoretical Methods in Physics*, ed. G. Gilmore, World Scientific, Singapore (1987).
P. Cvitanović, *Phys. Rev. Lett.* **61**, 2729 (1988).
P. Cvitanović, J. Myrheim, *Commun. Math. Phys.* **121**, 225 (1989).
B. Derrida, A. Gervois, Y. Pomeau, *J. Phys.* **12A**, 269 (1979).
R. L. Devaney, *An Introduction to Chaotic Dynamical Systems*, Benjamin/Cummings, Menlo Park (1986).
J. L. Doob, *Stochastic Processes*, Wiley, New York (1953).
M. Dörfle, *J. Stat. Phys.* **40**, 93 (1985).
P. G. Drazin, G. P. King (eds.), Interpretation of Time Series from Nonlinear Systems, *Physica* **58D**, 1 (1992).
F. J. Dyson, *Commun. Math. Phys.* **12**, 91, 212 (1969).
B. Eckhardt, *Physica* **33D**, 89 (1988).
J.-P. Eckmann, D. Ruelle, *Rev. Mod. Phys.* **57**, 617 (1985).
J.-P. Eckmann, I. Procaccia, *Phys. Rev.* **34A**, 659 (1986).
D. F. Escande, *Phys. Rep.* **121**, 165 (1985).
K. J. Falconer, *J. Phys.* **21A**, L737 (1988).
J. Feder, *Fractals*, Plenum Press, New York (1988).
M. J. Feigenbaum, *J. Stat. Phys.* **19**, 25 (1978).
M. J. Feigenbaum, *J. Stat. Phys.* **21**, 669 (1979).
M. J. Feigenbaum, *Commun. Math. Phys.* **77**, 65 (1980).
M. J. Feigenbaum, L. P. Kadanoff, S. J. Shenker, *Physica* **5D**, 370 (1982).
M. J. Feigenbaum, M. H. Jensen, I. Procaccia, *Phys. Rev. Lett.* **57**, 1503 (1986).
M. J. Feigenbaum, *J. Stat. Phys.* **46**, 919, 925 (1987).
M. J. Feigenbaum, *J. Stat. Phys.* **52**, 527 (1988).
M. J. Feigenbaum, I. Procaccia, T. Tél, *Phys. Rev.* **39A**, 5359 (1989).

M. I. Freidlin, A. D. Wentzell, *Random Perturbations of Dynamical Systems*, Springer, Berlin (1984).
U. Frisch, G. Parisi, in *Turbulence and Predictability of Geophysical Flows and Climate Dynamics*, eds. N. Ghil, R. Benzi, G. Parisi, North-Holland, Amsterdam (1985).
F.-J. Fritz, B. Huppert, W. Willems, *Stochastische Matrizen*, Springer, Berlin (1979).
H. Fujisaka, *Progr. Theor. Phys.* **70**, 1264 (1983).
H. Fujisaka, M. Inoue, *Prog. Theor. Phys.* **77**, 1334 (1987).
F. R. Gantmacher, *Applications of the Theory of Matrices*, Interscience, New York (1959).
P. Gaspard, G. Nicolis, *Phys. Rev. Lett.* **65**, 1693 (1990).
B. W. Gnedenko, *Lehrbuch der Wahrscheinlichkeitsrechnung*, Akademie Verlag, Berlin (1968).
I. S. Gradshteyn, I. M. Ryzhik, *Table of Integrals, Series, and Products*, Academic Press, New York (1965).
P. Grassberger, *J. Stat. Phys.* **19**, 25 (1981).
P. Grassberger, I. Procaccia, *Phys. Rev.* **28A**, 2591 (1983a).
P. Grassberger, I. Procaccia, *Phys. Rev. Lett.* **50**, 364 (1983b).
P. Grassberger, I. Procaccia, *Physica* **13D**, 34 (1984).
P. Grassberger, *Phys. Lett.* **97A**, 227 (1983).
P. Grassberger, *Phys. Lett.* **107A**, 101 (1985).
P. Grassberger, Estimating the Fractal Dimensions and Entropies of Strange Attractors, in *Chaos*, ed. A. V. Holden, Manchester University Press (1986a).
P. Grassberger, *Int. J. Theor. Phys.* **25**, 907 (1986b).
P. Grassberger, R. Badii, A. Politi, *J. Stat. Phys.* **51**, 135 (1988).
P. Grassberger, *Z. Naturforsch.* **43A**, 671 (1988).
P. Grassberger, H. Kantz, U. Moenig, *J. Phys.* **22A**, 5217 (1989).
C. Grebogi, E. Ott, J. A. Yorke, *Phys. Rev. Lett.* **57**, 1284 (1986).
C. Grebogi, E. Ott, F. Romeiras, J. A. Yorke, *Phys. Rev.* **36A**, 5365 (1987).
R. B. Griffiths, Rigorous Results and Theorems, in *Phase Transitions and Critical Phenomena* **1**, ed. C. Domb, M. S. Green, Academic Press, London (1972).
S. Grossmann, S. Thomae, *Z. Naturforsch.* **32a**, 1353 (1977).
J. Guckenheimer, P. Holmes, *Nonlinear Oscillations, Dynamical Systems, and Bifurcations of Vector Fields*, Springer, New York (1983).
G. Györgyi, P. Szépfalusy, *J. Stat. Phys.* **34**, 451 (1984).
T. C. Halsey, M. H. Jensen, L. P. Kadanoff, I. Procaccia, B. I. Shraiman, *Phys. Rev.* **33A**, 1141 (1986).

Hao Bai-Lin, *Chaos*, World Scientific, Singapore (1984).
G. N. Hatsopoulos, H. J. Keeman, *Principles of General Thermodynamics*, Wiley, New York (1965).
N. T. A. Haydn, *Ergod. Th. Dynam. Sys.* **7**, 119 (1987).
J. F. Heagy, *Physica* **57D**, 436 (1992).
M. Hénon, *Commun. Math. Phys.* **50**, 69 (1976).
H. G. E. Hentschel, I. Procaccia, *Physica* **8D**, 435 (1983).
J. E. Hirsch, M. Nauenberg, D. J. Scalapino, *Phys. Lett.* **87A**, 391 (1982).
B. Hu, J. M. Mao, *Phys. Rev.* **25A**, 3259 (1982).
B. Hu, J. Rudnick, *Phys. Rev. Lett.* **48**, 1645 (1982).
E. Ising, *Z. Physik* **31**, 253 (1925).
S. Iyanaga, Y. Kawada (eds.), *Encyclopedic Dictionary of Mathematics*, Vol. II, MIT Press, Cambridge (1977).
E. T. Jaynes, *Phys. Rev.* **106**, 620 (1957).
E. T. Jaynes, *Phys. Rev.* **118**, 171 (1961a).
E. T. Jaynes, *Ann. J. Phys.* **31**, 66 (1961b).
E. T. Jaynes, *Information Theory and Statistical Mechanics*, Brandeis Lectures **3**, 181 (1962).
M. H. Jensen, L. P. Kadanoff, I. Procaccia, *Phys. Rev.* **36A**, 1409 (1987).
R. V. Jensen, C. R. Oberman, *Phys. Rev. Lett.* **46**, 1547 (1981).
G. Julia, *J. Math. Pures et Appl.* **4**, 47 (1918).
H. Jürgens, H.-O. Peitgen, D. Saupe, Fraktale – eine neue Sprache für komplexe Strukturen, *Spektrum der Wissenschaft* (Sept.), 52 (1989).
L. P. Kadanoff, C. Tang, *Proc. Natl. Acad. Sci. USA* **81**, 1276 (1984).
H. Kantz, P. Grassberger, *Physica* **17D**, 75 (1985).
J. L. Kaplan, J. A. Yorke, Chaotic behaviour of multidimensional difference equations, in *Functional Differential Equations and Approximation of Fixed Points*, eds. H. O. Peitgen and H. O. Walther, Lecture Notes in Mathematics **730**, 204; Springer, Berlin (1979).
J. L. Kaplan, J. Mallet-Paret, J. A. Yorke, *Ergod. Th. Dynam. Syst.* **4**, 261 (1984).
D. Katzen, I. Procaccia, *Phys. Rev. Lett.* **58**, 1169 (1987).
Z. Kaufmann, P. Szépfalusy, *Phys. Rev.* **40A**, 2615 (1989).
Z. Kaufmann, *Physica* **54D**, 75 (1992).
A. I. Khinchin, *Mathematical Foundations of Information Theory*, Dover Publ., New York (1957).
M. Kohmoto, *Phys. Rev.* **37A**, 1345 (1987).

A. N. Kolmogorov, *Dokl. Akad. Sci. USSR* **119**, 861 (1958).

Z. Kovács, *J. Phys.* **22A**, 5161 (1989).

Z. Kovács, T. Tél, *Phys. Rev.* **45A**, 2270 (1992).

S. Kullback, *Annals Mathem. Statistics* **22**, 79 (1951a).

S. Kullback, *Information Theory and Statistics*, Wiley, New York (1951b).

O. E. Lanford, *Bull. Am. Math. Soc.* **6**, 427 (1982).

A. Lasota, M. Mackey, *Probabilistic Properties of Deterministic Systems*, Cambridge University Press, Cambridge (1985).

F. Ledrappier, M. Misiurewicz, *Ergod. Th. Dynam. Sys.* **5**, 595 (1985).

F. Ledrappier, L. S. Young. *Ann. Math.* **122**, 509, 540 (1985).

T.-Y. Li, *J. Approx. Theory* **17**, 177 (1976).

A. J. Lichtenberg, M. A. Liebermann, *Regular and Stochastic Motion*, Springer, Berlin (1983).

E. N. Lorenz, *J. Atmos. Sci.* **20**, 130 (1963).

S.-K. Ma, *Modern Theory of Critical Phenomena*, Benjamin, London (1976).

R. S. MacKay, *Physica* **7D**, 283 (1983).

R. S. MacKay, J. D. Meiss, I. C. Percival, *Physica* **13D**, 55 (1984).

B. B. Mandelbrot, *J. Fluid Mech.* **62**, 331 (1974).

B. B. Mandelbrot, *Ann. N. Y. Acad. Sci.* **357**, 249 (1980).

B. B. Mandelbrot, *The Fractal Geometry of Nature*, Freeman, San Francisco (1982).

M. Markosová, P. Szépfalusy, *Phys. Rev.* **43A**, 2709 (1991).

R. M. May, *Nature* **261**, 459 (1976).

D. H. Mayer, *The Ruelle–Araki Transfer Operator in Classical Statistical Mechanics*, Springer, New York (1980).

D. H. Mayer, G. Roepstorff, *J. Stat. Phys.* **31**, 309 (1983).

D. H. Mayer, *Phys. Lett.* **121A**, 390 (1987).

D. H. Mayer, G. Roepstorff, *J. Stat. Phys.* **47**, 149 (1987).

D. H. Mayer, G. Roepstorff, *J. Stat. Phys.* **50**, 331 (1988).

D. H. Mayer, *Commun. Math. Phys.*, **130**, 311 (1990).

N. Metropolis, M. L. Stein, P. R. Stein, *J. Combinat. Th.* **15**, 25 (1973).

M. R. Michalski, *Perturbation Theory for the Topological Pressure in Analytic Dynamical Systems*, Thesis, Virginia Polytechnic Institute and State University (1990).

J. Naas, H. L. Schmidt (ed.), *Mathematisches Wörterbuch*, Akademie Verlag, Berlin (1972).

E. Nelson, *Dyamical Theories of Brownian Motion*, Princeton University Press, Princeton (1967).

J. S. Nicolis, G. Mayer-Kress, G. Haubs, *Z. Naturf.* **36a**, 1157 (1983).
L. Onsager, *Phys. Rev.* **65**, 117 (1944).
V. I. Oseledec, *Moscow Math. Soc.* **19**, 197 (1968).
E. Ott, W. Withers, J. A. Yorke, *J. Stat. Phys.* **36**, 697 (1984).
G. Paladin, A. Vulpiani, *Phys. Rep.* **156**, 147 (1987).
P. Paoli, A. Politi, R. Badii, *Physica* **36D**, 263 (1989).
W. Parry, *Trans. AMS* **112**, 55 (1964).
K. Pawelzik, H. G. Schuster, *Phys. Rev.* **35A**, 481 (1987).
H.-O. Peitgen, P. H. Richter, *The Beauty of Fractals*, Springer, Berlin (1986).
I. C. Percival, D. Richards, *Introduction to Dynamics*, Cambridge University Press, London (1982).
Ya. Pesin, *Russ. Math. Surveys* **32**, 55 (1977).
H. Poincaré, *Les Méthodes Nouvelles de la Méchanique Celeste*, Gauthier-Villars, Paris (1892).
M. Pollicott, *J. Stat. Phys.* **62**, 257 (1991).
T. Prellberg, J. Slawny, *J. Stat. Phys.* **66**, 503 (1992).
D. Rand, S. Ostlund, J. Sethna, E. D. Siggia, *Phys. Rev. Lett.* **49**, 132 (1982).
D. Rand, *Ergod. Th. Dynam. Sys.* **9**, 527 (1989).
A. Rényi, On Measures of Entropy and Information, in *Proc. 4th Berkeley Symp. on Math. Stat. Prob.* 1960 Vol. 1, University of California Press, Berkeley, Los Angeles, 1961, 547 (1960).
A. Rényi, *Probability Theory*, North-Holland, Amsterdam (1970).
G. Roepstorff, *Pfadintegrale in der Quantenphysik*, Vieweg, Braunschweig (1991).
E. Ruch, *Theor. Chim. Acta* **38**, 167 (1975).
D. Ruelle, *Commun. Math. Phys.* **9**, 267 (1968).
D. Ruelle, *Statistical Mechanics: Rigorous Results*, Benjamin, New York (1969).
D. Ruelle, *Thermodynamic Formalism*, Addison-Wesley, Reading (1978a).
D. Ruelle, *Bol. Soc. Bras. Mat.* **9**, 83 (1978b).
D. Ruelle, *Ergod. Th. Dynam. Sys.* **2**, 99 (1982).
D. Ruelle, *Commun. Math. Phys.* **125**, 239 (1989).
G. Ruppeiner, *Phys. Rev.* **27A**, 1116 (1983).
P. Salamon, J. Nulton, E. Ihrig, *J. Chem. Phys.* **80**, 436 (1984).
P. Salamon, J. Nulton, R. S. Berry, *J. Chem. Phys.* **82**, 2433 (1985).
M. Sano, S. Sato, M. Sawada, *Progr. Theor. Phys.* **76**, 945 (1986).
C. Schaefer, *Einführung in die Theoretische Physik* II, p. 440, De Gruyter, Berlin (1929).

F. Schlögl, On the Statistical Theory of Entropy Production, in *Statistical Mechanics – Foundations and Applications*, ed. T. A. Bak, Benjamin Inc., New York (1967).

F. Schlögl, *Z. Physik* **248**, 446 (1971).

F. Schlögl, *Z. Physik* **253**, 147 (1972).

F. Schlögl, *Phys. Rep.* **62**, 267 (1980).

F. Schlögl, *Z. Physik* **B52**, 51 (1983).

F. Schlögl, *Z. Physik* **B58**, 157 (1985).

F. Schlögl, *J. Stat. Phys.* **46**, 135 (1987).

F. Schlögl, E. Schöll, *Z. Physik* **B72**, 231 (1988).

F. Schlögl, *Probability and Heat*, Vieweg, Braunschweig (1989).

L. S. Schulman, *Techniques and Applications of Path Integration*, Wiley, New York (1981).

H. G. Schuster, *Deterministic Chaos*, Physik-Verlag, Weinheim (1984).

C. Shannon, W. Weaver, *The Mathematical Theory of Communication*, University of Illinois Press, Urbana (1948).

T. Shimizu, *Physica* **164A**, 123 (1990).

W. Sierpinski, Sur une courbe cantorienne qui contient une image biunivoque et continue de toute courbe donné, *Comptes Rendus* (*Paris*) **162**, 629 (1916). More detailed version in W. Sierpinski, *Oeuvres Choisies*, eds. S. Hartman *et al.*, Vol. II, Editions scientifiques de Pologne, Warsaw (1974), p 107.

Ya. G. Sinai, *Dokl. Akad. Sci. USSR* **124**, 768 (1959).

Ya. G. Sinai, *Funct. Anal. Appl.* **2**, 245 (1968).

Ya. G. Sinai, *Russ. Math. Surveys* **166**, 21 (1972).

Ya. G. Sinai (ed.) *Dynamical Systems* II, Springer, New York (1986).

Ya. G. Sinai, *Modern Problems of Ergodic Theory*, Princeton University Press, Princeton, to appear (1993).

U. Smilansky, in *Chaos and Quantum Physics*, eds. M.-J. Giannoni *et al*, Elsevier, New York (1990).

C. Sparrow, *The Lorenz Equations: Bifurcations, Chaos, and Strange Attractors*, Springer, New York (1982).

H. E. Stanley, *Introduction to Phase Transitions and Critical Phenomena*, Clarendon Press, Oxford (1971).

W.-H. Steeb, *A Handbook of Terms used in Chaos and Quantum Chaos*, BI-Wissenschaftsverlag, Mannheim (1991).

R. Stoop, J. Parisi, *Phys. Rev.* **43A**, 1802 (1991).

R. Stoop, J. Peinke, J. Parisi, *Physica* **50D**, 405 (1991).

P. Szépfalusy, T. Tél, *Phys. Rev.* **34A**, 2520 (1986).

P. Szépfalusy, T. Tél, *Phys. Rev.* **35A**, 477 (1987).

P. Szépfalusy, T. Tél, A. Csordás, Z. Kovács, *Phys. Rev.* **36A**, 3525 (1987).

P. Szépfalusy, T. Tél, G. Vattay, *Phys. Rev.* **43A**, 681 (1991).

T. J. Taylor, *On Stochastic and Chaotic Motion*, preprint, Arizona State University (1991).

T. Tél, *Phys. Lett.* **119A**, 65 (1986).

T. Tél, *Phys. Rev.* **36A**, 1502, 2507 (1987).

T. Tél, *Z. Naturforsch.* **43a**, 1154 (1988).

T. Tél, Transient Chaos, in *Directions in Chaos* **3**, ed. Hao Bai-Lin, World Scientific, Singapore (1990).

S. M. Ulam, J. von Neumann, *Bull. Am. Math. Soc.* **53**, 1120 (1947).

S. Vaienti, *J. Stat. Phys.* **56**, 403 (1989).

S. Vaienti, *J. Phys.* **21A**, 2023, 2313 (1988).

J. P. van der Weele, H. W. Capel, R. Kluiving, *Physica* **145A**, 425 (1987).

N. G. van Kampen, *Stochastic Processes in Physics and Chemistry*, North-Holland, Amsterdam (1981).

R. Vilela Mendes, *Phys. Lett.* **84A**, 1 (1981).

P. Walters, *An Introduction to Ergodic Theory*, Springer, New York (1981).

M. Widom, D. Bensimom, L. P. Kadanoff, S. J. Shenker, *J. Stat. Phys.* **32**, 443 (1983).

N. Wiener, *Cybernetics or Control and Communication in the Animal and the Machine*, Wiley, New York (1949).

J. A. Yorke, C. Grebogi, E. Ott, *Phys. Rev.* **38A**, 3688 (1988).

L.-S. Young, *Ergod. Th. Dynam. Sys.* **2**, 109 (1982).

G. M. Zaslavsky, *Chaos in Dynamic Systems*, Harwood, Chur (1985).

# Index

allowed sequence 33, 37, 202
alphabet 38
aperiodic trajectory 2
Arnold's cat map 171, 176
astrophysics 83
asymmetric triangular map 153, 162, 206
attractor 2
availability 63

band splitting 10, 260
Barnsley's fern 109
bath 61
Bernoulli shift 34, 36, 37, 138
bifurcation 5, 8
binary shift map 33, 39, 152, 185, 217, 231
Birkhoff ergodic theorem 25, 169
bit-cumulants 127
bit-number 44, 91
bit-variance 129, 131, 132
bivariate phase transition 254
bivariate thermodynamics 226
Boltzmann constant 46, 59
Bowen–Ruelle formula 223
box 32
box counting algorithm 98, 120
box dimension 98, 116
Brownian motion 13, 271

canonical distribution 59
canonical dynamical coordinates 59
cantor set 96, 101
  classical 96, 99, 205
  two-scale 101, 123, 206, 232
capacity 98, 116
cat map 171, 176
cell 21, 32
chaos 4, 148, 170
chaotic chaos phase 251
characteristic function 25, 128
chemical potential 56, 60, 91, 181, 234, 240
coarse graining 32
coding 37, 39, 206
concavity 52, 65, 74, 92
conditional information 48
conditional probability 42, 91, 203, 240
conjugation 28, 34, 40
contact transformation 66
control parameter 4, 260
continued fraction map 194, 250
convexity 52, 65, 92
coordinate transformation 28
correlation dimension 116
correlation function 27, 261
correlation matrix 74
correlations between subsystems 131
cosmology 83
crisis 261
critical correlations 131, 261
critical density 259
critical exponents 76, 261
critical line 75, 256, 270
critical opalescence 76
critical point 5, 16, 67, 75, 236, 271
crowding index 114
cumulants 128
cusp map 195
cycle 2, 179, 238
cylinder 39, 147, 179, 204

density 23, 230
diffeomorphism 16, 173
dimension 90, 98, 100, 115
  box 98, 116
  correlation 116
  embedding 98
  fractal 98
  Hausdorff 100, 102, 104, 206, 253
  information 116, 269
  marginal 138
  partial 225
  Rényi 115, 123, 222, 244
Dirac delta function 13, 23, 26
dissipation rate 16
dissipative system 2
doubling operator 264
dynamical bit-number 155
dynamical Brownian motion 13, 271
dynamical free energy 150, 182
dynamical scaling index 156
dynamical system 1

eigenvalues 3, 86, 169, 196, 267
  of doubling operator 262, 267
  of Jacobi matrix 3, 169, 171, 173
  of Perron–Frobenius operator 196
  of transfer operator 86, 198, 203, 254
elementary event 45
elementary particle physics 83
embedding dimension 98
energy 59, 119, 227
energy shell 61
ensemble 21, 59
  canonical 56, 59, 119
  grand canonical 56, 60, 233
  microcanonical 56, 61
  pressure 56, 60, 91, 226
ensemble average 24
entropy 56, 62, 120
  Kolmogorov–Sinai 146, 148, 152, 220
  metric 146, 148, 152, 220
  Rényi 149, 218, 248
  Shannon 46
  topological 151
entropy function 221
equation of state 230, 255
equilibrium ensembles 58
ergodicity 21, 25, 196
escape 204
escape rate 207, 217, 220, 224
escort distributions 53, 88, 91
Euler homogeneity relation 72, 92

exergy 63
expanding dynamical system 175, 211
expansion rate 158, 184
expansion phase transition 251, 254
expectation value 24
exponentially mixing 197, 272
extensities 56
external phase transition 260

$f(\alpha)$ spectrum 118, 122, 248
Feigenbaum attractor 5, 121, 125, 208, 257, 262
Feigenbaum constants 7, 16, 103, 125, 209, 259, 261, 267
Feigenbaum's universal function 209, 262
ferromagnet 75, 78
finite size effects 136
fixed point 2, 173
  hyperbolic 173
  of the doubling operator 264
  of the Perron–Frobenius operator 193
  saddle type 173
  stable 3, 173
  unstable 3, 173, 179, 238
fluctuations 65, 73
forbidden sequence 33, 37, 202
fractal 4, 94
fractal dimension 98, 116, 206
free energy 58, 61, 89, 182
  dynamic 150, 182
  Gibbs 60, 73, 143, 228, 230
  Helmholtz 59, 70, 90, 141, 181, 230
  metric expansion 165, 182
  static 116, 182
  topological expansion 181, 182
fugacity 235, 240

Gaussian distribution 138, 271
generalized canonical distribution 56, 58
generalized free energy 58
generalized Liapunov exponent 165, 167, 172, 254
generalized Perron–Frobenius operator 197
generating function 127, 128
generating partition 35, 148
generator 127, 128
Gibbs distribution 58
Gibbs–Duhem equation 65, 71, 240
Gibbs free energy 60, 73, 143, 228, 230

Gibbs fundamental equation 65, 68
Gibbs measure 214, 217
global manifold 173
grammar 37
grand canonical ensemble 60, 91, 233
Grassberger–Procaccia algorithm 116
grid 32, 114

Hamilton function 79
Hamiltonian system 2, 16, 168, 171, 208
Hausdorff dimension 100, 102, 104, 206, 253
heat capacity 130, 132
Heisenberg model 79
Helmholtz free energy 59, 70, 90, 141, 181, 230
Hénon map 14, 39, 176, 195, 246
histogram 25, 30, 139
homoclinic points 175
homoclinic tangencies 175, 177, 247
homogeneous function 72, 92
homogeneous system 72, 118
hyperbolic dynamical system 175, 211
hyperbolic fixed point 173

IFS 109
information 44, 50
  conditional 48
  Kullback 51, 53, 63
  reduced Rényi 137, 145, 157
  Rényi 50, 136
  Shannon 46
information dimension 116, 269
information gain 51, 53, 63
information loss 149, 152, 158, 160
information production by quantum measurements 154
intensities 58, 69, 226
interaction 79, 184
interaction energy 84, 184
intermittency 8, 267
invariant density 23, 30, 193, 201, 207, 214
invariant measure 23, 30, 193, 201, 207, 214
invariant tori 17
invertible mapping 15
irregular scattering 208
Ising model 79, 155
iterated function system 109
iterates 1

Jacobi determinant 2, 30, 167, 169, 192, 225
Jacobi matrix 2, 170, 225
Jaynes principle 56
Julia set 104, 186, 253

KAM theory 17
Kaplan–Yorke conjecture 269, 271
Kaplan–Yorke type of map 11, 16, 21, 25, 28, 138, 246, 268
Khinchin Axioms 47, 50
kicked particle 13, 16, 138, 271
kicked rotator 19
kneading sequence 38
Koch curve 94, 99, 110
Kolmogorov–Sinai entropy 146, 148, 152, 220
KS entropy 146, 148, 152, 220
Kullback information 51, 53, 63

Lagrange multiplier 58
Langevin force 13, 271
lattice 78
lattice gas 79
Legendre transformation 65, 120
length scale interpretation 182
Liapunov exponent 3, 159, 169, 172, 218, 221
  generalized 165, 167, 172, 254
  local 164
linearized doubling operator 266
Liouville operator 191
Liouville theorem 59
local dimension 115
local Liapunov exponent 164
local manifold 173
logistic map 4, 37, 103, 134, 162, 176, 186, 253, 257, 261
Lorenz model 14
Loschmidt number 73

Mandelbrot set 102
manifolds 173
map 1
marginal dimension 138
marginal invariant density 138
Markov chain 41
Markov process 41, 241
Markov partition 41, 202, 212, 216
maximum entropy measure 42

maximum entropy principle 56
measure 22
  conditionally invariant 217
  Gibbs 214, 217
  invariant 23, 30, 193, 201, 207, 214
  natural invariant 26, 207, 214, 217
  SRB 214, 217
  of maximum entropy 42
metric entropy 146, 148, 152, 220
metric expansion phase transition 254
microcanonic distribution 61
microstates 21
Misiurewicz point 11, 25, 30
mixing 27, 196
moments 127
Monger sponge 97
multifractal 114
multiplicative ergodic theorem 169

natural invariant density 26, 207, 214, 217
natural invariant measure 26, 207, 214, 217
natural variables 62, 70
*N*-cylinder 39, 147, 179, 204
nearest-neighbour interaction 80, 82
negative temperature 89
nonlinear mapping 1
nonuniform fractal 115
nonuniformity factor 162

observable 24, 27, 197
orbit 2
order parameter 76, 261
Ornstein–Uhlenbeck process 271
oscillator 16

paramagnetic states 75
partial dimensions 225
particle density 230, 242, 257
particle number 60, 181
partition 32
  generating 35, 148
  into boxes of equal size 32, 114
  Markov 41, 202, 212, 216
partition function 59
  bivariate 227, 233
  dynamic 150
  expansion 165
  grand canonical 236
  static 115
  topological 180
  with conditional probabilities 241
path integral 241
period doubling scenario 5, 258, 266
periodic orbit 2
  stable 3, 38
  superstable 7, 38, 266
  unstable 3, 179, 238
Perron–Frobenius operator 190, 217, 244, 272
Pesin's formula 172, 220, 268
phase portrait 17
phase space 1
phase transition 67, 75, 82, 131, 243
  bivariate 254
  dynamical 248
  external 260
  first order 75, 245, 254, 267
  internal 260
  metric expansion 254
  of spin system 82
  second order 75, 260
  static 243
  topological expansion 251
  with respect to the volume 256
pitchfork bifurcation 5
Potts model 79, 83, 155
preimage 15, 22, 191
pressure 56, 60, 91, 141, 227
  for bivariate thermodynamics 227
  for finite volume effects 141
  topological 178, 188, 198, 214, 218, 221, 251
pressure ensemble 56, 60, 91, 226
principle of maximum entropy 56
principle of maximum topological pressure 211
principle of minimum free energy 61
probability 20
probability density 22
probability distribution 21, 45
probability measure 21
pure state 21, 47

quantum systems 154, 189

reduced Rényi information 137, 145, 157
regular chaos phase 251
relative frequency 20
relaxation time 197, 271

renormalization group 76, 262
Rényi dimensions 115, 123, 222, 244
Rényi entropies 149, 218, 248
Rényi information 50, 136
repeller 204, 207, 209, 217
repelling fixed point 173
rescaling 262
round-off errors 116, 258

saddle point method 119
saddle type fixed point 173
sample set 20
scaling 76, 261, 267
scaling function 184
scaling index 115, 121, 156, 184
Schwartz distribution 23
self-similar structure 4, 78, 99
sensitive dependence on initial conditions 4, 160
Shannon entropy 46
Shannon information 44
shrinking ratio 124
Sierpinski carpet 96, 99
Sierpinski gasket 97, 99
Sierpinski sponge 97, 99
specific heat 130
spectrum 118, 120, 125, 143, 156, 164, 184, 196
- of dynamical scaling indices 156
- of local cylinder expansion rates 184, 225
- of local dimensions 120, 225
- of local Liapunov exponents 164
- of the Perron–Frobenius operator 196
- of the transfer operator 86, 198, 203, 254

spin system 78, 155, 203
SRB measure 214, 222
stable manifold 173
stable periodic orbit 3, 38
standard map 16
state density 118
stationary process 24, 84
Stieltjes integral 23
stochastic chaos phase 251
stochastic force 13
stochastic matrix 200
stochastic process 41
strange attractor 4, 12, 15
strings of symbols 37
superconducting phase 75
superfluid phase 75
superstable orbit 7, 38, 266
susceptibility matrix 73
symbol sequence 33, 146
symbolic dynamics 32, 37, 202
symbolic stochastic process 40, 78, 83, 202

tangent bifurcation 8, 261
tangent space 168
temperature 59
tent map 26, 28, 34, 40, 193
test function 27, 91, 188
thermal variables 62, 69
thermally conjugated variables 69
thermodynamic limit 54, 61, 86, 257
thermodynamic potential 62, 70, 138, 242
time 2
time average 24
time series analysis 117
topological conjugation 28, 34, 40
topological entropy 151
topological expansion free energy 181, 182
topological expansion phase transition 251
topological Markov chain 41, 202
topological pressure 178, 188, 198, 214, 218, 221, 251
trajectory 1
transfer matrix 85, 199
transfer operator 197
transient chaos 208
triangular map 26, 153, 162, 206
two scale cantor set 101, 123, 206, 232

Ulam map 29, 37, 40, 143, 160, 166, 185, 189, 194, 218, 243, 251, 256
unbiased guess 56
uniform distribution 26, 47
universality 8, 77, 262
universality class 8, 259
unstable periodic orbit 3, 179, 238
unstable manifold 173

variance of the information loss 161
variational principle 56, 61, 211, 214, 221
volume 117

window 6
words 38

*XY* model 79

zeta function 210, 237

Printed in the United States
By Bookmasters